CRITICAL PERSPECTIVES ON URBAN REDEVELOPMENT

RESEARCH IN URBAN SOCIOLOGY

Series Editor: Ray Hutchison

Recent volumes:

Volume 1: Race, Class and Urban Change, 1989

Volume 2: Gentrification and Urban Change, 1992

Volume 3: Urban Sociology in Transition, 1993

Volume 4: New Directions of Urban Sociology, 1997

Volume 5: Constructions of Urban Space, 2000

RESEARCH IN URBAN SOCIOLOGY   VOLUME 6

CRITICAL PERSPECTIVES ON URBAN REDEVELOPMENT

EDITED BY

KEVIN FOX GOTHAM
Tulane University, New Orleans, USA

2001

JAI
An Imprint of Elsevier Science

Amsterdam – London – New York – Oxford – Paris – Shannon – Tokyo

ELSEVIER SCIENCE Ltd
The Boulevard, Langford Lane
Kidlington, Oxford OX5 1GB, UK

© 2001 Elsevier Science Ltd. All rights reserved.

This work is protected under copyright by Elsevier Science, and the following terms and conditions apply to its use:

Photocopying
Single photocopies of single chapters may be made for personal use as allowed by national copyright laws. Permission of the Publisher and payment of a fee is required for all other photocopying, including multiple or systematic copying, copying for advertising or promotional purposes, resale, and all forms of document delivery. Special rates are available for educational institutions that wish to make photocopies for non-profit educational classroom use.

Permissions may be sought directly from Elsevier Science Global Rights Department, PO Box 800, Oxford OX5 1DX, UK; phone: (+44) 1865 843830, fax: (+44) 1865 853333, e-mail: permissions@elsevier.co.uk. You may also contact Global Rights directly through Elsevier's home page (http://www.elsevier.nl), by selecting 'Obtaining Permissions'.

In the USA, users may clear permissions and make payments through the Copyright Clearance Center, Inc., 222 Rosewood Drive, Danvers, MA 01923, USA; phone: (+1) (978) 7508400, fax: (+1) (978) 7504744, and in the UK through the Copyright Licensing Agency Rapid Clearance Service (CLARCS), 90 Tottenham Court Road, London W1P 0LP, UK; phone: (+44) 207 631 5555; fax: (+44) 207 631 5500. Other countries may have a local reprographic rights agency for payments.

Derivative Works
Tables of contents may be reproduced for internal circulation, but permission of Elsevier Science is required for external resale or distribution of such material.
Permission of the Publisher is required for all other derivative works, including compilations and translations.

Electronic Storage or Usage
Permission of the Publisher is required to store or use electronically any material contained in this work, including any chapter or part of a chapter.

Except as outlined above, no part of this work may be reproduced, stored in a retrieval system or transmitted in any form or by any means, electronic, mechanical, photocopying, recording or otherwise, without prior written permission of the Publisher.
Address permissions requests to: Elsevier Science Global Rights Department, at the mail, fax and e-mail addresses noted above.

Notice
No responsibility is assumed by the Publisher for any injury and/or damage to persons or property as a matter of products liability, negligence or otherwise, or from any use or operation of any methods, products, instructions or ideas contained in the material herein. Because of rapid advances in the medical sciences, in particular, independent verification of diagnoses and drug dosages should be made.

First edition 2001

Library of Congress Cataloging in Publication Data
A catalog record from the Library of Congress has been applied for.

British Library Cataloguing in Publication Data
A catalogue record from the British Library has been applied for.

ISBN: 0-7623-0541-X
ISSN: 1047-0042 (Series)

∞ The paper used in this publication meets the requirements of ANSI/NISO Z39.48-1992 (Permanence of Paper).
Printed in The Netherlands.

CONTENTS

LIST OF CONTRIBUTORS — vii

URBAN REDEVELOPMENT, PAST AND PRESENT
Kevin Fox Gotham — 1

MONUMENTS OF TOMORROW: INDUSTRIAL RUINS AT THE MILLENNIUM
James Dickinson — 33

CITY REDEVELOPMENT POLICIES AND THE CRIMINALIZATION OF HOMELESSNESS: A NARRATIVE CASE STUDY
Adalberto Aguirre, Jr. and Jonathan Brooks — 75

INNOVATION, SPECULATION, AND URBAN DEVELOPMENT: THE NEW MEDIA MARKET BROKERS OF NEW YORK CITY
Michael Indergaard — 107

REGIME STRUCTURE AND THE POLITICS OF ISSUE DEFINITION: URBAN REDEVELOPMENT IN PITTSBURGH, PAST AND PRESENT
Gregory J. Crowley — 147

ON FRAGMENTATION, URBAN AND SOCIAL
Judit Bodnár — 173

HISTORIC PRESERVATION, GENTRIFICATION, AND TOURISM: THE TRANSFORMATION OF CHARLESTON, SOUTH CAROLINA
Regina M. Bures — 195

GENTRIFICATION, HOUSING POLICY, AND THE
NEW CONTEXT OF URBAN REDEVELOPMENT
  Elvin K. Wyly and Daniel J. Hammel 211

HOUSEHOLD SURVIVAL STRATEGIES IN A PUBLIC
HOUSING DEVELOPMENT
  Joel A. Devine and Petrice Sams-Abiodun 277

ABSTRACT SPACE, SOCIAL SPACE, AND THE
REDEVELOPMENT OF PUBLIC HOUSING
  Kevin Fox Gotham, Jon Shefner and Krista Brumley 313

WORLD CITY THEORY: THE CASE OF SEOUL
  Jamie Paquin 337

THE CITY AS AN ENTERTAINMENT MACHINE
  Richard Lloyd and Terry Nichols Clark 357

THE "DISNEYFICATION" OF TIMES SQUARE:
BACK TO THE FUTURE?
  Bart Eeckhout 379

REDEVELOPMENT FOR WHOM AND FOR WHAT
PURPOSE? A RESEARCH AGENDA FOR URBAN
REDEVELOPMENT IN THE TWENTY FIRST CENTURY
  Kevin Fox Gotham 429

LIST OF CONTRIBUTORS

Adalberto Aguirre, Jr.	Department of Sociology, University of California, California, USA
Judit Bodnár	CRCEES, Rutgers University, New Brunswick, USA
Jonathan Brooks	Department of Sociology, University of California, California, USA
Krista Brumley	Department of Sociology, Tulane University, New Orleans, USA
Regina M. Bures	School of Social Welfare, University at Albany, USA
Terry Nichols Clark	Department of Sociology, University of Chicago, USA
Gregory J. Crowley	Department of Sociology, University of Pittsburgh, Pittsburgh, USA
Joel A. Devine	Department of Sociology, Tulane University, New Orleans, USA
James Dickinson	Department of Sociology, Rider University, Lawrenceville, USA
Bart Eeckhout	Ghent Urban Studies Team (GUST), Ghent University, Ghent, Belgium
Kevin Fox Gotham	Department of Sociology, Tulane University, New Orleans, USA

Daniel J. Hammel	Department of Geography and Geology, Illinois State University, Normal, USA
Michael Indergaard	Department of Sociology, St. Johns University, New York, USA
Richard Lloyd	Department of Sociology, University of Chicago, USA
Jamie Paquin	Department of Sociology, York University, Toronto, Canada
Petrice Sams-Abiodun	Department of Sociology, Tulane University, New Orleans, USA
Jon Shefner	Department of Sociology, University of Tennessee, Knoxville, USA
Elvin K. Wyly	Department of Geography and Center for Urban Policy Research, Rutgers University, Piscataway, USA

URBAN REDEVELOPMENT, PAST AND PRESENT

Kevin Fox Gotham

Urban "redevelopment" has emerged, in recent years, as one of the key concerns of urban social science in both theoretical and empirical-based settings. This reflects a concomitant trend associated with urban studies more generally, toward specifying the economic, political, and cultural factors responsible for uneven metropolitan development. Indeed, even a causal look at our metropolitan areas reveals that they are composed of many different cities and spatial forms that are divided according to different land uses as well as related to patterns of race and class. One city is reserved for the rich and affluent; another is composed of working-class and middle-class neighborhoods; other areas cater to commercial interests, entertainment, tourists, and consumers; and still others languish in chronic disinvestment and decay, designated for the homeless, the poor, minorities, and the urban underclass. These contrasts are quite graphic, as anyone touring our metropolitan areas can attest, and represents an extreme crisis of inequality produced by the uneven nature of metropolitan development and growth. The city continues to provide the prime socio-spatial context within which economic and political elites and ordinary people construct and act out the processes of disinvestment, fiscal crisis, and inner city "renaissance." Scholars have begun to study redevelopment with an eye toward clarifying the links between macrostructural processes, specific urban redevelopment efforts, and locally lived realities. This review issue will therefore consider the significance of urban redevelopment as a focus for urban theory and urban research, outlining what such theoretical and methodological contributions and changes may mean for the future of urban scholarship.

The study of cities, urbanism, and urban change – redevelopment, disinvestment, and so on – has a rich tradition in urban scholarship. Europeans such as Marx and Engels, Weber, and Simmel devoted much thought to the importance of the city, for example, as a seat of the emerging capitalist economy, a site political and economic power, and force of cultural change that affects mental life. In the United States, the early Chicago School urban sociologists focused their empirical attention on the spatial distribution of people and organizations, the causes and consequences of neighborhood racial succession, and ethnic and racial group "adaptation" to the urban environment. Robert Park (1925), Ernst Burgess (1925), Lewis Wirth (1938), all commented on community structure and local institutions, often drawing analogies to biological systems. From these writings there emerged a theory of urban and neighborhood change as a "life-cycle" beginning with investment and growth and ending with inevitable decline. By the 1930s, social scientists around the nation were employing the insights, models, and analyses developed by the Chicago School to study cities, as well as influence public policy. Yet a lacuna of American urban scholarship in general, and the Chicago School in particular, was the lack of specificity in identifying the webs of interconnections between urban life and wider macro-level processes. Early urban sociologists, in short, were primarily concerned with the internal organization and dynamics of cities, while ignoring the larger macrostructures that linked urban change to extra-local processes.

In the early 1970s, several Marxist social scientists including Manuel Castells (1977), David Harvey (1973), and Henri Lefebvre (1991), among other scholars began to revise Karl Marx's ideas to explain uneven metropolitan development, urban industrial decline, and other urban trends. Castells proposed that urban scholars focus on the collective consumption characteristic of urbanized nations and way in which political and economic conflicts within cities generate urban social movements for change. David Harvey, in contrast, argued that the central issue in making sense of cities was not collective consumption but the more basic Marxist concern with capital accumulation. Influenced by Lefebvre, Harvey (1973) argued that investment in land and real estate is an important means of accumulating wealth and a crucial activity that pushes the growth of cities in specific ways. Processes as diverse as urban disinvestment and decay, suburbanization, deindustrialization, urban renewal, and gentrification are part and parcel of the continuous reshaping of the built environment to create a more efficient arena for profit making. According to Harvey (1989), powerful real estate actors invest, disinvest, and reshape land-uses in a process of "creative destruction" that is continually accelerating, destroying communities and producing intense social conflicts and struggles over meanings and uses of urban space. Despite their different emphases, the work of Marxists helped focus scholarly attention on the capitalist system of for-profit production generally, and class struggle and capital accumulation specifically, as analytical starting points for understanding the nature of urban redevelopment and disinvestement (for overviews, see Jaret, 1983; Tabb & Sawyers, 1984).

By the late 1970s and continuing into 1980s, a new critical approach to the study of cities and urban redevelopment had developed. Usually called the "critical political-economy" or "sociospatial approach," this perspective emphasized several major dimensions of cities: (1) the importance of class and racial domination (and, more recently, gender) in shaping urban development; (2) the primary role of powerful economic actors, especially those in the real estate industry, in building and redeveloping cities; (3) the role of growth-assisted government actors in city development; (4) the importance of symbols, meanings, and culture to the shaping of cities; (5) attention to the global context of urban development (for overviews see Feagin, 1998; Gottdiener & Feagin, 1988; Hutchison, 2000; Savage & Ward, 1993; Smith, 1995). Gottdiener (1994) and Hutchison (2000) prefer the term "sociospatial" perspective to describe the critical political economy paradigm, a term that accents the society/space synergy, and emphasizes that cities are multifaceted expressions of local actions and macrostructural processes. They also use the term to distance themselves from older Marxist approaches of Gordon (1984), Dear and Scott (1981), and Storper and Walker (1983) and highlight the diversity of theory and method within the broad paradigm. Molotch (1999, 1976) and Logan and Molotch (1987) prefer the term urban political economy and have developed their own "growth machine" theory to explain urban redevelopment. Other critical scholars have embraced a more eclectic and multi-perspectival focus in their empirical work, attempting to develop middle range variants of general theories as an expedient to moving toward constructive dialogue and common ground in the direction of synthesized theory building (see e.g. Smith, 1995; Orum, 1995; Squires, 1994; for an overview, see Gotham & Staples, 1996).

Despite ambitious theoretical and analytical contributions in critical urban scholarship, discussions of "redevelopment" have been slow to deal with the complex nature of the interrelationships that exist between the macro-micro, class/race/gender, the economic and cultural, and the dialectic of space and social, among others. Indeed, these relationships manifest themselves in urban settings making urban redevelopment an auspicious process fraught with many dilemmas, conflicts, and contradictions regarding the links between space, capital, and power. A number of edited collections on urban redevelopment have appeared in the 1980s and 1990s, including the work of Rosenthal (1980) and contributors on "urban revitalization" in the 1960s and 1970s, the work of Squires (1989) and colleagues on "unequal partnerships," Fainstein and

Fainstein's (1986) first and second editions on the political economy of urban redevelopment, Cumming's (1988) edited volume on the role of business elites, Kearns and Philo's (1993) collection of essays on culture and the "selling of cities," Stone and Sander's (1987) edited book on the "politics" of urban development, and King's (1996) edited volume on transformations in "world cities." As these collections show, the social context, strategies, and meaning of urban redevelopment have changed dramatically throughout the twentieth century. Moreover, areas such as law, transportation, housing, policy, and local financing devices are important mechanisms for promoting urban redevelopment and exacerbating urban disinvestment. In addition, conflicts over disinvestment and redevelopment reflect broader struggles over an array of social issues, for example, poverty, race, and schools, among others.

The contributors to this volume add to this growing literature but also tackle several unresolved issues and unanswered questions: What are the historical sources of continuity and variation in the structure, influence, and composition of public-private partnerships and related growth coalitions in U.S. cities? What impact have global level changes had on the transformation of cities into places of leisure and entertainment in the 1970s and later? How are these changes related to the dynamics of race, class, and gender? How have changes in federal funding and bureaucratic arrangements affected the content and process of local redevelopment measures? What ideologies, themes, symbols, and motifs undergird current and past redevelopment strategies? What forms of conflict, opposition, and collective mobilization have occurred in response to local redevelopment efforts? What new theories, methods, and data sources have proved more useful than others in helping scholars understand the causes and consequences of urban redevelopment?

It is the purpose of this volume to help clarify conceptual and theoretical issues, furnish relevant empirical data, and illuminate empirical relationships between the phenomenon of redevelopment and broader process of urban and global change. In the first section of this introduction, I present a broad history of urban redevelopment in the United States through the 1960s. Second, I discuss trends in urban redevelopment from the end of the 1960s through the end of the 1990s. In the final section, I consider some of the significant ways in which urban redevelopment has emerged as a key focal point of urban scholarship in general, and of urban sociology more specifically. As the contributors to the volume show, the factors that are responsible for urban redevelopment do not operate in a uniform way and there is a noticeably uneven pattern of urban redevelopment and disinvestment across cities. Questions about the full extent, meaning, significance, and empirical study of urban redevelopment remain a source of much debate and controversy. While the contributors to this volume may not offer definitive answers, they do provide additional evidence for assessing the impact of urban redevelopment and present new ideas, models, and methods for future research.

URBAN REDEVELOPMENT FROM THE INDUSTRIAL REVOLUTION THROUGH THE 1960s

The economic and social circumstances of urban areas have stimulated calls for redevelopment and renewal since the Industrial Revolution transformed the economic base and spatial makeup of America cities during the second half of the nineteenth century. Industrialization not only brought about a fundamental transformation in the nature of work, but also systems of transportation, the structure and operations of city government, the experience of poverty, and the demographic movement of racial and ethnic groups on an unprecedented scale. Henry Mayhew, Charles Booth, and Lewis Mumford, among others, documented in graphic detail the pathological nature of uneven development characterizing urban growth under industrial capitalism. The erosion of subsistence production and the rise of mass production, the emergence of the factory system dominated by hierarchical lines of authority, urbanization of wage labor, and the gradual spatial separation of home from work were the hallmark of the industrialization during the late nineteenth century (Saunders, 1981; Taylor, 1993; Sugrue, 1996, pp. 88–92; Katz, 1986, chapter 6). During this time, attempts at redeveloping and planning the city were typically haphazard, uncoordinated, and chaotic. Indeed, early twentieth century planning schemes such as zoning laws, building codes, land-use ordinances, annexation drives, and subdivision regulations were attempts to segregate land-uses, class and ethnic groups, and other social activities to facilitate profitable investment and economic development (for an overview, see Sies & Silver, 1996). Economic and political elites and planners offered many blueprints for change, ranging from the utopian solutions proposed by advocates of the "garden city," to the development of parks and recreational spaces by the City Beautiful movement (Wilson, 1989), to the class and racially homogeneous residential environments marketed by the emerging real estate industry (Gotham, 2000; Weiss, 1987).

By the 1930s, many American cities were experiencing increasing physical deterioration of their core neighborhoods and commercial districts, forced concentration of inner city blacks into crowded areas, and loss of population and industry (Banfield & Wilson, 1963; Silver, 1984; Mollenkopf, 1983; Teaford, 1990). The population of central cities essentially stopped growing after 1920 and the collapse of economy because of the Great Depression stimulated calls for federal and local governments to attack the problems of urban deterioration. In New York, Chicago, Cincinnati, Philadelphia, Buffalo, Atlanta, and many other cities, real estate elites and political officials recited the same litany of ills – dilapidated neighborhoods, eroding property values, declining revenues from commercial and industrial sites, snarled traffic, and increasingly "drab appearance" (for overviews, see Mohl, 1993; Teaford, 1990). In several cities, the Franklin Roosevelt's "New Deal" provided federal grants to urban officials to revitalize of their deteriorating cores. In Kansas City, the Public Works Administration (PWA), the Civil Works Administration (CWA), and the Works Projects Administration (WPA) brought thousands of new jobs to the city and paid the construction of costs for several new downtown skyscrapers, a country courthouse, police station, a new City Hall, a municipal auditorium, and a new convention center and baseball stadium (Gotham, 2001; Reddig, 1947). In Detroit and New Orleans, federal funding through the WPA helped establish a nascent urban renewal program, a housing authority with the legal power to acquire and clear slum areas, and money to rebuild city parks (Darden, Jill, Thomas & Thomas, 1987; Mahoney, 1985). Yet New Deal federal programs and subsidies were insufficient in eradicating urban blight and facilitating large-scale urban redevelopment. By the early 1940s, local and national real estate elites openly called for federal and local government assistance in revitalizing downtown business districts and eliminating blighted areas (Gotham, 2000b; Weiss, 1987).

Beginning in the early 1930s, the National Association of Real Estate Boards (NAREB) took the lead in developing and lobbying for a national urban redevelopment policy. During this time, the NAREB devised a series of policy proposals that could facilitate public acquisition of land in blighted areas for clearance and resale to private builders (Davies, 1958, pp. 182–185). These proposals included state acts empowering municipalities to redevelop slum areas, close public-private coordination of urban land-use and control, long-term federal loans to cities at low interest rates, and generous tax subsidies and write-offs for local redevelopers. In 1939, the NAREB formed a research agency, the Urban Land Institute (ULI), to study and research the causes and consequences of urban blight and to identify the policy and financial tools needed to revitalize cities. The ULI undertook as its first major research project a study of real estate values in 221 cities (Urban Land Institute, 1940). During the next two years the ULI published case studies of Boston, Cincinnati, Detroit, Louisville, Milwaukee, New York, and Philadelphia, recommending a plan by which cities could condemn land in the blighted areas near the CBD and then sell or lease the land to private developers for replanning and rebuilding (i.e. eminent domain) (Gotham, 2001). Government aid and subsidies, as the ULI maintained, would be necessary to revitalize the central city, eliminate slums, and maintain profitable land sales and real estate markets.

During this time, officials of the NAREB along with downtown business elites agreed that state action – through eminent domain and public subsidies for private revitalization – was necessary to counter the specter of urban blight and obsolescence. Yet a number of problems forestalled private efforts to harness the legal and financial power of the local state to undertake slum clearance. First, private interests proved incapable of acquiring land in large enough parcels to permit large-scale revitalization. Second, the most desirable inner city land sought by downtown businesses tended to be residential land-use. Hence, private business did not have the organizational capacity or legal right to acquire such properties. Financing also proved to be a major problem since few private firms and redevelopers possessed the huge cash reserve necessary to pay for clearance and revitalization (Davies, 1958, pp. 183–184). Thus, while real estate elites wholeheartedly supported public-private action for urban revitalization, the existing political structure in many local areas of the country did not possess the legal or fiscal capacity to undertake large-scale slum clearance and redevelopment during the 1930s and 1940s (Hirsch, 1993, pp. 87–88; Gelfand, 1975, pp. 151–156, 275–276; Hays, 1985, chapter 5; Kleniewski, 1984; Weiss, 1980; Wilson, 1966).

In many cities, public housing provided a partial answer to the problem of freeing inner city land for private redevelopment. The Housing Act of 1937 empowered local communities to create local housing authorities with the legal power of eminent domain to acquire privately-owned land for slum clearance and rehousing (McDonnell, 1957; Quercia & Galster, 1997, p. 537; Bauman, 1987, pp. 40–42). By the mid-1940s, a number of cities and states had passed legislation to enable local governments to designate, acquire, and clear "slum" areas and sell the land to private developers. Yet early on there were opposing views over whether public housing legislation was supposed to be legislation that authorized local housing authorities to build low-income housing or whether it was legislation to allow the clearing of land and the subsidization of the local redevelopment industry (Hoffman, 1996, p. 425; Marcuse, 1986; Jackson, 1985, chapter 12). Proponents of public housing, including progressive housing advocates, social workers, and union officials argued for the building of low-rent housing for working- and middle-class workers by the federal government (Bauman, 1987, 1981). On the other hand, real estate industry and home building officials rejected a strong federal role and embraced a privatist vision of slum clearance with no government regulations on private redevelopment. Moreover, the NAREB and real estate elites attacked the view of public housing shared by labor unions, social workers, and housing activists as "creeping socialism" and opposed any federal building of low-income housing on the grounds that it would put the government in competition with private housing construction and real estate (Keith, 1973, pp. 30–38, 94–95; Bratt, 1986, pp. 336–337; Bauman, 1981, p. 8). The NAREB favored a trickle down model of housing distribution, where excess production of suburban housing would free up dwelling units at the lower end of the housing market for low-income families displaced by slum clearance (Davies, 1958, pp. 180–182; Gelfand, 1975, pp. 184–204; Checkoway, 1984).

The Housing Acts of 1949 and 1954 appropriated the model of federal funding and local decision-making provided in the 1937 Housing Act and empowered localities to create urban redevelopment authorities to designate and clear "blighted" areas.[1] Title I of the Housing Act of 1949 provided federal subsidies for property acquisition and slum clearance while requiring federal and local governments to finance one-to-two thirds of the site preparation.[2] The Housing Act of 1954 changed the name of the program from "urban redevelopment" to "urban renewal" and required each city to submit a Workable Program detailing its plan of clearance, demolition, and redevelopment. The 1954 Act intended projects to be "predominantly residential" but in later years the federal government waived this requirement for projects that would benefit universities, hospitals, business parks, and industrially depressed districts (Barnekov, Boyle & Rich, 1989, pp. 38–39; Gelfand, 1975, pp. 151–156, 275–276; Hays, 1985, chapter 5; Kleniewski, 1984; Weiss, 1980; Wilson, 1966). While the stated goal of Housing Acts of 1949 and 1954 was "To provide a decent home and suitable environment for every American family" urban leaders and real estate elites considered the Acts less as a "housing" program and more of urban "redevelopment" program (Gotham and Wright 2000). According to housing activist, Charles Abrams (1971, p. 244), "No sooner had the federal law [Housing Act of 1949] been enacted then two facts became plain: (1) there were no houses available for the slum-dwellers to be displaced from the sites; (2) these slum-dwellers were largely minorities to who housing in new areas was banned."

Over the next two decades, local redevelopment authorities used the legal mechanisms and public subsidies of the urban renewal program to acquire residential and commercial land, clear the residents and buildings, and rebuild their aging downtown business districts (Adams et al., 1987; Hirsch, 1983; Squires, Bennett, McCourt & Nyden, 1987; Feagin & Parker, 1990, chapter 9; Wilson, 1966). Although urban renewal officials justified the urban renewal program based on attacking substandard housing conditions, very few officials used federal resources and subsidies to improve deteriorated or dilapidated housing or "renew" slum neighborhoods (Teaford, 2000; Greer, 1965, p. 3;

Jacobs, 1961; Hartman & Kessler, 1978; National Commission on Urban Problems, 1969, p. 153). Nationally, the program destroyed thousands more units than it replaced, dislocated tens of thousands of small business and residents, and became the target of intense civil rights protest from leaders who labeled it "black removal" due to the large number of African American residents and neighborhoods cleared under the guise of "urban renewal" (Bayor, 1989; Friedland, 1982, pp. 81, 85, 195; Gans, 1962; Jacobs, 1961; Weiss, 1980; Kleniewski, 1984). According to Scott Greer (1965), blacks occupied nearly 70% of the dwelling units condemned for urban renewal projects. This was primarily due, Greer felt, to their "central locations and deteriorated conditions, but the effects [were] the same as they would be if dehousing Negroes were the goal." As a program for "pushing people around," Greer argued, urban renewal was about, at best, "slum-shifting"; and, at worst, increasing slums by eliminating housing (pp. 55–56). Herbert Gans's (1962) case study of the "urban villagers" in Boston's West End, Arnold Hirsch's (1983) study of Chicago, Ronald Bayor's (1989) analysis of Atlanta, and John Bauman's (1987) examination of Philadelphia show that the urban renewal program involved long delays, cost overruns, painful dislocation of thousands of people, and long periods where cleared land remained vacant.

A vast literature has focused on the mobilization of downtown real estate elites, the formation of growth coalitions, and the reshaping of business-government relations in the post-World War II era as a result of urban renewal (Hoffman, 2000; Barnekov & Boyle, 1989; Fainstein & Fainstein, 1986; Mollenkopf, 1978). In many cities, urban coalitions of downtown real estate and business elites worked to establish and legitimize a "public-private partnership" (Squires, 1989) of private actors and government redevelopment officials dedicated to redeveloping the downtown. In the 1980s and later, critical scholars such as Nancy Kleniewski (1984), Heywood Sanders (1980), and R. Allen Hays (1985), among others, showed that urban renewal was never a single, national program, as researchers and policy analysts contended during the 1960s and 1970s. Many cities did not participate in the urban renewal program (e.g. Houston and New Orleans) and the meaning of urban "renewal" (e.g. slum clearance, neighborhood rehabilitation, etc.) changed considerably from its inception in 1949 through the early 1970s. The physical shape of the program, its economic benefits and liabilities, levels of funding, site selection, and project duration displayed much local variation. Arnold Hirsch (1983, 1993) and Marc Weiss (1980) have shown how urban renewal programs were largely initiated and carried out by local business and real estate oriented groups with the assistance of federal subsidies and local government. Downtown real estate and other business elites not only helped the frame the laws under which renewal proceeded, but they controlled the character, location, and pace of redevelopment efforts (for overviews, see Barnekov, Boyle & Rich, 1989; Barnekov & Rich, 1989; Squires, 1989).

URBAN REDEVELOPMENT, FISCAL AUSTERITY, AND FEDERAL RETRENCHMENT, 1968–2000

In the 1960s and later, the character of urban redevelopment underwent a shift in emphasis away from large-scale clearance, unilateral taking of private property, and urban displacement. During the middle and late 1960s, the nation experienced five consecutive summers of racial unrest in its cities. The riots in Chicago, Cleveland, Detroit, Kansas City, Los Angeles, and many other cities followed a decade of massive displacement of people and neighborhoods because of urban renewal slum clearance, large-scale expressway building, and public housing. Critics such as Jane Jacobs (1961) and Herbert Gans (1962) blasted urban renewal for destroying neighborhoods while Martin Anderson (1964) and other conservative critics castigated urban renewal as symptomatic of "big government," a "federal bulldozer," and a "housing" program imposed on cities by the federal government. According to John Mollenkopf (1978, p. 141), "[m]ost of the community turbulence of the 1960s was firmly directed against urban renewal, highway construction, the declining availability of decent inexpensive housing, expansion of dominant institutions, and city bureaucracies tightly dominated by ethnic groups being displaced in urban population by minority newcomers." As the costs of urban displacement became apparent, neighborhood constituencies that had previously supported urban renewal began to swing into opposition thus fanning the flames of urban protest and provoking mobilization for reform (Piven & Cloward, 1974). In Detroit, Kansas City, Chicago, San Francisco, and other cities, resident dissatisfaction with displacement helped spawn combative coalitions of housing activists, civil rights organizations, and historical preservationist groups dedicated to halting slum clearance (Adams et al., 1987, pp. 120–121; Gotham, 1999; Hirsch, 1983; Bennett et al., 1987; Hartman & Kessler, 1978). In 1974, the federal government discontinued the urban renewal program amid widespread urban protest and neighborhood discontent surrounding the destabilizing effects of the program on inner city communities (Hoffman, 2000; Teaford, 2000; Hirsch, 1983; Weiss, 1980).

By this time the NAREB, central city businessmen, and local economic and political elites had apparently not fulfilled urban renewal's stated goal of providing a "decent home" for every American family. Urban renewal was supposed to reverse central city decline, counteract decentralization trends, and revitalize blighted neighborhoods. However, by the 1970s urban renewal had not saved the central city but had exacerbated disinvestment and decay, and aggravated racial tensions and neighborhood unrest. David Bartelt (1993, p. 149) has pointed to two faulty assumptions embedded within the urban renewal program concerning the causes and consequences of urban disinvestment. First, the program assumed that private capital would flow into renewal areas once local officials had cleared the blighted structures, relocated the previous occupants, and rebuilt infrastructure. However, what usually happened was that large-scale clearance produced vacant lots and abandoned dwellings rather than revitalize neighborhoods and generate new jobs (Weiss, 1980; Sanders, 1980; Gotham, 2000a). Second, urban renewal assumed that residents within affected areas would support large-scale clearance of their impoverished neighborhoods as evidence of improvement and progressive change. In fact, however, neighborhoods often opposed renewal efforts because of their destabilizing and segregative consequences (Hartman, 1966; Bennett, 1986). Nancy Kleniewski (1984) and Gregory Squires (1989) also pointed out that neighborhoods opposed urban renewal because local officials often earmarked tax subsidies and other federal resources for the expansion of universities, hospitals, large corporations, and other elite institutions that tended to employ middle- and upper-income residents rather than lower-income inner city residents (Kleniewski, 1984; Bartelt, 1993; Squires, 1989).

In 1974, the federal government replaced the urban renewal program with the Community Development Block Grant Program (CDBG) that gave cities more control and discretion over the use of federal funds. With the new CDBG program, the amount of money assured solely for urban renewal areas was now competing with demands for federal assistance from all areas of the city. By this time, many cities had embraced a "triage" strategy in which government officials divided federal money into three categories: deteriorated areas, transitional areas, and healthy areas (Downs, 1975; Metzger, 2000). In this way, cities could use federal money to revitalize moderately deteriorating areas, little money would be spent in the most deteriorated areas, or in healthy areas that could survive without federal assistance. Despite this emphasis on a more efficient use of federal resources, two contradictory goals plagued the CDBG program: on the one hand, the Carter Administration intended CDBG dollars to go to low- and moderate-income areas; on the other hand, the Administration encouraged cities to use CDBG money to stimulate private sector investment. Not surprising, in many cities, local officials and elites sacrificed the former goal to accommodate the later goal (Judd & Swanstrom, 1998, p. 230).

Another important federal subsidy for private development was the 1977 Housing and Urban Development Act, which created the Urban Development

Action Grant (UDAG) program to dispense monies to local governments to encourage private investment in depressed areas. Unlike the CDBG program, the UDAG program required applicants to propose and accomplish specific projects, and to leverage private money to supplement the federal grant. Between 1977 and 1987, the federal government spent nearly $4.5 billion on nearly 3,000 UDAG projects in 1,180 cities (Feagin & Parker, 1990, p. 135). Yet evidence from several cities showed that UDAGs were frequently used for the streets, utilities, and other services needed for various types of private residential, commercial, and industrial projects that directly benefitted businesses and affluent suburbanites, rather than needy residents and poor neighborhoods (for an overview, see Hubbell, 1979). For example, local officials and elites in various cities used UDAG to convert Union Station in St. Louis to a festival mall, rebuild Harborplace in Baltimore, and expand the Bartell Hall convention center in Kansas City. In the 1980s, the federal government folded UDAGs earmarked for specific purposes into block grants. This shift in funding strategies gave local political authorities considerably more control over the allocation of federal funds during a period in which interurban competition for federal monies and private capital rose sharply in the United States (Blair & Nachimas, 1979; Helen & Yinger, 1989; Feagin & Parker, 1990; Ladd & Yinger, 1989).

Shrinking federal resources, devolution of policy implementation to state governments, and increased reliance on market-centered strategies have been the core features of federal urban policy since the 1970s (Kincaid, 1999; Cummings, 1988; Fainstein & Fainstein, 1986; Gotham, 1998; Gotham & Wright, 2000; Squires, 1989). Barnekov, Boyle and Rich (1989, pp. 100–101) pointed out that from the 1930s through the 1970s, the view that cities were important to national economic prosperity and growth had guided U.S. urban policy. Yet by the 1980s, policy analysts at the federal level no longer accepted this assumption as valid and, as a result, gave little attention or priority to the redevelopment of central cities (see also Bartelt, 1993; Block, Cloward, Ehrenreich & Piven, 1987; Piven & Cloward, 1982). In the 1980s, the Reagan and Bush administrations eliminated general revenue sharing and public works monies. Moreover, funding for the CDBG program dropped from more than $6.2 billion to $2.8 billion from 1980 to 1990, a decline of 54% in constant 1990 dollars. Other budget cuts also affected cities. Economic development assistance went from $717 million in 1980 to $160 million by 1990 (a 78% loss), and funding for the UDAG program was cut by 41%, from $357 million in 1980 to $209 million by 1990. In 1980, federal government monies made up 14.3% of city budgets in the nation. In 1992, federal government dollars account for less than 5%. That same year, federal assistance as a percentage of city budgets had dropped to 64% below what it had been in 1980. Devolution of federal policy accompanied these budget cuts. State governments spread federal funds across many jurisdictions and gave little priority to distressed urban areas (Kincaid, 1999; Cole, Taebel & Hissong, 1990; Judd & Swanstrom, 1998, pp. 238–240; Wolman, 1986).

A report released in 1980 by the President's Commission for a National Agenda for the Eighties (1980) foreshadowed this shift in funding and thinking about cities. Appointed by President Carter to review urban policy and recommend policy changes, the Commission essentially adopted a "hands-off" approach to cities recommending that the "best interest of the nation" would be to promote "locationally neutral economic and social policies rather than spatially sensitive urban policies that either explicitly or inadvertently seek to preserve cities in their historical roles." According to the Commission, suburbanization of people and industry was an inevitable outcome of broad-based economic change and the "historical dominance of more central cities will be diminished as certain production, residential, commercial, and cultural functions disperse to places beyond them" (pp. 66–67). The function of the federal government is not to aid deteriorating cities but to allow the process of decay to run its natural course. Cities can adapt and change in response to economic and social forces and this process of adaptation can be encouraged, but not altered, by the federal government. Glickman (1984, p. 471) referred to this non-urban or anti-urban federal approach as a policy of "spatial trickle-down." Numerous scholarly studies and prominent institutions such as the National Research Council and the Brookings Institute supported much of the Commission's analysis. In a widely cited book, *The New Urban Reality*, Paul Peterson (1981) arrived at conclusions similar to those of the Commission and proclaimed that the "industrial city has become an institutional anachronism" (p. 1). The implications of much of Peterson's work suggested a Pollyannaish outcome where the transition to a technologically dynamic, post-industrial city would benefit all cities and regions equally, a view sharply criticized by John Logan and Todd Swanstrom (1990), Mark Gottdiener (1994), Ray Hutchison (1993), and other critical urban scholars.

Within the intellectual and policy environment defined by President's Commission's report, and carried on during the Reagan, Bush, and Clinton Administrations, federal officials and policy analysts have endorsed the view that cities must make substantial monetary and policy concessions to attract capital if they are to survive. The advocacy of Enterprise Zones and Empowerment Zones reflects this view and represents the latest in a series of market-centered federal efforts to revitalize cities by promoting economic competitiveness between places. Both the Reagan/Bush Enterprise Zone program and the Clinton/Gore Empowerment Zone (EZ) initiative embrace a supply-side policy approach that provides economic incentives to the private sector to invest in central city communities. However, under Clinton/Gore, officials attempt to target economic incentives to resident-based "empowerment zones" rather than industry-based businesses as the Reagan/Bush initiative. Whereas the Reagan/Bush program included no direct, citizen-based participation, a hallmark of the Clinton/Gore initiative is the creation of partnerships between government, business, and community organizations to encourage and sustain grassroots participation in the policy making and implementation process. Yet, as Baum (1999) has recently noted, the structure of partnership and success of community participation varies widely throughout the United States. Case studies of Detroit and Chicago suggest that government/business/community partnerships of the EZ process tend to reflect the prevailing power structures and political alignments of the particular cities and, as a result, community participation varies considerably, from high to almost nonexistent (Bockmeyer, 2000; Herring et al., 1998).

To compensate for declining federal resources and tax revenues, many cities have attempted to redevelop themselves as entertainment destinations, devoting enormous public resources to the construction of large entertainment projects, including professional sports stadiums, convention centers, museums, redeveloped riverfronts, festival malls, and casinos and other gaming facilities (Eisinger, 2000; Fainstein & Judd, 1999; Hannigan, 1998; Zukin, 1997; Kearns & Philo, 1993). The justification for spending huge sums of money for engineering economic growth through entertainment lies in the hope that (1) economic revival can bring the middle classes back to the city, not as residents but as visitors and spenders; and (2) the benefits of urban entertainment facilities will "trickle down" into the local economy, generating ancillary investment, high employment in the hospitality and retail sectors, and bring in needed tax revenue. As noted most recently by Peter Eisinger (2000), while the expenditure of local public resources for entertainment is not entirely new, the current pattern of local government spending and investment is very different from earlier eras in U.S. history. First, the pace, scale, cost, and variety of construction have markedly increased since the 1970s with huge investments in casino gaming, sports stadiums, convention centers, and other forms of entertainment. Moreover, unlike previous periods in U.S. history, "in the contemporary period, many local governments have been making large public investments in entertainment facilities at the same time that the municipal tax base is declining and social welfare needs are rising" (p. 321). Lastly, during the nineteenth century local elites designed and built urban recreational facilities for the local population. Today, however, large-scale tourist promoters market entertainment facilities vigorously, to bring visitors to the city. As much research has shown, building cities for the interests of "visitors" rather than the concerns of "residents" translates into a skewed public agenda, declining quality of municipal services for residents, and increasing social division and conflict (Baade, 1996; Euchner, 1993; Sanders, 1992; Hudson, 1999; Rosentraub, 1999; Felsenstein, Littlepage & Klacik, 1999).

In short, since the 1970s, U.S. cities have developed new strategies and tools for engineering urban redevelopment in response to broad economic changes and shifts in the nature of federal intervention. Market-centered initiatives such as enterprise zones and empowerment zones have accompanied the proliferation of new forms of local tax subsidies such as tax abatements, industrial revenue bonds, business improvement districts (BIDS), tax increment financing (TIF), and the rise of community development corporations (CDCs) (Gotham, 1998; Gotham & Wright, 2000; Harrison & Bluestone, 1988; Squires, 1991, 1989; Barnekov & Rich, 1989; Barnekov, Boyle & Rich, 1989; Gaffikin & Warf, 1993; Marcuse, 1993; Cowan, Rohe & Baku, 1999; Gittell & Wilder, 1999). These new strategies and policy tools reflect a wholesale shift in the priorities of local governments, which are increasingly less concerned with issues of social redistribution, provision of public services, and so forth, and the more concerned with promoting economic competitiveness, attracting investment capital, and the creating a favorable "business climate." In the 1980s, Logan and Molotch (1987) and Squires (1989, 1991) showed how the increasing use of tax subsidies and other redevelopment tools to attract capital had weakened democratic processes and safeguards, strengthened the power of private capital in both private and public life, and "depoliticized" local decision-making and policy implementation. In the 1990s, much research has shown that the building of urban entertainment facilities and the traditional public subsidization of the private sector has done little to improve living conditions for the majority of urban dwellers and, in fact, has exacerbated inequality and the fiscal problems of local governments (Felsenstein, Littlepage & Klacik, 1999; Feagin & Parker, 1990; Gotham, 1998; Hudson, 1999; Lord & Price, 1992). As a result, today many local governments find themselves deeply in debt, forced to slash funding for public schools, infrastructure, and other public services while financially pressured to expend greater funds to leverage capital investment.

URBAN REDEVELOPMENT IN THE TWENTY-FIRST CENTURY

The contributors to this volume of *Research in Urban Sociology* document the diverse urban redevelopment strategies, different land-use techniques, and novel market-based funding tools cities are using to redevelop themselves. There is much diversity in the ways cities attempt to engineer economic growth and the key actors and organized interests involved in planning and redeveloping the city have changed considerably over the last two decades. Fainstein (1991, p. 79) notes that as urban planners have become more directly involved in economic development, "market rationality and local competitiveness have replaced comprehensiveness and equity as the primary criteria by which planning projects are judged." Hannigan's (1998) discussion of the "maverick" developers of the new entertainment-based "fantasy city" points to how national- or international-scale corporations today dominate urban redevelopment and have the economic and political power to take their investments elsewhere should local officials not prove compliant, a trend observed by Bennett (1986), Feagin and Parker (1990), Smith (1996), and others. Today, economic shifts on global level, transformations in federal policy, the development of new modes of consumption, and the accompanying processes of privatization and commodification have helped to create a new context for urban redevelopment (Featherstone, 1991; Gottdiener, 1997; Gottdiener, Collins & Dickens, 1999; Judd & Fainstein, 1999; Urry, 1995). The articles in this volume provide illustration of some the key expressions of urban redevelopment, seen from different methods and theory. Collectively, their endeavor highlights the multifaceted relationship between urban redevelopment, public policy, and extra-local forces.

The chapter by James Dickinson, "Monuments of Tomorrow: Industrial Ruins at the Millennium," is concerned with how past structures and buildings – e.g. pyramids, temples, warehouses, and so on – become cherished monuments, relics that add historical depth and continuity to a civilization. As the era of industrial production and concentrated urban life winds down, a new class of structures is appearing in the landscapes of the Europe and the United States: abandoned and derelict factories, warehouses, steel mills, refineries, office buildings, railway stations, and dock facilities, as well as unused prisons, workhouses, asylums, and housing projects. Dickinson suggests that these obsolete commercial and industrial structures of the industrial era are potentially the monuments of tomorrow, contemporary rivals to the antiquities admired in other, older cultures. He uses the ideas of Austrian philosopher and art historian Alois Riegl to identify four possible fates for today's large-scale remnants of the industrial age: (1) demolition and disappearance; (2) recycling into new (primarily commercial) uses; (3) transformation into historical monuments; or (4) persistence in the landscape as conventional ruins. He illustrates this typology with examples of urban and architectural change drawn from the recent history of the United States and Europe. His purpose is to clarify how local people, economic elites, and public officials negotiate and assign new meanings and values to recently antiquated utilitarian structures. Moreover, Dickinson's purpose is to increase our understanding of the role of meaning and discourse in the framing of urban redevelopment and the role of derelict structures in the urban landscapes of the future.

Aldaberto Aguirre and Jonathan Brooks ("City Redevelopment Polices and the Criminalization of Homeless: A Narrative Case Study") examine the relationship between city redevelopment policies and the treatment of the homeless in the city of Riverside, California. They illustrate how elites have used redevelopment policies to dislocate the homeless from the principle business district of the city. In particular, they focus on how local law enforcement has worked alongside with city government to design redevelopment policies that criminalized the homeless. The city's efforts to revitalize the downtown, for example, have resulted in the police department creating new ordinances that prohibit loitering, panhandling, sitting on a park bench, and sitting in grassy park areas. In short, city redevelopment policies and law enforcement procedures have turned public space into contested terrain in which the homeless have become the victims, defined as deviants and impediments to local reform. Thus, local crime control measures impose hierarchical control over the homeless and urban space, transforming public space into privatized space reserved for middle class consumption, a theme echoed by Davis (1992), Judd (1995), Marcuse (1993), Wright (1997), and others who link contemporary urban redevelopment efforts with the militarization of urban space.

The development of new electronic and computer technologies are becoming a major topic of urban research in the United States and elsewhere, especially the role of the Internet in fashioning new (cyber)spaces of community identity and economic growth. For example, in "Innovation, Speculation, and Urban Development: The New Media Brokers of New York City," Michael Indergaard draws attention to how new media districts in major cities are becoming focal points for innovation related to the Internet, drawing in billions of dollars in speculative investment into many cities. Indergaard uses the case of New York's "Silicon Alley" to examine the role of new media districts in organizing speculation and in diffusing its risks across the city. He draws on theories of innovative districts and more general network models to explain how the same ties that aid innovation are now supporting speculation. He proposes that "new media districts" are arenas for various kinds of market brokers: new media entrepreneurs who use their networks to create markets, and venture capitalists who weave new media firms into networks of risk capital. Recently, many entrepreneurs and their venture capitalist allies have shifted their emphasis to positioning themselves vis-à-vis the stock market. Moreover, these networks of speculation are extending into the rest of the urban economy as new media suppliers and other interests accept new media stock as a sort of currency. Thus, while the new media has emerged as the leading producer of new jobs in New York and has helped fuel a business boom, especially in the real estate market, the role of speculation casts doubt on the sustainability of the development achieved.

Greg Crowley ("Regime Structure and the Politics of Issue Definition: Urban Redevelopment in Pittsburgh, Past and Present") examines the politics of urban redevelopment in Pittsburgh to demonstrate the relationship between urban regime structure and issue-definition strategies available to actors for influencing the policy process. His study compares two phases of urban redevelopment in Pittsburgh, each pursued under a different regime structure. At mid-century, a "corporatist" partnership between an elite business community and the Democratic political machine formulated and carried out urban renewal in Pittsburgh. The stability of this public-private alliance depended upon reciprocity, cooperation, and trust between regime partners. Under this corporatist regime, insiders to the regime monopolized the process of policy issue-definition. Regime outsiders, and opponents of urban renewal generally, were only minimally successful at exploiting strategies of issue redefinition to shift the balance of power in their favor. According to Crowley, the corporatist regime in Pittsburgh has gradually become more open and pluralistic than it was at mid-century. With declining trust and predictability among members of the policy community, and with coalitions more likely to form around specific issues rather than long-term interest affiliations, the politics of issue-definition are more important now than ever before in determining policy outcomes. Crowley finds that marginal political actors are increasingly successful at redefining issues and shifting policy debates beyond elite policy circles and onto the public agenda. Thus, the composition of community power has changed since mid-century. Crowley's analysis corroborates the work of Zukin (1997, 1998), Croucher (1997), and Beauregard (1993), among others, who attempt to integrate structural analysis with a social constructionist approach to research the politics of local economic development.

In his widely cited book, the *Condition of Postmodernity*, David Harvey (1989, p. 355) proposed that scholars direct their attention to how the "production of images and discourses is an important facet of activity that has to be analysed as part and parcel of the reproduction and transformation of any social order." In the decade since the publication of Harvey's seminal book, a vast literature has examined the role of imagery, symbols, motifs, and discourse in representing the city. Beauregard's (1993) examination of the "discourse of urban decline," Smith's (1996) focus on the frontier motif and the revanchist city, and Sorkin's and contributors (1992) discussions of the "theme park" city reflect a major trend in urban research toward examining how the production of imagery is integral to the process of urban redevelopment, growth, and decay (see also Holt, 2000). "On Fragmentation, Urban and Social," by Judit Bodnar, is illustrative, concentrating, as it does, on the discourse of "fragmentation" in explanations of urban growth and decline. She suggests that there is a growing anxiety among urban people that the city is becoming increasingly fragmented, uncontrollable, and unmanageable. Her paper points out that, on the one hand, social fragmentation is not altogether new, as the discourse of fragmentation (and related synonyms) has long been a recurring theme of modernity (for an overview, see Berman, 1981). On the other hand, our present millennial concern with urban and social fragmentation is quite different from the advent of modernity, a theme discussed at length by David Harvey (1989) and Steven Best and Douglas Kellner (1997), among others. Here her analysis departs from Georg Simmel's thoughts on fragmentation, individual freedom, and the city, and examines public space as an exemplary instance of the modern city. She goes on to present a critical analysis of today's urban and social fragmentation and highlights empirical and theoretical links between fragmentation and globalization.

Regina Bures examines the relationship between changing patterns of segregation and the development of the historic district in Charleston, South Carolina. Before 1950, people often cited Charleston as an anomaly, a Southern city with a historically low rate of segregation. Yet, this is no longer the case. Between 1950 and 1990, segregation increased significantly within the city at the same time law makers declared segregationist land-use practices were illegal. The establishment and ongoing expansion of the Old Historic District has played an important role in this increase in segregation. Over time, the expansion of Charleston's historic district has led to gentrification, increased property values and taxes, and increased racial and economic segregation. Recently, economic and political support for historic preservation has contributed to the growth of tourism as a major industry in the city.

The relationship between gentrification and urban redevelopment has become a hot topic in the late 1990s. An entire issue of *Research in Urban Sociology* (1992) focused on gentrification and the contributors offered a wealth of insights on the meaning, causes and consequences of gentrification for U.S. cities from the perspective of the late 1980s and early 1990s. As we move into the next century, scholars are again focusing their attention on gentrification asking how much gentrification is there? How much of the city does it cover? Where is it? And, finally, how is it related to macrostructural changes in the state and economy? In their paper, "Gentrification, Housing Policy, and the New Context of Urban Redevelopment," Elvin Wyly and Daniel Hammel argue that the resurgence of gentrification in many cities emanates from recent transformations in federal regulatory policy and mortgage financing. First, they point out that the economic expansion of the last decade has revived inner-city housing markets, abet in a spatially uneven and unequal fashion. More important, they argue that the force, depth, and focus of resurgent capital investment have been sufficient to invalidate the predictions of "degentrification" voiced by many scholars in the early 1990s. Second, they suggest that decades of uneven development have helped create a new socio-spatial context for the reinvention of low-income housing policy. This shift is apparent in many local policies and public-private redevelopment efforts, but is most starkly illustrated by local endeavors under the federal HOPE VI program which, as they point out, attempts to recast the disinvested core as a profit opportunity and a testing ground for the devolved and privatized social policy of the federal government. Third, Wyly and Hammel hypothesize that restructuring of the national system of housing finance has altered key facets of the gentrification process itself, opening new markets for low-income and minority borrowers and neighborhoods, increasing access to conventional mortgage capital through automation and standardization, and accelerating class turnover in cities.

While Wyly and Hammel aim for breadth and generalizability in their analysis, the article by Joel Devine and Petrice Sams-Abiodun, and my article with Jon Shefner and Krista Brumley use ethnographic field observations and in-depth interviews with residents in a single public housing development in New Orleans currently undergoing redevelopment through the HOPE VI program. HOPE VI pioneers a new model of public policy that links the redevelopment of public housing with the twin goals of lessening the concentration of poverty and creating stable inner city neighborhoods. As a major component of the federal government's attempt to "reinvent" public housing authorities (PHAs) and rebuild inner cities, HOPE VI permits expenditures for PHAs to cultivate "public-private partnerships" with local non-profit organizations, private businesses, and other agencies (Quercia & Galster, 1997). In the process, the program encourages PHAs to transform their public housing developments into for-profit developments that are competitive within the broader inner city housing markets (Salama, 1999). Finally, HOPE VI initiatives include a variety of innovative financing mechanisms, involving combinations of Low-Income Housing Tax Credit (LIHTC) funds, private lenders, HUD grants, and local contributions (Epp, 1996). Overall, federal officials have designed the HOPE VI program to induce PHAs to develop creative programs to remedy public housing's problems and become more "entrepreneurial" in attracting private investment to rebuild inner cities.

Devine and Sams-Abiodun use historical, census, survey, ethnographic, and focus group methods to provide a case study of the economic survival strategies employed by residents of the C. J. Peete public housing development (formerly Magnolia). While their motivating research question and ultimate focus is on how residents of this impoverished, inner-city community meet their material needs, they initially explore prevailing conceptions of households and families and scrutinize their applicability concerning a poor, inner-city, African-American population. Devine and Sams-Abiodun then discuss the history of C. J. Peete and interrogate the critical policy decisions that affected its development. Next, they examine tract-level census data and demographic and economic information gathered in a series of community surveys conducted over the past four years. Faced with the constraints imposed by these information gathering techniques, Devine and Sams-Abiodun subsequently undertook an ethnographically-based research strategy and a series of focus group conversations to investigate and more fully appreciate the formal and informal strategies people in a low-income community use to provide for their survival.

In the context of economic deprivation and social isolation, the redevelopment of C. J. Peete has become a potential resource for improving the lives of residents and a harbinger of insecurity and social instability. My article with Jon Shefner and Krista Brumley uses Henri Lefebvre's (1991) conceptual tools of "abstract space" and "social space" to critically analyze the dilemmas, and conflicts surrounding the relocation of public housing residents. We argue that the redevelopment of public housing is an attempt to recommodify public housing space, to increase its exchange-value through privatization while diminishing its use-value for the inner city poor. The data gathered by Wyly and Hammel show that the derelict spaces surrounding many inner city public housing developments are currently being selectively re-valorized as a resurgence of gentrification, in some cities, has created "islands of decay in seas of renewal." Yet this process is fraught with major inconsistencies and difficulties. Our ethnographic field notes of public meetings held over the last two years indicate substantial resident opposition to the redevelopment with residents condemning the housing authority for being insensitive to their concerns, disseminating inaccurate and conflicting information, and refusing to work with them to find affordable housing. We investigate how housing authority officials and other local elites use various rhetorical devices, imagery, and motifs to define public housing as an undesirable area of overcrowding and social pathology and convince the public and housing residents of the necessity of redevelopment. The dominant discourse of urban redevelopment views public housing as a symbol of urban "decline" (Beauregard, 1993), a "failure" of federal policy, and a rampant ghetto of crime, drugs, and violence. This vision is a key to understanding the cultural and racial politics that supports the redevelopment effort and, at the same time, discourages alternative views. Following Lefebvre's insights, we wish to understand the conflicts between public housing residents and housing authority officials over the redesign of public housing, broad policy shifts involving public housing space, and the role of the state in using space for social control.

The article by Jamie Paquin, "'World City' Theory: The Case of Seoul," examines the global dimensions of urban redevelopment by applying world city theory to examine the contemporary economic restructuring of Seoul, South Korea. As elaborated by King (1990), Knox and Taylor (1995), and Sassen (1991, 1998), world city theory suggests that the processes driving urban change are global rather than national or local. A related argument is that a hierarchy of cities performing different functions within a global economy has been emerging in recent decades. Cities such as London, Toyko, and New York have become the major command centers for the global economy and are considered the preeminent "world" cities. Other cities struggle to achieve world city status by redeveloping their economic bases, others face stagnation and decay, and still others languish in chronic misery and despair. Paquin points to the significance of globalization, the problems scholars have with theorizing global processes, and the conceptual and empirical links between global processes and local redevelopment. His intent is to highlight how a country outside the United States and Europe experiences "globalization" and to contribute to a more comprehensive understanding of urban economic change and its consequences.

Reflecting the increasing scholarly focus on the redevelopment of cities as sites of fun, leisure, and entertainment, Richard Lloyd and Terry Nichols Clark argue in the "City as an Entertainment Machine," that cities are now in the business of leveraging culture to enhance their economic competitiveness. The components of the entertainment machine heuristic are not altogether new, as cities have long been sites of entertainment, consumption, and aesthetic innovation. In "Paris: Capital of the Nineteenth Century," Walter Benjamin (1978) discussed how "art is brought in to the service of commerce" with the emergence of a capitalist entertainment economy as displayed in the World Exhibition of 1867 in Paris. Both Thorsten Veblen and Georg Simmel discussed the intersection of culture and commerce, focusing on the commodification of fashion, art, leisure, in early twentieth century Europe. The Frankfurt School theorists discussed the powerful impact of the "culture industry" in the production and transmission of capitalist ideology and the massification of society. In the 1980s, both Zukin (1981) and Whitt (1987) noted the emerging importance of the arts industry and related art-oriented growth coalitions as an element of the urban growth machine. Today, urban scholars have identified ethnicity, culture, and historic preservation as important elements of a new postmodern (Harvey, 1989) or symbolic (Zukin, 1995) economy in which cultural strategies drive the production of commercialized urban spaces geared toward entertainment and tourism (Bassett, 1993; Boyer, 1992; Hannigan, 1998; Lin, 1995, 1998; Reichl, 1997; Strom, 1999; Urry, 1995; Whitt, 1988).

Lloyd and Clark follow the lead of John Walton (1993) and Todd Swanstrom (1993) in rejecting the conceptual and analytical separation of "culture" and the "economy" in urban theory and research. Their paper points to the growing importance of aesthetic production and consumption in redeveloping urban spaces as sites of consumption and entertainment. Echoing Zukin (1997) and Harvey (1989), Lloyd and Clark suggest that the business of cities has become the business of marketing culture and entertaining consumers, both residents and tourists (see also Wright and Hutchison 1997). They link the city as an "entertainment machine" to post-industrial occupational categories, and the competition among cities for tourists and expert labor.

In recent years, scholars have begun to examine the profound socio-spatial transformation of Times Square in New York City over the last two decades. Bart Eeckhout ("The 'Disneyification' of Times Square: Back to the Future?") builds upon the insights of Hannigan (1998), Reichl (1999) and other scholars to investigate the conversion of Times Square into themed and revitalized space of hype, advertising, and cultural consumption. Eeckhout's purpose is to interrogate prevailing explanations of the dramatic changes that have affected Times Square, critically assess the analytical merit of the concept "Disneyification," and finally, suggest future avenues for urban research. In particular, Eeckhout provides a historical overview of Times Square and identifies seven core components or meanings of "Disneyification" that currently frame urban research on the transformation of urban public spaces into commodified spaces. His chapter brings together a number of recent themes developed within the urban literature on the arts, entertainment, and urban tourism: the construction of themed fantasy spaces geared toward pleasure and consumption, the transition to a postindustrial urban economy, the hegemony of consumerism and globalized tourism, the commodification of history and culture in preservation- and arts-based redevelopment, and the privatization and militarization of urban space. For Eeckhout, a "back-to-the-future" ideology is driving the redevelopment of Times Square where economic and political elites are superimposing images of a nostalgic small town Main Street – images developed by Walt Disney in the 1950s – upon the cityscape to foster urban redevelopment primarily for urban tourists, a process that is occurring in many cities today as recently noted by Sharon Zukin, John Hannigan, David Harvey, and others.

As should be clear from these brief summaries, the aim of this collection is to highlight ambiguities, inconsistencies, and contradictions in the nature, meaning, and consequences of urban redevelopment. The contributors seek not to be definite about urban redevelopment but rather the explore the complexity of the major issues. Hence this edited volume brings together some of the latest ideas and research on the topic. The aim is to present to the reader a range of perspectives on the various aspects of urban redevelopment set in the context of the evolving political economy and uneven development.

NOTES

1. The major outlines of the 1949 and 1954 Housing Acts appeared at least as early as 1941 in the NAREB's plans and reports for developing the means of teaming state action and private enterprise to carry out large-scale clearance of slum areas (Davies, 1958, pp. 182–185; Weiss, 1980). As early as 1932, the NAREB and its affiliated organizations called for government assistance to simplify and coordinate local building codes to promote new building and urban revitalization. In 1936, the NAREB's Committee on Housing recommended that local government acquire land, undertake demolition, and sell or lease it back to private enterprise for the construction of housing or business facilities. By the time the bills that were to become the Housing Acts of 1949 and 1954 had reached Congress, leading officials within the real estate industry had already set the basic agenda and legislation. There were disagreements over administrative issues but not basic policy goals (Gotham, 2000a).

2. Title II of the 1949 Housing Act raised by $500 million the amount the Federal Housing Administration (FHA) was allowed to offer as mortgage insurance. Title III authorized the federal government to build 810,000 new public housing units over the next ten years and required local public housing authorities demolish or renovate one slum dwelling for every public housing unit they built (Hoffman, 2000). Due to opposition from the real estate industry and conservative members of Congress, public housing never came close to the construction levels provided in the 1949 Act (810,000 units) (Gotham & Wright, 2000). The abandonment of the commitment to 810,000 units of public housing signaled a movement away from urban renewal as a "housing" program. By 1960, only 250,000 units had been made available and by 1979 only about 1 million total units had been built across the nation (Mitchell, 1985, pp. 9–11).

REFERENCES

Abrams, C. (1971). *The City is the Frontier*. New York: Harper and Row.

Adams, C., Bartelt, D., Elesh, D., Goldstein, I., Kleniewski, N., & Yancey, W. (1991). *Philadelphia: Neighborhoods, Division, and Conflict in a Postindustrial City*. Philadelphia: Temple University Press.

Anderson, M. (1964). *The Federal Bulldozer: A Critical Analysis of Urban Renewal, 1949–1962*. Cambridge: MIT Press.

Baade, R. A. (1996). Professional Sports and Economic Development. *Journal of Urban Affairs*, *18*(1), 1–18.

Barnekov, T., & Rich, D. (1989). Privatism and the Limits of Local Economic Policy. *Urban Affairs Quarterly*, *25*(2), 212–238.

Barnekov, T., Boyle, R., & Rich, D. (1989). *Privatism and Urban Policy in Britain and the United States.* New York: Oxford University Press.

Bartelt, D. (1993). Housing the 'Underclass.' In: M. B. Katz (Ed.), *The Underclass Debate: Views from History* (pp. 118–157). Princeton: Princeton University Press.

Bassett, K. (1993). Urban Cultural Strategies and Urban Regeneration: A Case Study and Critique. *Environment and Planning A, 25,* 1773–1788.

Baum, H. S. (1999). Education and the Empowerment: Ad Hoc Development of an Interorganizaitonal Domain. *Journal of Urban Affairs, 21*(3).

Bauman, J. F. (1981). Visions of a Postwar City: A Perspective on Urban Planning in Philadelphia and the Nation, 1942–1945. *Urbanism Past and Present, 6*(11), Winter/Spring 1980–1, 1–11.

Bauman, J. F. (1987). *Public Housing, Race, and Renewal: Urban Planning in Philadelphia, 1920–1974.* Philadelphia: Temple University Press.

Bayor, R. H. (1989). Urban Renewal, Public Housing, and the Racial Shaping of Atlanta. *Journal of Policy History, 1*(4), 419–439.

Beauregard, R. A. (1993). *Voices of Decline: The Postwar Fate of U.S. Cities.* Blackwell.

Benjamin, W. (1978). *Reflections, Essays, Aphorisms, Autobiographical Writing.* New York: Harcourt Brace and Company.

Bennett, L. (1986). Beyond Urban Renewal: Chicago's North Loop Redevelopment Project. *Urban Affairs Quarterly, 22*(2), December 1986, 242–260.

Bennett, L., McCourt, K., Nyden, P. W., & Squires, G. D. (1988). Chicago's North Loop Redevelopment Project: A Growth Machine on Hold. In: S. Cummings (Ed.), *Business Elites and Urban Development: Case Studies and Critical Perspectives* (pp. 183–202). Albany: State University of New York Press.

Berman, M. (1988). *All That Is Solid Melts into Air: The Experience of Modernity.* (2nd ed.). New York: Penguin Books.

Best, S., & Kellner, D. (1997). *The Postmodern Turn.* New York: Guilford.

Blair, J. P., & Nachmias, D. (1979) (Eds). *Fiscal Retrenchment and Urban Policy.* Beverly Hills, CA: Sage.

Block, F., Cloward, R. A., Ehrenreich, B., & Fox Piven, F. (Eds). (1987). *The Mean Season: The Attack on the Welfare State.* New York: Pantheon Books.

Bockmeyer, J. L. (2000). The Culture of Distrust: The Impact of Local Political Culture on Participation in the Detroit EZ. *Urban Studies, 37*(13), 2417–2440.

Boyer, C. (1992). Cities for Sale: Merchandising History at South Street Seaport. In: M. Sorkin (Ed.), *Variations on a Theme Park* (pp. 181–204). New York: Hill and Wang.

Bratt, R. G. (1986). Public Housing: The Controversy and Contribution. In: R. G. Bratt, C. Hartman & A. Meyerson (Eds), *Critical Perspectives on Housing* (pp. 362–377). Philadelphia: Temple University Press.

Burgess, E. (1925). The Growth of the City: An Introduction to a Research Project. In: R. Park, E. Burgess & R. McKenzie (Eds), *The City.* Chicago: University of Chicago Press.

Castells, M. (1977). *The Urban Question.* Cambridge, MA: MIT Press.

Checkoway, B. (1984). Large Builders, Federal Housing Programs and Postwar Suburbanization. In: W. K. Tabb & L. Sawyers (Eds), *Marxism and the Metropolis: New Perspectives in Urban Political Economy* (2nd ed., pp. 152–173). New York: Oxford University Press.

Cole, R. L., Taebel, D. A., & Hissong, R. V. (1990). America's Cities and the 1980s: The Legacy of the Reagan Years. *Journal of Urban Affairs, 12*(4).

Cowan, S. M., Rohe, W., & Baku, E. (1999). Factors Influencing the Performance of Community Development Corporations. *Journal of Urban Affairs, 21*(3).

Croucher, S. L. (1997). Constructing the Image of Ethnic Harmony in Toronto, Canada: The Politics of Problem Definition and Nondefinition. *Urban Affairs Review, 32*(3), Jan.

Cummings, S. (1988) (Ed.). *Business Elites and Urban Development: Case Studies and Critical Perspectives.* Albany: State University of New York Press.

Darden, J. T., Hill, R. C., Thomas, J., & Thomas, R. (1987). *Detroit: Race and Uneven Development.* Philadelphia: Temple University Press.

Davies, P. J. (1958). *Real Estate in American History.* Washington, DC: Public Affairs Press.

Davis, M. (1992). *City of Quartz: Excavating the Future in Los Angeles.* New York: Vintage Books.

Dear, M., & Scott, A. (Eds) (1981). *Urbanization and Urban Planning in Capitalist Society.* New York: Methuen.

Downs, A. (1975). Using the Lessons of Experience to Allocate Resources in the Community Development Program. In: *Recommendations for Community Development Planning: Proceedings of the HUD/RERC Workshops on Local Urban Renewal and Neighborhood Preservation,* 1–28. Chicago: Real Estate Research Corporation.

Eisinger, P. (2000). The Politics of Bread and Circuses: Building the City for the Visitor Class. *Urban Affairs Review, 35*(3), January, 316–333.

Engels, F. (1967). *The Condition of the Working Class in England: From Personal Observations and Authentic Sources.* Moscow: Progress Publishers.

Epp, G. (1996). Emerging Strategies for Revitalizing Public Housing Communities. *Housing Policy Debate, 7*(32), 563–588.

Euchner, C. C. (1993). *Playing the Field: Why Sports Teams Move and Cities Fight to Keep Them.* Baltimore: Johns Hopkins University Press.

Fainstein, S. (1991). Promoting Economic Development: Urban Planning in the United States and Great Britain. *Journal of the American Planning Association, 57,* 22–33.

Fainstein, S. S., & Fainstein, N. I. (1986). Economic Change, National Policy, and the System of Cities. In: S. S. Fainstein, N. I. Fainstein, R. C. Hill, D. R. Judd & M. P. Smith (Eds), *Restructuring the City: The Political Economy of Urban Redevelopment* (2nd ed., pp. 1–26). New York: Longman.

Feagin, J. R., & Parker, R. (1990). *Building American Cities: The Urban Real Estate Game* (2nd ed.). Englewood Cliffs, NJ: Prentice Hall.

Feagin J. R. (1998). *The New Urban Paradigm: Critical Perspectives on the City.* New York: Rowman and Littlefield.

Featherstone, M. (1991). *Consumer Culture and Postmodernism.* London: Sage.

Felsenstein, D., Littlepage, L., & Klacik, D. (1999). Casino Gambling as Local Growth Generation: Playing the Economic Development Game in the Reverse. *Journal of Urban Affairs, 21*(4), 409–421.

Friedland, R. (1982). *Power and Crisis in the City.* London: Macmillan

Gaffikin, F., & Warf, B. (1993). Urban Policy and the Post-Keynesian State in the United Kingdom and the United States. *International Journal of Urban and Regional Research, 17*(1), 67–84.

Gans, H. J. (1962). *The Urban Villagers: Group and Class in the Life of Italian-Americans.* New York: Free Press.

Garreau, J. (1991). *Edge City: Life on the New Frontier.* New York: Anchor Books.

Gittell, R., & Wilder, M. (1999). Community Development Corporations: Critical Factors that Influence Success. *Journal of Urban Affairs, 21*(3).

Glickman, N. (1984). Economic Policy and the Cities: In Search of Reagan's Real Urban Policy. *Journal of the American Planning Association. 50,* 471–478.

Gordon, D. (1984). Capitalist Development and the History of American Cities. In: W. K. Tabb & L. Sawyers (Eds), *Marxism and the Metropolis: New Perspectives in Urban Political Economy* (2nd ed., pp. 21–53). New York: Oxford University Press.

Gotham, K. F. (1998). Blind Faith in the Free Market: Urban Poverty, Residential Segregation, and Federal Housing Retrenchment, 1970–1995. *Sociological Inquiry, 68*(1), Winter.

Gotham, K. F. (1999). Political Opportunity, Community Identity, and the Emergence of a Local Anti-Expressway Movement. *Social Problems, 46*(3). August.

Gotham, K. F. (2000). Urban Space, Restrictive Covenants, and the Origin of Racial Residential Segregation in a U.S. City, 1900–1950. *International Journal of Urban and Regional Research, 24*(3), 616–633, September.

Gotham, K. F. (2001). A City Without Slums: Urban Renewal, Public Housing, and Downtown Revitalization in Kansas City, Missouri. Forthcoming in *American Journal of Economics and Sociology, 60*(1), January.

Gotham, K. F., & Staples, W. G. (1996). Narrative Analysis and the New Historical Sociology. *Sociological Quarterly, 37*(3), 481–501.

Gotham, K. F., & Wright, J. D. (2000). Housing Policy. In: J. Midgley, M. Tracy & M. Livermore (Eds), *Handbook of Social Policy*, Chapter 15. Sage Publications.

Gottdiener, M. (1994). *The Social Production of Urban Space* (2nd ed.). Austin: University of Texas Press.

Gottdiener, M., Collins, C. C., & Dickens, D. (1999). *Las Vegas: The Social Production of an All-American City*. New York: Blackwell.

Gottdiener, M., & Feagin, J. R. (1988). The Paradigm Shift in Urban Sociology. *Urban Affairs Quarterly, 24*(2), December, 163–187.

Gottdiener, M. (1997). *Theming of America: Dreams, Visions, and Commercial Spaces*. Boulder, CO: Westview Press.

Greer, S. (1965). *Urban Renewal and American Cities: The Dilemma of Democratic Intervention*. New York: Bobbs-Merril.

Hannigan, J. (1998). *Fantasy City: Pleasure and Profit in the Postmodern Metropolis*. New York: Routledge.

Harrison, B., & Bluestone, B. (1988). *The Great U-Turn*. Basic Books.

Hartman, C., & Kessler, R. (1978). The Illusion and Reality of Urban Renewal: San Francisco's Yerba Buena Center. In: W. K. Tabb & L. Sawyers (Eds), *Marxism and the Metropolis: New Perspectives in Urban Political Economy* (1st ed., pp. 153–178). New York: Oxford University Press.

Harvey, D. (1973). *Social Justice and the City*. Baltimore: Johns Hopkins University Press.

Harvey, D. (1989). *The Condition of Postmodernity*. New York: Free Press.

Hays, R. A.. (1985). *Federal Government and Urban Housing: Ideology and Change in Public Policy*. State University of New York Press.

Herring, C., Bennett, M., Gills, D., & Jenkins, N. T. (Eds) (1998). *Empowerment in Chicago: Grassroots Participation in Economic Development and Poverty Alleviation*. Urbana and Chicago: University of Illinois Press.

Hirsch, A. (1993). With or Without Jim Crow: Black Residential Segregation in the United States. In: H. Arnold & R. H. Mohl (Eds), *Urban Policy in Twentieth Century America*. Rutgers University Press.

Hirsch, A. R. (1983). *Making the Second Ghetto: Race and Housing in Chicago, 1940–1960*. Cambridge: Cambridge University Press.

Hoffman, A. von. (2000). A Study in Contradictions: The Origins and Legacy of the Housing Act of 1949. *Housing Policy Debate, 11*(2), 299–326.

Holt, W. G. (2000). Distinguishing Metropolises: The Production of Urban Imagery. In: R. Hutchison (Ed.), *Constructions of Urban Space, Vol. 5 Research in Urban Sociology* (pp. 225–252). JAI Press.

Hubbell, L. K. (1979). *Fiscal Crisis in American Cities: The Federal Response.* Cambridge, MA: Ballinger.

Hudson, I. (1999). Bright Lights, Big City: Do Professional Sports Teams Increase Employment? *Journal of Urban Affairs, 21*(4), 397–407.

Hutchison, R. (1993). The Crisis in Urban Sociology. *Urban Sociology in Transition. Research in Urban Sociology, 3,* 3–26. JAI Press.

Hutchison, R. (2000). Introduction to Constructions of Urban Space. In: R. Hutchison (Ed.), *Constructions of Urban Space Research in Urban Sociology* (pp. ix-xvii). Stamford, CT: JAI Press.

Jackson, K. T. (1985). *Crabgrass Frontier: The Suburbanization of the United States.* New York: Oxford University Press.

Jacobs, J. (1961). *The Death and Life of Great American Cities.* New York: Random House.

Jaret, C. (1983). Recent Neo-Marxist Urban Analysis. *Annual Review of Sociology, 9,* 499–525.

Judd, D., & Fainstein, S. S. (Eds) (1999). *The Tourist City.* New Haven: Yale University Press.

Judd, D. R. (1995). The Rise of the New Walled Cities. In: H. Liggett & D. C. Perry (Eds), *Spatial Practices: Critical Explorations in Social/Spatial Theory* (pp. 144–166). Thousand Oaks, CA: Sage.

Judd, D. R., & Swanstrom, T. (1998). *City Politics: Private Power and Public Policy* (2nd ed.). New York: Longman.

Katz, M. (1986). *In the Shadow of the Poorhouse: A Social History of Welfare in America.* New York: Basic Books.

Kearns, G., & Philo, C. (1993). *Selling Places: The City as Cultural Capital, Past and Present.* Oxford: Pergamon Press.

Kincaid, J. L. (1999). De Facto Devolution and Urban Defunding: The Priority of Persons Over Places. *Journal of Urban Affairs, 21*(2).

King, A. (Ed.) (1996). *Re-Presenting the City: Ethnicity, Capital, and Culture in the 21st Century Metropolis.* NYU Press.

King, A. D. (1990). *Global Cities: Post-Imperialism and the Internationalization of London.* London: Routledge.

Kleniewski, N. (1984). From Industrial to Corporate City: The Role of Urban Renewal. In: W. K. Tabb & L. Sawyers (Eds), *Marxism and the Metropolis: New Perspectives in Urban Political Economy* (2nd ed., pp. 205–222). New York: Oxford University Press.

Knox, P. L., & Taylor, P. J. (1995). *World Cities in a World-System.* Cambridge: Cambridge University Press.

Ladd, H. F., & Yinger, J. (Eds) (1989). *America's Ailing Cities: Fiscal Health and the Design of Urban Policy.* Baltimore: Johns Hopkins University Press.

Lefebvre, H. (1991). *The Production of Space.* Oxford: Basil Blackwell.

Lin, J. (1995). Ethnic Places, Postmodernism, and Urban Change in Houston. *Sociological Quarterly, 36*(4), 629–647.

Lin, J. (1998). Globalization and the Revalorization of Ethnic Places in Immigrant Gateway Cities. *Urban Affairs Review, 34*(2). November, 313–339.

Logan, J., & Molotch, H. (1987). *Urban Fortunes: The Political Economy of Place.* Berkeley: University of California Press.

Logan, J. R., & Swanstrom, T. (Eds) (1990). *Beyond the City Limits: Urban Policy and Economic Restructuring in Comparative Perspective.* Philadelphia: Temple University Press.

Lord, G., & Price, A. C. (1992). Growth Ideology in a Period of Decline: Deindustrialization and Restructuring, Flint Style. *Social Problems, 39*(2), May, 155–169.

Marcuse, P. (1993). What's So New About Divided Cities? *International Journal of Urban and Regional Research, 17*(3), September, 355–365.

Marcuse, P. (1986). The Beginning of Public Housing in New York. *Journal of Urban History, 12*(4), August, 353–390.

McDonnell, T. (1957). *The Wagner Housing Act: A Case Study of the Legislative Process.* Chicago: Loyola University Press.

Metzger, J. T. (2000). Planned Abandonment: The Neighborhood Life-Cycle Theory and National Urban Policy. *Housing Policy Debate, 11*(1).

Mitchell, J. P. (Ed.) (1985). *Federal Housing Programs: Past and Present.* Rutgers: State University of New Jersey Press.

Mohl, R. (1993). Shifting Patterns of American Urban Policy Since 1900. In: A. Hirsch & R. H. Mohl (Eds), *Urban Policy in Twentieth Century America.* Rutgers University Press, New Brunswick, NJ.

Mollenkopf, J. H. (1983). *The Contested City.* Princeton University Press.

Mollenkopf, J. H. (1978). The Postwar Politics of Urban Development. In: W. Tabb & L. Sawers. *Marxism and Metropolis* (1st ed., pp. 117–152). New York: Oxford University Press.

Molotch, H. (1976). The City as a Growth Machine: Toward a Political Economy of Place. *American Journal of Sociology, 82,* 309–330.

Molotch, H. (1999). Growth Machine Links: Up, Down, and Across. In: A. E. G. Jonas & D. Wilson (Eds), *The Urban Growth Machine: Critical Perspectives Two Decades Later* (pp. 247–266). Albany: State University of New York (SUNY) Press.

National Commission on Urban Problems (1969). *Building the American City. Report of the National Commission on Urban Problems.* Paul A. Douglas Chairman. New York: Frederick A. Praedger Publishers.

Orum, A. (1995). *City Building in America.* Boulder, CO: Westview Press.

Park, R. (1925). The City: Suggestions for the Investigation of Human Behavior in the Urban Environment. In: R. E. Park, E. Burgess & R. D. McKenzie (Eds), *The City* (pp. 1–46). Chicago: University of Chicago Press.

Peterson, P. (Ed.) (1981). *The New Urban Reality.* Washington, DC: Brookings Institute.

Piven, F. F., & Cloward, R. (1974). *The Politics of Turmoil.* New York, NY: Vintage Books.

Piven, F. F., & Cloward, R. (1982). *The New Class War: Reagan's Attack on the Welfare State and Its Consequences.* New York: Pantheon.

President's Commission for a National Agenda for the Eighties (1980). *Urban American in the Eighties.* Washington, DC: Government Printing Office.

Quercia, R. G., & Galster, G. (1997). The Challenges Facing Public Housing Authorities in a Brave New World. *Housing Policy Debate, 8*(3), 535–569.

Reddig, W. M. (1947). *Tom's Town: Kansas City and the Pendergast Legend.* Columbia, MO: University of Missouri Press.

Reichl, A. (1997). Historic Preservation and Progrowth Politics in U.S. Cities. *Urban Affairs Review, 32*(4), March, 513–535.

Rosentraub, M. S. (1999). Are Public Policies Needed to Level the Playing Field Between Cities and Teams? *Journal of Urban Affairs, 21*(4).

Rosenthal, D. (Ed.) (1980). Urban Revitalization. *Urban Affairs Annual Reviews,* 18. Sage Publications.

Salama, J. J. (1999). The Redevelopment of Distressed Public Housing: Early Results from HOPE VI Projects in Atlanta, Chicago, and San Antonio. *Housing Policy Debate, 10*(1).

Sanders, H. T. (1980). Urban Renewal and the Revitalized City: A Consideration of Recent History. In: D. Rosenthal (Ed.), *Urban Revitalization* (pp. 103–126). Urban Affairs Annual Reviews. Vol. 18. Sage Publications.

Sanders, H. T. (1992). Building the Convention City: Politics, Finance, and Public Investment in Urban America. *Journal of Urban Affairs, 14*, 135–159.

Sassen, S. (1991). *The Global City: New York, London, and Tokyo*. Princeton, NJ: Princeton University Press.

Sassen, S. (1998). *Globalization and Its Discontents: Essays on the New Mobility of People and Money*. New York: The New Press.

Saunders, P. (1981). *Social Theory and the Urban Question*. London: Hutchinson.

Savage, M., & Ward, A. (1993). *Urban Sociology, Capitalism, and Modernity*. New York: Continuum.

Sies, M. C., & Silver, C. (Eds) (1996). *Planning the Twentieth Century American City*. Baltimore: Johns Hopkins University Press.

Silver, C. (1984). *Twentieth Century Richmond: Planning, Politics, and Race*. Knoxville: University of Tennessee Press.

Smith, D. A. (1995). The New Urban Sociology Meets the Old: Rereading Some Classical Urban Ecology. *Urban Affairs Review, 30*(3), January, 432–457.

Smith, N. (1996). *The New Urban Frontier: Gentrification and the Revanchist City*. New York: Routledge.

Sorkin, M. (Ed.) (1992). *Variations on a Theme Park: The New American City and the End of Public Space*. New York: Hill and Wang.

Squires, G. D., Bennett, L., McCort, K., & Nyden, P. (1987). *Chicago: Race, Class, and the Response to Urban Decline*. Philadelphia: Temple University Press.

Squires, G. D. (Ed.) (1989). *Unequal Partnerships: The Political Economy of Urban Redevelopment in Postwar America*. New Brunswick: Rutgers University Press.

Squires, G. D. (1991). Partnership and Pursuit of the Private City. In: M. Gottdiener & C. Pickvance (Eds), *Urban Life in Transition, 39*. Urban Affairs Annual Reviews (pp. 196–211). Newbury Park: Sage Publications.

Squires, G. D. (1994). *Capital and Communities in Black and White: The Intersections of Race, Class, and Uneven Development*. State University of New York Press.

Stone, C. N., & Sanders, H. T. (Eds) (1987). *The Politics of Urban Development*. Lawrence, KS: University Press of Kansas.

Storper, M., & Walker, R. (1983). The Theory of Labor and the Theory of Location. *International Journal of Urban and Regional Research, 7*, 1–41.

Strom, E. (1999). Let's Put on a Show! Performing Acts and Urban Revitalization in Newark, New Jersey. *Journal of Urban Affairs, 21*(4), 423–435.

Sugrue, T. J. (1996). *The Origins of the Urban Crisis: Race and Inequality in Postwar Detroit*. Princeton University Press, Princeton, NJ.

Swanstrom, T. (1993). Beyond Economism: Urban Political Economy and the Postmodern Challenge. *Journal of Urban Affairs, 15*(1), 55–78.

Tabb, W. D., & Sawyers, L. (Eds) (1984). *Marxism and the Metropolis: New Perspectives in Urban Political Economy* (2nd ed.). New York: Oxford University Press.

Taylor, H. L. (Ed.) (1993). *Race and the City: Work, Community, and Protest in Cincinnati, 1820–1970*. Chicago: University of Illinois Press.

Teaford, J. C. (1990). *The Rough Road to Renaissance: Urban Revitalization in America, 1940–1985*. Baltimore: Johns Hopkins University Press.

Teaford, J. C. (2000). Urban Renewal and its Aftermath. *Housing Policy Debate, 11*(2), 443–466.

Urban Land Institute (1940). *Decentralization: What is it Doing to Our Cities?* New York, NY: Urban Land Institute.

Urry, J. (1995). *Consuming Places.* London and New York: Routledge.

Walton, J. (1993). Urban Sociology: The Contribution and Limits of Political Economy. *Annual Review of Sociology, 19,* 301–320.

Weiss, M. A. (1987). *Rise of the Community Builders: The American Real Estate Industry and Urban Land Planning.* New York: Columbia University Press.

Weiss, M. A. (1980). The Origins and Legacy of Urban Renewal. In: P. Clavel, J. Forester & W. W. Goldsmith (Eds), *Urban and Regional Planning in an Age of Austerity* (pp. 53–79). New York: Pergamon Press.

Whitt, J. A. (1987). The Arts Coalition in Strategies of Urban Development. In: C. N. Stone & H. T. Sanders (Eds), *The Politics of Urban Development* (pp. 144–158). Lawrence, KS: University Press of Kansas.

Whitt, J. A. (1988). The Role of the Performing Arts in Urban Competition and Growth. In: S. Cummings (Ed.), *Business Elites and Urban Development: Case Studies and Critical Perspectives* (pp. 49–70). Albany, NY: State University of New York.

Wilson, W. H. (1989). *The City Beautiful Movement.* Baltimore: Johns Hopkins University Press.

Wilson, J. Q. (1966). (Ed.). *Urban Renewal: The Record and the Controversy.* M. I. T. Press.

Wirth, L. (1938). Urbanism as a Way of Life. *American Journal of Sociology, 44*(July), 1–38.

Wolman, H. (1986). The Reagan Urban Policy and Its Impacts. *Urban Affairs Quarterly, 21*(3), 311–335.

Wright, T., & Hutchison, R. (1997). Socio-Spatial Reproduction, Marketing Culture, and the Built Environment. In: R. Hutchison (Ed.), *New Directions in Urban Sociology,* Vol. 4 Research in Urban Sociology (pp. 187–214). Greenwhich, CN: JAI Press.

Wright, T. (1997). *Out of Place: Homeless Mobilizations, Subcities, and Contested Landscapes.* Albany, NY: State University of New Press.

Zukin, S. (1981). *Loft Living.* New Brunswick: Rutgers University Press.

Zukin, S. (1997). Cultural Strategies of Economic Development and the Hegemony of Vision. In: A. Merrifield & E. Swyngedouw (Eds), *Urbanization of Injustice* (pp. 233–242). New York: New York University Press.

Zukin, S. (1995). *The Cultures of Cities.* Cambridge, MA: Blackwell.

MONUMENTS OF TOMORROW: INDUSTRIAL RUINS AT THE MILLENNIUM

James Dickinson

> I saw huge buildings rise up faint and fair,
> and pass like dreams.
> H. G. Wells, *The Time Machine* (1895)

> The shape of a city changes more quickly, alas!
> than the heart of a man.
> Baudelaire, *The Flowers of Evil* (1857)

> Look on my works, ye Mighty, and despair!
> Nothing beside remains. Round the decay
> Of that colossal wreck, boundless and bare
> The lone and level sands stretch far away.
> Shelley, "Ozymandias" (1817)

INTRODUCTION

In any culture, structures from the past commingle with the contemporary and the new. Some survivals from the past – pyramids, henges, temples – become monuments, relics that add historical depth and continuity to the civilization. Now, however, as the era of accumulation based on industrial production and concentrated urban life winds down, a new class of structures is appearing in the landscapes of the United States, Europe, and elsewhere: abandoned and derelict factories, warehouses, steel mills, refineries, office buildings, railway stations, and dock facilities, as well as unused prisons, workhouses, asylums, and housing projects. What is the fate of these remnants of the industrial age? Can these giant relics of the immediate past become the monuments of tomorrow, authentic rivals to the royal tombs, medieval castles, and Gothic cathedrals of other civilizations and earlier times?

In earlier articles I considered how new meanings are negotiated for old structures in the entropic zones of the contemporary industrial city (Dickinson, 1996, 1999b). Here I look more systematically at industrial ruins, identifying attributes and contextual factors that account for the varied fates they experience today. Lacking the imposing, timeworn monuments found in other cultures, the United States has based its national identity on appreciation of natural wonders such as the Grand Canyon, Monument Valley, and Niagara Falls, or on engagement with what David Nye (1994, p. 43) calls the "technological sublime", in which the experience of technology rather than nature or history "transformed the individual's experience of immensity and awe into a belief in national greatness". I suggest here that America's obsolete, abandoned, and increasingly derelict commercial and industrial structures are potentially the monuments of tomorrow, contemporary rivals to the antiquities admired in other, older cultures.

To begin, I briefly discuss how European modernists appreciated industrial structures, particularly those erected in the United States, as elementary forms of an appropriately modern, up-to-date aesthetic, and as essential expressions of the potentialities of the new age. I then review ideas about entropy and architecture put forward by Austrian philosopher and art historian Alois Riegl, especially his understanding of historical change as a process in which structures undergo continual decay and thus become the ruins and monuments of each successive age. In the main part of the paper I apply Riegl's theory of monuments to an interpretation of industrial ruins. I identify four possible fates for today's large-scale remnants of the industrial age: Demolition and disappearance; recycling into new (primarily commercial) uses; transformation into historical monuments; or persistence in the landscape as conventional ruins. I illustrate this typology with examples of urban and architectural change drawn from the recent history of the United States and Europe. My purpose is to clarify how new meanings and values are negotiated for recently obsolete, derelict, and abandoned utilitarian structures, and to increase the understanding of their possible role in the landscapes of our own future. I conclude that recent changes in culture and technology may have made the industrial structures of the immediate past the last great source of ruins and monuments for future epochs.

THE "SILO DREAMS" OF ERICH MENDELSOHN

Industrial buildings have always possessed value and meaning beyond the merely commercial and functional. In the first decades of this century, for example, European modernists were attracted to large-scale industrial structures, particularly those in the United States, because they thought these utilitarian buildings supplied the vocabulary of pure architectural forms necessary for developing an appropriately modern, up-to-date aesthetic. Walter Gropius compared the monumentality of American industrial buildings to the "work of the ancient Egyptians"; and Le Corbusier, inspired by photographs of American industrial structures assembled by Gropius and published in the 1913 edition of *Jahrbuch des Deutschen Werkbundes*, celebrated the inherent monumentalism of the grain silo and the reinforced concrete factory in his book, *Towards a New Architecture* (Banham, 1986, p. 6).[1]

Erich Mendelsohn was one of the first European modernists to travel to the United States to see at first hand the wonders of American industrial construction.[2] After visiting Buffalo, New York, he enthusiastically proclaimed in a letter to his wife: "Mountainous silos, incredibly space-conscious but creating space. A random confusion amidst the chaos of loading and unloading of corn ships, of railways and bridges, crane monsters with live gestures, hordes of silo cells in concrete, stone and glazed brick. Then suddenly a silo with administrative buildings, closed horizontal fronts against the stupendous verticals of fifty to a hundred cylinders ... (E)verything else so far seemed to have been shaped interim to my silo dreams" (quoted in Banham, 1986, p. 6).[3]

In *Amerika Bilderbuch eines Architekten*, his book of photographs of the industrial and commercial wonders of New York, Detroit, Chicago, and Buffalo, Mendelsohn claimed to "perceive in the midst of this magma the first solid foundations of a new era". He described the grain elevators he had seen as "(c)hildhood forms, clumsy, full of primeval power, dedicated to purely practical ends ... a preliminary stage in a future world that is just beginning to achieve order". With the appearance of these new, uncompromising structures, Mendelsohn announced, "a bare practical form becomes an abstract beauty" (Mendelsohn [1926] 1993, pp. xi, 45, 47).

What struck these modernists about the "real world of engineering" was that industrial builders were now utilizing ("for 'objective', functional reasons that supposedly did not involve rhetoric or aesthetics" [Banham, 1986, p. 214]) precisely the same geometrical solids of cube, cone, cylinder, sphere, and pyramid as had been employed to great effect in antiquity – for example, by Egyptian funereal architects and Greek temple builders. Subscribing to a Platonic idealism, modernists such as Le Corbusier also believed that everybody

– "the child, the savage and the metaphysician" – accepted such architectural forms as "always beautiful in their very nature" (quoted in Banham, 1986, p. 223). Consequently they made architecture, in Le Corbusier's memorable definition, "the masterly, correct and magnificent play of masses brought together in light" (Le Corbusier [1926] 1986, p. 29). When this dialogue was established between European modernists and American industrial building, Banham writes, "[t]he stage was set for the legitimization, so to speak, of industrial forms as the basic vocabulary of modernism" (Banham, 1986, p. 215). As Le Corbusier put it, "American grain elevators and factories [are] the magnificent *first fruits* of the new age" (Le Corbusier [1926] 1986, p. 31).

As a result of deindustrialization, many buildings celebrated by these eager modernists have been cruelly deprived of the meaning and value that originally sustained them. Formerly prosperous – even imposing – manufacturing and industrial districts in older cities have become vast areas of dereliction and decay. Signaling the passing from one historical epoch to another, these remaining industrial structures face a varied, uncertain future.

RIEGL'S THEORY OF MONUMENTS

The appearance of a new class of ruins comprising the detritus of the industrial age has revived interest in the ideas of Austrian philosopher and art historian Alois Riegl (1858–1905). According to Kurt Forster (1982, p. 15), Riegl is important for contemporary scholarship because he advocated the "historical contingency of all aesthetic values", such that "contemporary concerns, the *Kuntswollen* of our epoch, profoundly determine our perception of the past". Thus for Forster, Riegl teaches that "there is no objective past, constant over time, but only a continual refraction of the absent in the memory of the present".

In *The Modern Cult of Monuments* ([1903] 1982) Riegl identified three principal ways in which structures from the past are incorporated into the present. First is what Riegl calls the "intentional monument", a surviving structure that "recall(s) a specific moment or complex of moments from the past" (p. 24), which largely retains the same meaning for the present generation as established by its original builder or creator. Indeed, every epoch produces intentional monuments as elites decide to commemorate this or that event or person. These are familiar to us as war memorials and tombs or as commemorative structures such as the Lincoln, Jefferson, or Franklin D. Roosevelt memorials in Washington, D.C., the Eiffel Tower in Paris, or the monument to the Great Fire in London.[4] As long as factory, mill, and office are home to activities for which they were originally conceived and constructed, the buildings that house these activities, as unchanged survivals, might be considered intentional monuments of the industrial age.

Second, when the specific moment from the past that a structure commemorates "is left to our subjective preference", the result is what Riegl describes as an "historical monument" (p. 24). According to Riegl, each generation scans the past and selects certain surviving artifacts as monuments on the basis of how that generation conceives of, values, and interprets the past. In such cases, the meaning assigned to a structure may deviate significantly from that originally intended. The Great Wall of China, the Liberty Bell and Independence Hall in Philadelphia, Ellis Island, Stonehenge, and Alcatraz are historical monuments. As cities and regions seek to remake themselves as essential destinations in the landscape of consumption, structures associated with the industrial era can be a rich source of future historical monuments.

Historical monuments are often "unintentional" monuments in that they emerge as such when new value and meaning replace the meaning or utility that sustained their earlier existence. In such a change, a structure is typically transformed through a process of ruination, rediscovery, and preservation. Often the new meaning is supplied by cultural movements such as romanticism or by science or historical scholarship. (The rise of modern archaeology was particularly important for assigning new meaning to the detritus of the past, as evidenced, for example, by Schliemann's sensational discoveries at Troy in 1871.) Today other, more democratic values may play a role: Witness the transformation of Paul McCartney's boyhood home in Liverpool into a property of Britain's National Trust.

When industrial buildings find new uses for which they were not originally intended – educational and tourist sites, art galleries, museums, restaurants, sports bars, or condos – they become contemporary forms of the historical or unintentional monument. For Riegl, the discourse by which structures achieve monument status is determined historically; consequently the category of unintentional monument is more broadly based than that of intentional monument. Thus the transformation of structures into unintentional monuments is more difficult to predict.

Finally, survivals from the past may be appreciated for the way their surface patina and eroded structure record earlier stages in the life of the artifact; Riegl calls these monuments with "age value". Age value derives less from scholarly knowledge or historical education (as in the case of the historical monument) than from an "appreciation of the time which has elapsed since (the work) was made and which has burdened it with traces of age" (p. 24). Any kind of decayed or worn structure, whatever its original function or purpose, can become an age value monument under appropriate conditions. Because of their large size, specialized function, and solid construction, many industrial structures are well equipped for a lengthy sojourn in the ruination mill. Perhaps in this way, the factory, skyscraper, or prison can survive long enough to acquire new meaning as an age value monument.

THE FATES OF INDUSTRIAL STRUCTURES

In later life, industrial structures deviate from their original uses because they are subject to the peculiarly modern phenomenon of technological obsolescence: physical persistence despite loss of designated function. According to James Fitch, doyen of architectural preservation, the idea that objects could become "useless economically without reference to any residual physical utility" originated with the industrial revolution. As science and technology came to dominate economic development, "the old artifact (tool, factory, town) began to be regarded as an intolerable restriction upon increased productivity" (Fitch, 1990, p. 30). Thus plants and equipment were consistently junked (often in favor of more wasteful and less efficient technologies) long before they wore out physically. Marx called this process "moral depreciation" of the means of production; he pointed out that in a commodity-driven economy the machine constantly loses "exchange-value, either by machines of the same sort being produced cheaper than it, or by better machines entering into competition with it" (Marx, n.d., p. 381). In this way, "the corpses of machines, tools (and) workshops . . . are always separate from the products they helped to turn out" (p. 197).

Below I identify four possible future scenarios for the "corpses" of the industrial era: (1) destruction or demolition, whereby possibilities for the building or structure as new utility, historical monument, or elegant ruin are canceled by its elimination from the landscape and by the transformation of its site, at best, into a cleared lot for some future development project; (2) adaptive recycling of the structure as new utility under whatever commercial or other regime prevails. The structure is inserted into the prevailing landscape of consumption, in which the building's architectural envelope is preserved, if not its original internal program of spaces; (3) transformation of building, structure, or site into an historical monument. This fate favors highly specialized structures that are difficult to recycle directly into commercial use in the new economic regime; and, finally, (4) persistence of the structure in the landscape as an age value monument or ruin, a fragment of some past whole which poignantly recalls a lost universe of activity and meaning. I discuss each of these scenarios in turn, bearing in mind that the categories are necessarily crude and impure, and are likely to overlap considerably in the untidiness of the real world.

DESTRUCTION OR DEMOLITION: THE NEW *SVENTRAMENTI*

Most structures rot or collapse long before they have a chance to be preserved and recycled into a new universe of meaning. In the past, some structures survived (even when local elites wanted them razed) because the means of demolition were underdeveloped; this explains the survival, into our own age, of structures that are now cherished monuments and heritage sites. Today, however, several factors conspire to remove unused, unwanted, or unoccupied structures from the landscape at an unprecedented rate. Indeed, according to one commentator, there may be no point in directing people to industrial ruins "because it would be the strangest freak if they lasted until those visitors came" (Harbison, 1991, p. 122).

For one thing, Western cultures periodically speed up the process of natural decay by pursuing programs of urban renewal that actively eviscerate cities, undertaking the massive surgery thought necessary to save the "badly infected urban organism" by clearing away congested older neighborhoods in the interests of the rational organization of urban space (Kostof, 1982, p. 33). Baron Haussmann used the term *eventrement* to refer to his reorganization of Paris in the 1850s where a rationalized circulatory network of wide and straight boulevards, continuous frontages, and "spatially profligate piazzas" functioned to "sweep away the dross of the community's promiscuous life through time", so creating the first modern city (p. 34). In the 1920s and 1930s, Mussolini personally involved himself in massive clearance projects called *sventramenti*. He praised the role of *il piccone risanatore*, "the healing pick", in clearing condemned neighborhoods in Rome, Florence, and elsewhere (p. 33). In this ambition he was not alone; he merely put into practice the Futurists' visceral hatred of old cities. Indeed, he acted as Marinetti had urged: "Get hold of picks, axes, hammers and demolish, demolish without pity, the venerated cities" (quoted in Kostof, 1982, p. 33).

Such *sventramenti* were not confined to Europe. In the 1940s and 1950s, Robert Moses' empire in New York directed wrecking machines that leveled houses and whole urban blocks in a campaign "no less destructive to culture than if they had been the tanks and artillery of an attacking army". The fruits of war – the highways and parks – were "actually monuments to the victory of autocratic authority over the fragile lives of mere people" (Woods, 1995, p. 50). Virtually no city escaped being remade by superhighways that bored through its neighborhoods until Jane Jacobs organized a counterstrike with her enormously influential book *The Death and Life of Great American Cities* (1961).

It was fashionable to destroy then, and so it is now. Demolition technology has now reached the point where virtually no structure exists that cannot be brought down quickly and efficiently.[5] Development of various powerful hydraulic attachments including giant grapples, mobile pulverizers, and "huge, swiveling shears ... that can snip steel support beams like scissors cutting cardboard" have transformed excavators into "state-of-the-art wrecking machine(s)", capable of pulling apart most structures (Johnson, 1996, p. 16). The spectacular but less frequently used implosion method of demolition – a staple of local TV news – not only can reduce the proudest architecture to rubble in seconds but also provides recurring opportunities for public entertainment. It transforms "working stiffs into rock stars" and even increases spectators' "consciousness of their own mortality" (Samuels, 1997, p. 37). (Indeed, several spectators were killed or injured when an incompetently imploded hospital building in Australia showered the area with debris.) Moreover, modern buildings are generally not as "over-engineered" as those of antiquity (the walls of the keep at Kenilworth Castle in England, for example, are 16 feet thick), and thus succumb more economically to the wrecker's ball.

Vandalism and fire also play their part. Fire, a particularly frequent and destructive agent of decay, remains a major factor in the remaking of the urban landscape. Not all fires are accidental: In Philadelphia, for example, arson now accounts for more than 40% of building fires – double the rate of a few years ago. Buildings are torched by persons engaged in domestic disputes, teenagers seeking thrills, and drug dealers protecting their turf. (The Quaker Lace factory in North Philadelphia was burned down by a local drug dealer who thought narcotics detectives were using its upper floors to monitor illicit street activity.) Indeed, on Devil's Night (the night before Halloween) in cities such as Camden and Detroit, youths set hundreds of fires, gutting many abandoned properties. Arson also offers a way for owners to extract value from unmarketable, tax-burdened properties – in police jargon "selling them back to the insurance company" (McCoy & Sataline, 1996, p. 24). In the Bronx, a new spate of fires threatens economic revival (Halbfinger, 1998, p. 3). Burning of buildings, coupled with municipal policies of demolishing unsafe structures that remain, has become so widespread as to hasten the return of the industrial city to the open fields and weed-filled lots described by Camilo Vergara (1995a) as the "green ghetto".

The elimination of unwanted industrial structures from the landscape is also encouraged by speculative development. This process is reinforced by municipal authorities' failure to protect historic or other important structures. Typically developers, unenthusiastic about the restrictions that historic district and landmark building designations place on their activities, challenge in court the legal

Monuments of Tomorrow: Industrial Ruins at the Millennium

Demolition as street theater. A housing project is imploded in Philadelphia.
(Photo by James Dickinson).

basis of such ordinances, or otherwise subvert them by political manipulation or corruption of local officials.[6] Corporations cynically give vacant properties to underfunded community and nonprofit groups in exchange for tax credits. These structures, too large to be developed successfully by such groups, are later demolished at taxpayers' expense.

Local authorities, moreover, often fail to see that vacant or abandoned structures are adequately sealed. Such failure facilitates occupation by addicts, scavengers, and the homeless. As these structures acquire reputations as eyesores and problem buildings, local politicians and the public lobby for their demolition. Thus they encourage that mind-set which favors the wholesale razing of structures – urban clear-cutting to create the vacant land parcels desired by today's giant development corporations. Communities even actively promote the dereliction of empty but otherwise sound local buildings, encouraging vandalism and minor building degradation in the hope that the structure will be condemned as unsafe and demolished (Parmley, 1997). The clearing of the Camden and Philadelphia waterfronts, which involved the demolition of the majestic, eminently recyclable RCA buildings as a prelude to the rise of a gambling and entertainment economy, demonstrates the destructiveness of these new development policies.

Moreover, because new construction on cleared sites is often tacky and insubstantial, the city loses its traditional solidity and density, and increasingly acquires the look and feel of suburbia. For example, one proposal for a two-acre entertainment complex on the Camden waterfront included structures no more substantial than tents or others buildings requiring less than a week to erect. When a vacant lot in the densest part of the city is commercially redeveloped, it is often reborn as a mini-mall: a drugstore or fast food outlet set in the middle of a parking lot (Johnson, 1999). One development in northeast Philadelphia proposes the literal suburbanization of the city by turning an abandoned 214-acre industrial site into a golf course. This transformation will require the demolition of an assembly building "so vast that five football teams could practice inside it and never rub shoulder pads with each other" (Davis, 1998).

Suburbanization also affects the city in other, more profound ways. In Philadelphia, population loss (the city has lost 500,000 residents over the past 50 years) has rendered obsolete much of the city's stock of traditional row houses; indeed, that city's Office of Housing and Community Development (1997) anticipates that most of the 27,000 houses currently vacant in the city will have to be demolished because, even if they were made habitable, there would be no one to rent or buy them (p. 62). In combination with its aggressive policy of demolishing up to 1,500 vacant residential properties a year,

Monuments of Tomorrow: Industrial Ruins at the Millennium

Ruins near South Street, Philadelphia.
(Photo by James Dickinson).

Philadelphia proposes to rebuild residential neighborhoods "based on land-use plans and building designs that are more relevant for the future" (Weyrich, 1998, p. 11). These plans include housing at a much lower density than in the past (13 units per acre as opposed to 45), the closure of back streets to create larger lots, replacement of narrow row houses with detached and semidetached houses with driveways, and reorganization of streets as cul-de-sacs. In a *sventramento* of its own, Detroit plans to demolish all its derelict and abandoned homes, some 10,000 structures, at one time. Such a move, in the absence of rebuilding, is likely to extend the desolate, surreal landscape for which that city is already famous (Katz, 1998).

The "new urbanism" proposes to cure the ills of the industrial city by extending the form and feel of suburbia into its most derelict and deprived neighborhoods, principally by calling for a downsizing of public housing. For example, one federally-funded report recommends demolishing nine Philadelphia Housing Authority high-rise apartment buildings and replacing them with low-rise townhouses. This plan will force 700 families to move and will do nothing for the 13,000 families on the waiting list for public housing. In other cities, similar policies are proposed or already being put into effect. In Newark, New Jersey, most high-rise public housing projects have been demolished. Under a $1.5 billion federal program, Chicago will raze 51 high-rise projects including the Cabrini-Green and Robert Taylor Homes, threatening homelessness for thousands of impoverished residents who currently occupy the structures without leases (Nicholas, 1998; Belluck, 1998; Cawthon, 2000).

Today, urban clear-cutting to make way for new development is not limited to the United States In Sheffield, England, for example, virtually every structure in the historic Don Valley steelmaking district has been swept away, creating space for American-style shopping malls, superstores, parking lots, and an airport. The Sheffield Development Corporation, the organization charged with facilitating this transformation, sees its mission primarily in terms of "land acquisition and assembly" so as to remove "stagnation in the land market", followed by "land disposal and development" to secure "regeneration" (Sheffield Development Corporation, n.d., p. 8). In its reports the Corporation proudly displays photographs of the huge vacant lots it has assembled and cleared of buildings, people, and history.

London's Docklands, a spectacular example of urban clear-cutting, is a project of staggering ambition and destructiveness – as much an attempt to wipe out the past as to build a future. An expression of the aggressive monetarist policies pursued by the Thatcher government in the 1980s, Docklands is the product of the London Docklands Development Corporation (LDDC), an urban development corporation (UDC) controlled by business and property interests

Monuments of Tomorrow: Industrial Ruins at the Millennium

The City Eviscerated. Don Valley, Sheffield, U.K.
(Photo by James Dickinson).

directly appointed by the Secretary of State. Because of this appointment, LDDC can bypass controls and regulations on urban land use and redevelopment traditionally exercised by elected bodies such as local councils. Vested with almost unlimited powers, LDDC has appropriated some 5,000 acres of publicly-owned land as well as 55 miles of waterfront property, which it sells off to private developers as cleared parcels.

To take a trip through Docklands (on the new driverless, computer-controlled Light Railway), according to Jon Bird (1993), is "to experience the global postmodern as building site. Here consortia and multinationals swallow up the generous offers of land ... and spew out varieties of architectural postmodernism and high-tech in paroxysms of construction that are as incoherent as they are unregulated" (p. 124).[7] Developers enjoy a simplified planning process (no pesky local interference), relief on property taxes, and the ability to offset construction costs against other taxes owed. With local government and citizen groups silenced or sidelined, LDDC relentlessly promotes the fantasy of a harmonious and well-functioning Utopia on the Thames.

The most visible component of Docklands is Canary Wharf on the Isle of Dogs (called the "Pompeii of London" before its seizure by developers), a $7 billion development of offices, apartments, restaurants, shopping malls, and other facilities said to be one of the largest building projects ever undertaken, comprising an unrivalled concentration of late twentieth-century architecture.[8] A recent guide describes this triumph of LDDC policy as "an architectural circus" where, in the rush "between balance sheet and building", the "necessary intermediate step" of urban design unfortunately "got missed". The resulting buildings are more suited to 1930s Chicago than to London's tradition of Regency squares and muted skylines. Moreover, the centerpiece building by Cesar Pelli, an 800-foot tower clad in stainless steel, is obviously misproportioned: the building's large footprint was dictated by LDDC rules, but its top five stories had to be sacrificed so that the flight path at the nearby City of London Airport would not be compromised. The outcome was a skyscraper that is simultaneously squat and overbearing (Hardingham, 1996a, pp. 296–300).

Docklands – along with other UDC-type projects in Sheffield, Manchester, Newcastle, and elsewhere – are United Kingdom examples of the "property-led approach to urban regeneration" vigorously pursued by the government in the 1980s. The idea was to link citizens' well-being more firmly to the private economy and to flows of global capital, rather than tempering market inequities with traditional social spending. In policy terms, this meant a shift from support for broad-based welfare programs in aid of poor and underprivileged urban families "towards infrastructure and building projects directly linked to the supply-side ethos of monetarism" (Imrie, 1997, p. 98). Overall the Conservative government became less concerned about the needs of local populations marginalized by uneven development; instead it "target[ed] resources towards reproducing the most economically active groups" (p. 99).

The result has been a rapid transformation of the built environment through "gentrification in working-class neighborhoods and the development of new business centers, while sweeping away older industrial areas with skyscrapers devoted to the emergent service economy" (Imrie, p. 99). In cities where these policies have been put into effect, new landscapes have emerged "richly symbolic of the social cleavages engendered by neo-liberalism" (p. 100). These landscapes are marked by social and economic division, privatization of public space, and a mentality favoring fortress architecture: razor wire, walled estates, and bunkerlike buildings. Such landscapes are much in evidence in the United States as well, leading to the notion that late capitalism increasingly takes on the appearance of a new feudalism – at least with respect to its organization of space.[9]

ADAPTIVE RECYCLING

Obsolete and abandoned industrial structures may be recycled or reinvented for new uses, and thus survive to become a functioning part of the new economic regime. As Fitch (1990) reminds us, "[T]he reworking of extant structures to adapt them to new uses is as old as civilization itself"; indeed, until a century ago, adaptive reuse was the "characteristic mode of energy conservation" (p. 165). Since World War II, however, because of the introduction of low-cost industrial construction techniques and tax breaks favoring demolition and new construction, the American landscape has largely been made and remade with little recourse to adaptive recycling.

Today, however, the sheer scale of building obsolescence and abandonment (as well as the fallout from financial disasters such as the saving and loan scandal) has placed adaptive recycling back on the agenda. As Sharon Zukin notes, capitalism is a process of "creative destruction" that continually "re-organizes space and time, reformulates economic roles, and revalues cultures of production and consumption" (Zukin, 1991, p. 29). Old structures and spaces of the industrial city are reinvented as "emporia of mass consumption". Harbors, docks and shipbuilding sites are turned into marinas, aquariums, waterfront parks, casinos, and sports bars; warehouses and factories become offices, condominiums, or artists' studios; banking halls become restaurants; railway stations are reborn as shopping malls. Numerous smaller, less glamorous transformations occur daily: A store becomes a church; the neighborhood bank becomes a photo-processing lab or a pizza parlor; an old workshop becomes a methadone clinic. In this way, Zukin suggests, the old landscape of production is converted into a new landscape of consumption.[10]

Adaptive reuse is connected with the rise to prominence of what Zukin (1995) calls the "symbolic economy". As traditional manufacturing and commercial activities have declined, many cities have experienced an expansion of the part of the economy devoted to symbolic transactions: abstract financial exchanges (speculative markets in stocks, derivatives, currency, and real estate) as well as the mass consumption of food, art, fashion, music, and tourism. The gradual realization that "culture is more and more the business of cities – the basis of their tourist attractions and unique competitive edge" (p. 2) motivates shifts in building use, causing the conversion of older commercial and industrial spaces into hotels, restaurants, art galleries, and design studios. Thus, not only do many buildings metamorphose but, in addition, public spaces such as parks and streets are remade by powerful interests that benefit from the transformation of the city into a symbolic economy. In this makeover, a critical role is played by "business improvement districts", organizations of business and property owners in commercial districts which voluntarily take over maintenance and control of public spaces (p. 33). They take up the slack left by government cutbacks and organize the redevelopment of downtown areas into malls devoted to Disney-style cultural consumption, especially entertainment-based retail shopping areas orchestrated by giant out-of-town corporations such as Warner Brothers and Sony (p. 3). Enterprise and tax abatement zones are designed to play a similar role in the commercial revitalization of less glamorous parts of town.

Often commercial development helps pay for renovation and preservation of historic buildings and public spaces, now undertaken increasingly by public-private partnerships. For example, the plan for restoring Grand Central Station called for almost four acres of commercial space to help finance this multi-million-dollar refit, packing in familiar chain and "boutique" stores that threaten to turn this once-proud "Gateway to a Continent" into the "Gateway to a Caviarteria" (Dunlap, 1998). In New York, restoration has led to the almost total privatization of Bryant Park, a nine-acre public space. A private agency, the Bryant Park Restoration Corporation, "raises most of the park's budget, supervises maintenance, and decides on design and amenities" (Zukin, 1995, p. 30). By providing a variety of upscale drinks and snacks as well as kiosks selling Broadway theater tickets, the Corporation encourages "normal" middle-class use as a way to crowd out vagrants and undesirables, initiating what Zukin calls "a model of pacification by cappuccino" (p. 28). As Benjamin Barber (1998) warns, however, such privatization of public space comes at a price: It threatens not only to reduce commercial civilization to the manufacture of private, individual needs, but also to turn citizen into consumer. Yet so intense is competition within the landscape of consumption that shopping malls are failing at an alarming rate; over the next decade, some 8,600 malls are expected to go bankrupt or be adapted for another use (Barry, 1999).

Older industrial cities in the United States and Europe, desperate to reinvent themselves for the new symbolic economy, often devise a "plan" aimed at redeveloping and reworking existing assets into centers of leisure and mass tourism. Examples of this process include London's Covent Garden, Baltimore's Inner Harbor, New York's South Street Seaport, and Philadelphia's Avenue of the Arts project (as well as that city's transformation of the Delaware River waterfront into a strip of sports bars and nightclubs).

Often, however, adjacent structures necessary for a full historical and aesthetic appreciation of the site are swept away. On the Liverpool waterfront, for example, only one warehouse complex, Albert Docks (1846), survives from what was once an uninterrupted seven-mile stretch of monumental industrial architecture. Predictably, redevelopment of this site as an urban/industrial leisure park has produced "a continuous arcade of card and candle shops, phony American restaurants and grotesque candy stores punctuated by an excess of unlet space" (Hardingham, 1996b, p. 68). Developments focusing on the conversion of warehouses and dock facilities to residential and commercial use, such as Salford Quays in Manchester (as well as London's Docklands), produce strangely deserted and sterile landscapes. For one thing, shops and pubs are generally missing (the enclave character of the development limits the number of potential customers). In addition, no family with young children could contemplate living next to so much open water.

Because they take so long to implement, redevelopment plans are often modified in the light of changing economic circumstances. Thus Cooper's Ferry Development Association, a private agency formed in 1983 to "bring new life to the Camden waterfront", originally envisaged redevelopment based on office buildings. But because the office market collapsed in the early 1990s, the agency now hopes to attract tourists and suburbanites to a regional family entertainment complex consisting of a baseball park, a children's garden, nightclubs, restaurants, and recreational facilities that complement the existing aquarium and outdoor music center. This plan, should it ever be implemented, is not likely to have any real impact on this city, the fifth poorest in the country (Mulvihill, 1998).[11]

If many of these grand adaptive recycling projects seem to be surprisingly similar, it is in part because they are often undertaken by the same development corporation. In the 1970s, for example, the Rouse Company was responsible

Monument of Tomorrow. USS *United States* on the Delaware River awaits conversion into floating casino. (Photo by James Dickinson).

for Boston's Faneuil Hall Marketplace, Baltimore's Inner Harbor, and Philadelphia's Gallery on Market Street. Soon every city had to have its "festive marketplace", but in the 1980s, as innovation gave way to imitation, there was a "spate of less than successful formula projects" (Central Philadelphia Development Corporation, 1997). Homogeneity is now global: Cities the world over offer uniform eating and shopping experiences in restored and pedestrianized "historic" downtowns filled with international chain stores and formula boutiques.

One particularly interesting example of large-scale adaptive recycling is the sprawling Roebling wire rope factory in Trenton, New Jersey. In its heyday, this factory employed 12,000 workers and spun the cables for many suspension bridges, including the George Washington and Golden Gate bridges (Zink & Hartman, 1992). The site recently has been converted into a multifunctional complex comprising a supermarket and shopping center, housing for senior citizens, and government offices, with educational facilities and museum and art spaces to come. This is one of the most comprehensive and most effective transformations of an old industrial site to date; the largely intact working-class dwellings that surround it increase its appeal.

Even in Trenton, however, despite evidence of the many and varied possible uses for specialized old industrial buildings and their potential in urban regeneration, the philosophy of adaptive recycling and historic preservation has only a tenuous hold. Several important structures, including a pre-Civil War foundry, ironworking and machine shops from the 1880s, and other mill buildings, have been demolished to make way for a new sports arena despite the developers' promises to preserve and reuse these historic buildings. Although the city might acquire a shiny new building, the future economic value of the city's rich architectural heritage will be lost forever (Zink, 1998).

Adaptive recycling is not always a benign process that produces glittering centerpieces for the official redevelopment plan. Appropriation and conversion of abandoned structures by squatters, drug dealers, and unlicensed businesses produce problems that become simply another reason for communities to pursue a policy of urban clear-cutting. State and local authorities may recycle older structures into the shelters for the homeless, the halfway houses for delinquents and the mentally ill, and the drug rehabilitation clinics needed to contain and warehouse the redundant and increasingly recalcitrant population that has no place in the new economy. (For a survey of the building types characteristic of the "institutional ghetto", see Vergara [1995a].) A particularly striking example of institutional recycling is found in Johannesburg, South Africa, where the once-prestigious, 54-story Conde skyscraper is about to become the world's tallest prison. The tower will be divided into 11 miniprisons by inserting double decks every five stories to contain recreation, visiting, and guardroom areas. Cells with steel mesh windows will ring the outside. With courtrooms, police station, and hospital on the ground floor, arrested suspects will never leave the building until acquitted or transferred to a long-term prison (McNeil, 1998).

Adaptive reuse is fundamentally a function of the relationship between a building's architectural envelope (its outer structure) and its internal program (the layout of rooms and spaces); this relationship must be flexible and dynamic if reuse is to occur. Indeed, the reuse of many warehouses, factories and workshops – including the concrete-and-steel wonders so greatly admired by European modernists – reflects the fact that many of these structures originally were designed and built as "universal spaces" able to accommodate a variety of activities and processes. Similarly, the boxlike character of the modern skyscraper – masterfully realized, for example, in Mies van de Rohe's stacking of universal, functionally nonspecific spaces into glass-curtain-walled towers – reflects modern architects' desire to accommodate the still-unknown uses to which the building may be put during its lifespan (Blake, 1996, pp. 236–237).

For this reason, office buildings and skyscrapers, less usually regarded as victims of deindustrialization and globalization, also can be converted into condos and hotels. In this way, these recent victims of corporate downsizing can be recycled into the expanding consumption economy. The Philadelphia Saving Fund Society building (built 1930), for example, one of the earliest International Style skyscrapers erected in the United States, will find new life as a hotel in support of the recently opened $600-million Pennsylvania Convention Center. This is only one of many preservation-based real estate developments currently under way in the city in which "older buildings are being rethought, redesigned and rebuilt" (Preservation Alliance for Greater Philadelphia, 1998).

In general, if the relationship between envelope and program is flexible, then conflict between what Rem Koolhaas (1986) calls the fixity of architecture and the fluidity of the city, between infrastructure and economy, can be resolved, and the city can be reproduced through time. This conflict is difficult to resolve within the industrial city because particularities of the manufacturing process often dictate a less flexible, more rigid relation between a structure's envelope and its program. Gradually the physicality of the city, its built environment, becomes divorced from prevailing systems of meaning; structures exist but have no use; abandonment and dereliction ensue. However, for cities dedicated to abstract functions such as government, administration, or market transactions rather than to the manufacture and assembly of physical objects, the relation between envelope and program can be negotiated more easily. As Koolhaas put it, New York survives "through a mutant architecture that combines the aura of monumentality with the performance of instability. Its interiors accommodate 'compositions' of programs and activity that change constantly and independently of each other – without ever affecting what is called, with accidental profundity, the envelope" (Koolhaas, 1986, p. 448). Thus New York's skyscrapers are secure as long as the city remains the administrative and calculating center of global capitalism.

Inflexibility between envelope and program, however, does not always mean that specialized industrial structures such as chemical plants, oil refineries, steel mills, power stations, and floating exploration platforms cannot find commercial viability or reuse within the ascendant regime, and thus face abandonment, ruination, or demolition. Indeed, a grain elevator adjacent to downtown Philadelphia has been converted into fine offices and design studios. Nuclear missile bunkers in the Midwest have been recycled as family dwellings, mushroom farms, or storage facilities. A rocket silo in Illinois has been turned into a corporate swimming pool; another, in Texas, into a scuba diving school. In Maryland, a salt storage dome has been converted into a school building, providing 7,600 square feet of prefabricated floor for only $106,000 (Brooke, 1997; "Missile Base Living", 1999).

Bankside Power Station in London, a massive industrial building "so powerful as a symbol that no one can bear the thought of its being demolished" (Harbison, 1991, p. 126), is now transformed into the Tate Gallery of Modern Art. (Unfortunately the same could not be said for New York's magnificent Pennsylvania Station designed by McKim, Mead and White in 1910 which was demolished between 1963 and 1966.) Elsewhere architectural students have imagined recycling an obsolete power station cooling tower in Washington State as an amphitheater ("for choral productions the acoustics are great"), a spa and watergarden (taking advantage of the plant's powerful water system), a motocross track, and a museum of energy complete with a Guggenheim Museum-style ramp running around the inside of the tower (Goldberg, 1997). With imagination, an abandoned sugar refinery in Philadelphia might find new life as "the world's most sinister casino" (Vergara, 1997).

TRANSFORMATION INTO HISTORICAL MONUMENTS

Monuments, intentional or otherwise, are structures whose meaning transcends, to some degree, the strictly commercial and functional. The industrial age might be regarded as antithetical, or at least indifferent, to monuments, to symbolic and nonutilitarian architecture. After all, many modernists, enamored of the spirit of strict functionalism, were committed to producing a timeless

Reading Co. Grain Silo, Philadelphia. Converted into offices and design studio.
(Photo by James Dickinson).

architecture in which buildings, devoid of historical and symbolic ornamentation, were conceived of as "machines" to be constructed as efficiently as possible. After 1945, however, a reassessment occurred. In the essay "Nine Points on Monumentality", discussed at CIAM (Congrès Internationaux d'Architecture Moderne) conferences in 1947 and 1951, leading figures in the modern movement admonished contemporary architecture for its failure to satisfy the popular desire for symbolic structures: "Monuments are human landmarks which men have created as symbols for their ideals . . . They are intended to outlive the period which originated with them, and constitute a heritage for future generations. As such they form a link between the past and the future . . . The people want the buildings that represent their social and community life to give more than functional fulfillment" (quoted in Frampton, 1997, p. 26). The heritage movement is a belated legacy of this sentiment and, of course, is beset by its own contradictions and pitfalls (Lowenthal, 1985; Loewen, 1999).

Obsolete industrial structures constitute an important stock of potential symbolic architecture and thus are prime candidates for transformation into historical monuments: sites and buildings that are visited for their alleged cultural or educational value or for the new meanings (often related to their original, distinctive function) that private and public advocates can negotiate and establish for them. Indeed, extravagant size, peculiar architecture, and expressed function are precisely the attributes that attracted commentators to industrial structures in the past and now excite advocates seeking new meaning and value to sustain them into the future.

By the middle of nineteenth century, visitors to the new industrial sites were caught up in the power and excitement of the technological sublime. According to Nye (1994), for example, early factory towns in Massachusetts were considered awesome in their "combination of complexity and order on a massive scale". The factories themselves, driven by a central power source (first water, then steam), "presented a system of linked machines that many ordinary observers could barely grasp, virtually compelling their admiration for the ingeniousness of its mechanical contrivances" (pp. 113–114). The famous Corliss steam engine, exhibited at the 1876 Philadelphia Exposition, proved a "mesmerizing vision" for fairgoers, its gigantic flywheel spinning in "rapid, silent, and seemingly effortless rotation". According to one commentator, the movement of its exposed rods, cranks, and levers "in plain sight, provid[ed] overtones and counterpoints to the steady rhythm of the great machine's chief members". The engine enthralled even Walt Whitman, who reportedly sat before "this colossal and mighty piece of machinery for half an hour in silence . . . contemplating the ponderous motions of the greatest machinery man has built" (quoted in Nye, 1994, pp. 120–122). Marx too was impressed with modern

industry, describing factory production as "(a)n organised system of machines
... a mechanical monster whose body fills whole factories, and, whose demon
power, at first veiled under the slow and measured motions of his giant limbs,
at length breaks out into the fast and furious whirl of his countless working
organs" (Marx, n.d., pp. 360–361).[12]

Enthusiasm for technology often underlies efforts to convert obsolete industrial sites into museums and historic places. In Pennsylvania, for example, the Lackawanna Coal Mine attracts 70,000 visitors a year, and the $66-million federally-funded Steamtown Museum in Scranton promises bliss for devotees of nostalgia and the mechanical sublime (Infield 1997). Plans also exist to convert part of the defunct Bethlehem Steel works in the Lehigh Valley (an extraordinary array of structures beautifully documented in Garn [1999]) into a museum of the Smithsonian Institution. A somewhat more Disneyesque remaking of the Mesabi Iron Range, once the largest open-pit iron mine in the world, is now under way in Minnesota (Goin & Raymond, 1999). Among more recently constructed sites, Cape Canaveral already anticipates its own future as a monument of the Space Age (Wilford 1998); others, however, such as the Hanford Nuclear Reservation and atomic bomb test sites around the world, cannot be enjoyed directly because of environmental contamination (Goin, 1991; United States Department of Energy, 1992, 1996). Other sites have yet to be recognized for their potential as monuments. As one prescient visitor asked when touring the Bayway Refinery in New Jersey, famous for harboring the world's largest catalytic cracker, a gigantic wet-gas scrubber, and an immense tank farm, "Is there a registry of national historic places that this is on?" (Newman, 1997).

In Europe, the transformation of industrial sites into historical monuments is well advanced. Coal mines and steel mills, like their obsolete counterparts in the United States, are now museums and tourist attractions. Particularly striking is the "ecological symbiosis between flora and industrial ruins" which is at the heart of a plan for transforming a steel mill in Duisburg, Germany into a multi-purpose recreation destination (Steinglass, 2000, p. 127). The entire Ironbridge Gorge district in Shropshire, England, center of the Industrial Revolution, is now a United Nations World Heritage Site, a multisite museum complex requiring days for a comprehensive visit. Earlier industrial and factory sites such as Abbeydale Industrial Hamlet in Sheffield, Quarry Bank Mill, and the Black Country Living Museum, are now preserved for the edification and amusement of current and future generations.

Steam engines and other industrial contrivances, many of enormous size and complexity, are installed at various sites such as Manchester's Museum of Science and Technology and the Railway Museum in York. The Trencherfield Mill at Wigan Pier Heritage Centre, a complex of historic industrial structures which assiduously exploits its pop cultural and literary associations (George Formby endlessly sings his famous song, and there is a pub called "The Orwell"), boasts a particularly impressive machinery hall housing what is claimed to be the largest working mill engine in the world. Demonstrations of this Goliath in action – surely every bit as impressive as the mighty Philadelphia Corliss engine – are followed, curiously, by a lecture on the wretchedness of the conditions for workers in the vast textile mill it once powered. This suggests that the interpretation of the past has a critical edge, at least at this "heritage industry" site.

Institutional structures such as prisons, workhouses, and company towns associated with the large-scale management, control, and processing of people are often good candidates not only for adaptive recycling, but also for transformation into historical monuments. In the United States, Ellis Island National Monument and Alcatraz prison in San Francisco Bay are popular tourist attractions. In England, a nineteenth-century workhouse in Nottinghamshire is about to become "the only fully-restored demonstration of the Poor Law system which cast a shadow of terror over the poor well into this century" (Kennedy, 1997; Morrison, 2000). In Clerkenwell, London, the remains of the House of Detention hint at the horrors of incarceration, often before inmates were transported to Australia. In that country, prison structures in Tasmania and on Norfolk Island, so central to its founding and history, are major tourist attractions. In South Africa the Robben Island prison has been reopened as a museum and draws thousands of tourists, many of them former prisoners of the apartheid regime (Brodie & Davis, 1999).

The history of Ellis Island, the U. S. Immigration Station through which more than 12 million immigrants passed, illustrates the complexity and unpredictability of the process by which such obsolete sites become monuments. The wooden buildings that constituted the original immigrant receiving station burned down in 1897, and new fireproof buildings opened in 1900. The largest and most impressive building is the Immigrant Receiving Station with the Registry Hall on the second floor. Designed to handle 5,000 immigrants a day, the interior until 1911 was divided by metal pipes into a series of pens resembling cattle runs. Here federal inspectors processed immigrants, weeding out the physically and mentally handicapped, paupers, criminals, and those with infectious diseases. Other buildings include the General and Contagious Diseases Hospitals, the Immigrant Building (to house detained aliens separately from criminal deportees), the Baggage and Dormitory Building, and restaurant and laundry facilities. In 1954 the buildings were closed and put up for sale. Over the years various proposals were made to turn Ellis Island into a luxury "Pleasure Island", a bible college, a gambling casino, and a mental institution.

In the meantime, neglect and decay transformed the site into a ruin. Declared a national monument in 1965, the site was opened to visitors in 1976 (von Pressentin Wright, 1983, pp. 75–76). To date, the main receiving buildings have been lavishly restored, but most of the site is still in ruins.

The Eastern State Penitentiary in Philadelphia is another interesting example of a specialized people-processing structure from the early industrial age transformed into an historical monument (Dickinson, 2000). Built in the 1820s and not closed until 1971, the penitentiary stood empty and exposed to the elements for several years. More recently, preservationists, historians, and community groups have mounted a major effort to establish the penitentiary as an important historical site, an outstanding tourist destination in the new landscape of consumption.

The historical importance of Eastern Penitentiary lies in its contribution to the development of modern penal systems, particularly to the Quaker belief that proper treatment and rehabilitation required incarceration of the prisoner under a regime of contemplative solitude. This regime was organized on the basis of solitary confinement and isolation of the inmate at all times (even during work and exercise), the moral benefits of hard labor in the cell, and religious instruction and prison visitation. Because of the way the penitentiary's distinctive architecture of cells and blocks radiating from a central observation rotunda expresses this peculiarly cruel but modern philosophy, it has been placed on the World Monuments Fund's list of the 100 most endangered monuments (Johnston, 1994, 2000).

PERSISTENCE AS RUINS: MEMENTI MORI OF THE INDUSTRIAL AGE

Ruins – age value monuments in Riegl's terminology – occur when a structure survives long enough to pass from one system of meaning, one cultural valuation, to another. Because of their large size and superior construction, their specialized function, which precludes easy recycling into the new economy, and their occupation of sites without significant redevelopment value, newly obsolete industrial structures can persist in the landscape for decades, even centuries. Thus they gradually acquire the worn patina and fragmented, eroded structure that give familiar survivals from the past, such as castles, temples, and pyramids, their distinctive allure. Once they fall into disuse, industrial buildings enter a period of abandonment and decay. Succumbing to the ravages of vernacular use, such buildings are stripped of all usable materials. In the absence of maintenance, exposure to the elements increases. Eventually the roof collapses; walls

Unintentional Monument, Eastern State Penitentiary, Philadelphia. (Photo by R. C. Horsch).

crumble; vines and bushes take over (Dickinson, 1996). The resulting ruins thus "lie upon the land like so much detritus", fragments appreciated for their ability to recall lost worlds of action and meaning (Frampton, 1997, p. 24).

Ruins possess a special appeal because they manifest the universal process by which nature destroys what human cultures create. As Riegl ([1903] 1982) put it: "From man we expect accomplished artifacts of a necessary human production; on the other hand, from nature acting over time, we expect their disintegration as the symbol of an equally necessary passing" (p. 32). Age value monuments typically appear at this intersection of culture and nature. Heavily dependent on the perceiver's attitude, ruins are thus a deeply pessimistic "way of seeing . . . counting more ancestors than descendants . . . heralds of the disintegrating mind and collapsing principles of the age after the end of stable belief" (Harbison, 1991, p. 99). According to Buck-Morss (1989, p. 170), in the Arcades project, Walter Benjamin wanted to make the point that the "debris of industrial culture teaches us not the necessity of submitting to historical catastrophe, but the fragility of the social order that tells us this catastrophe is necessary". However, for J. B. Jackson, ruins are necessary because they help us see history "not as a continuity but as a dramatic discontinuity". Only after a period of neglect and decay is it possible for a civilization to recover its "golden age . . . of harmonious beginnings", thus restore the world to "something like its former beauty". In this "cosmic drama", retrieval of the past necessitates "an interim of death or rejection before there can be renewal and reform". In Jackson's words, "[T]he old order has to die before there can be a born-again landscape" (Jackson, 1980, pp. 101–102).

Ruins, however, are neither pure works of art nor unmodified nature, but a combination of "created man-made forms and organic nature". On the one hand, "ruins can no longer be considered genuine works of art, since the original intention of the builder has been more or less lost"; on the other, they cannot simply be taken as "an outgrowth of nature, since man-made elements continue to exist as the basis for that which has been contributed or taken away by Time, its vindictiveness toward human creations" (Zucker, 1968, p. 3). Because of this dialectical relationship, the building fragment thus can reveal the past wholeness of the structure in ways that are meaningful to the present. In Georg Simmel's words, "[O]ut of what of art still lives in the ruin and what of nature already lives in it, there has emerged a new whole, a characteristic unity" (Simmel, 1965, p. 260).

Perhaps, strangely, ruins are a strictly modern phenomenon, relevant only in a world of rapid change. In antiquity, decayed structures were either reused or, if they possessed special religious significance, reburied. In the Middle Ages, unearthed ruins were recycled or incorporated into contemporary structures so as to establish a connection between the Christian present and its pagan Roman past. Not until the Renaissance did the idea of the "irresistible decay of a ruin become differentiated from dilapidation" (Roth, 1997, p. 79). Beginning at that time, meanings gradually were secularized. Archaeology, for example, established the systematic study of ruins as the basis for acquiring scientific knowledge of the past. To the Romantics, ruins signaled not only mortality but also the deep embedding of humanity in the natural world, evidence of nature's "capacity to integrate different stages of human life into a balanced whole" (Roth, 1997, p. 5). With the rise of photography, ruins became "scenes of moral instruction", warnings against the tragic foolishness of war and conflict (Roth, 1997, p. 15).

Ruins retain a strong temporal sense, evidence of the futility of seeking immortality in the false permanence of stone. This sense of the evanescence of things is captured by the term "graceful degradation", which Edward Tenner (1997) relates to "the ability of a system under stress to lose its vital functions in increasing order of importance, instead of behaving wildly or unraveling too quickly" (p. 76). Applied to architecture, this idea suggests that as "[a] place degrades gracefully, it acquires the fascinating complexity that comes with a succession of ownership, each age leaving its own subtle marks" (p. 77). Thus, in cities as varied geographically, historically and economically as Jacksonville (Oregon), Luneberg (Germany), and Shibam (Yemen), places that are neither very popular nor highly elitist, Tenner finds that "arrested growth" has given way to "glorious stagnation" (p. 78).

In the modern age, however, time is not the sole creator of ruins. In fact, contemporary culture generates ruins quickly and profusely. Ruins can be created artificially, as by the construction of follies or the celebration of cracked facades and obsolete styles in some postmodern architecture (SITE, 1980). Ruins also can be created institutionally: by warfare (Beirut, Sarajevo, and Grozny are three recent examples [Squiers, 1986]), by municipal policies of "development" centered on the wholesale razing of structures (discussed above), or by the dynamics of property speculation manifested in the dereliction of the "zone in transition", formed at the edge of downtown as speculators anticipate and seek to profit from expansion of the urban core (Park, Burgess & McKenzie, 1925, p. 50). Terrorist bombings instantly transform buildings into ruins, thus inspiring the massive fortification of surviving structures as well as new construction. Ruins created in these ways rarely become cherished monuments because they stand as reminders of immense suffering or failed and despised policies; usually they are cleared away as soon as possible. An exception is the preservation, as an historic ruin, of the French village of Oradour-sur-Glane, site of a World War II atrocity (Tindall, 1997).[13]

Modern buildings are particularly susceptible to rapid deterioration and hence to ruination. As Frampton (1997) points out, many modernist buildings, far from constituting a "timeless white architecture ... have not withstood the test of time particularly well". For example, canonical twentieth-century buildings including Gerrit Rietveld's house in Utrecht (1923) and Mies van der Rohe's house in Brno (1930), "have ... been extensively restored, not to say largely rebuilt" (pp. 22–28). Le Corbusier's Villa Savoye (1931) was nearly demolished after it fell into disrepair just a few years after being built; it was saved at the last minute by an international petition.[14] However, efforts to prevent demolition of the Tricorn Centre (1964) in Portsmouth, "one of the most brutal of all Britain's concrete buildings", failed in part because the *béton brut* style in which it was executed had fallen completely out of fashion (Glancey, 1997).[15]

Premature deterioration is also evident in many so-called postmodern buildings such as the Pompidou Center in Paris, Louis Kahn's library at the Phillips Academy in Exeter, and the Lloyds building in London. Indeed, Harbison (1991, p. 125) claims this latter building is already a ruin because its "service structures are revealed deliberately as a ruin's might be by accident". All of these buildings have undergone, or are undergoing, expensive refits and modifications. In Japan, ephemerality is celebrated in a financial speculation so fierce that "buildings are designed with the expectation that they will not last and will be torn down sooner rather than later". Thus Tokyo is continually reproduced as a "brand new city"; indeed, most of its architecture dates from after World War II, and "more than 30% of all its structures ... has been built since 1985" (Bongar, 1997, p. 33).

Despite their tendency to premature ruination, Nazi architect Albert Speer thought modern buildings constructed of steel and concrete poorly suited to form that "bridge of tradition" to future generations; indeed, be believed it "hard to imagine that rusting heaps of rubble could communicate these heroic inspirations which Hitler admired in the monuments of the past" (Speer, 1970, p. 56). In his "Theory of Ruin Value", Speer went so far as to suggest that "by using special institutions and by applying certain principles of statics, we should be able to build structures which even in a state of decay, after hundreds or ... thousands of years, would more or less resemble Roman models". Speer even prepared drawings of Zeppelin Field to show what the structure would look like after generations of neglect " ... overrun with ivy, its columns fallen, the walls crumbling here and there, but the outlines still clearly recognizable". Many in Hitler's entourage found this anticipation "blasphemous", conceiving as it did of a period of decline for the thousand year Reich. Hitler accepted it, however, and ordered future important buildings "to be erected in keeping with this law of ruins" (p. 56).

Whereas Speer rejected modern architecture because of its inability to produce a good ruin, its lack of durability has led one commentator to ask if "there is something inherently flawed in modern architecture's conception of a good building". Many modernists, less interested in architecture's traditional obsession with immortality in brick and stone, believed firmly in the power of technology to shape the future. The Italian Futurists, for example, eagerly embraced machine technology, speed, violence, and youth, imagining a world of art and architecture without the dead hand of the past on the throttle of taste and culture. "We combat ... patina and the obscurity of false antiques ... superficial archaism ... ", declared one manifesto; "(A)rchitecture, exhausted by tradition, begins again, forcibly, from the beginning", declared another (quoted in Banham, 1967, pp. 109, 128). Indeed, in the view of Antonio Sant'Elia and others, architecture was to become fluid, plastic, and ephemeral: Each generation, rejecting historical continuity and receiving little from the past, expected to build its structures anew.

Premature deterioration, however, is not only the product of speculative commodification or of an aesthetic that worships the new. It also stems from recent developments in construction technology. For example, a whole generation of modern architects was influenced by the minimalist doctrine which held that good construction was synonymous with a strict economy of materials. Buildings were conceived of as fine-tuned machines constructed of precisely the right amount of the latest materials; as strict expressions of function, buildings needed constant maintenance like machines, something which rarely was supplied.

More fundamentally, the rise of steel-frame construction has effectively replaced the traditional load-bearing wall with the curtain wall, and thus permits buildings to be enclosed with panels of glass or other nonstructural materials. This movement "away from the solid and monolithic wall toward the layered and the veneered" in modern construction, coupled with the need for evermore sophisticated environmental systems, makes "far greater demands on the building envelope than has traditionally been the case". With the decline of the load-bearing wall come many of the problems associated with modern buildings: water leaks, gasket failures, and, especially, deterioration of the exterior panels, producing what has been called "cladding's ticking time bombs" (Ford, 1997, pp. 12–18). These detonations are increasingly familiar to residents of cities such as New York and Philadelphia as buildings catastrophically lose portions of their facades.[16] The marble veneers often used to face buildings are particularly vulnerable to bowing; they necessitate, for example, the massively expensive recladding of the Amoco Building in Chicago and Finlandia Hall in Helsinki.[17]

In England, modern buildings became notorious for their shortcomings, in particular "icons of the machine age polemic" such as James Stirling's Engineering Building at Leicester University (1959) and his History Faculty Building at Cambridge University (1965). Stirling always stressed the functional rationale behind his designs, arguing that "glass buildings are . . . appropriate for the English climate . . . where it is seldom too hot or too cold, and on a normal cloudy day there is a high quality of diffused light in the sky. A glass covering keeps the rain out and lets the light through". As users of these buildings soon learned, however, "[S]loped glass was not ideal to keep the rain out and could let in too much light, cold, and heat" (Curtis, 1987, pp. 320–324; Stamp, 1997). Victims of premature deterioration are not limited to rich corporations or cultural institutions: Some council house tenants in Britain who chose to participate in the Conservative government's scheme to privatize public housing bought houses that the local authorities knew were obsolete and in danger of "collapse without warning" (Pigott, 1997).

By the time buildings are recognized as historical monuments, many have acquired characteristics of a ruin. The massive exterior walls protected Eastern State Penitentiary from the more destructive forms of scavenging and vernacular use, and from fire, the particular scourge of old buildings. The physical isolation of Ellis Island kept the complex relatively undisturbed, allowing its gradual evolution into an evocative ruin. The interiors of both Eastern Penitentiary and Ellis Island dramatically display the action of nature and time on human creations: Rooms and corridors are covered with abstract seascapes of peeling paint, collapsed plaster reveals the geometry of the hidden stonework, plaster dust and debris pile up, and vegetation erupts everywhere. As ruins, these two structures offer a rare chance to confront directly, in an urban environment, the unfathomable forces of entropy – to make contact, as Simmel put it, with "that depth where human purposiveness and the working of non-conscious natural forces grow from their common root" (Simmel, 1965, p. 260).

Relentless restoration, however, threatens to dilute this experience. Many modern ruins are fragile structures "even more vulnerable to environmental attrition than enclosed and inhabited buildings" (Fitch, 1990, p. 295). Hence, like their ancient cousins, they need consolidation and continuing maintenance if they are not to disappear from the landscape. This raises the paradox of attempting "to preserve a site in its decay" (Roth, 1997, p. 8). The ruin stands as a conduit to a past, but as fragment it cannot provide full knowledge of that past. If fully restored, the structure then provides knowledge of that past, but it would cease to be a ruin "since it would have lost nothing over time" (Roth, 1997, p. 9).[18]

Here the value of structures as ruins, as age value monuments, may be incompatible with their claimed status as historical monuments. As David Lowenthal

(1985) writes, "[P]reservation itself reveals that permanence is an illusion. The more we save, the more aware we become that such remains are continually altered and reinterpreted. We suspend their erosion only to transform them in other ways" (p. 410). Recycling old structures such as the Eastern Penitentiary into some future landscape of consumption necessarily involves a policy of preservation in which decay is checked, the structure is stabilized, and the site is cleaned up. Such new meaning can be imposed only at the expense of the structure as a ruin, of its emancipating potential as a liminal zone. Indeed, as Zucker (1962, p. 2) notes, "If structurally adapted to the needs of later centuries, ruins lose their character as ruins".[19]

Ominous signs of preservation, stabilization, and reconstruction are already evident at the penitentiary. Vegetation is being removed, new roofs are being installed, and some cells will be restored to their original condition. Similarly, the "businesslike" restoration of Ellis Island in 1991, its straitjacketed insertion into an officially constructed national mythology, seems to have extinguished its more imaginative properties as a ruin. As age value monuments, utilitarian buildings can transcend their historical origins and meanings. They can become refuges from, and critiques of, the prevailing system, an interstitial zone where dominant values, accepted norms, and conventional modes of behavior – even perception – are suspended or reversed, and where all sorts of experimentation can occur. Indeed, in 1970 Native Americans tried to occupy Ellis Island in an effort to draw attention to the destruction of indigenous peoples by European immigration. Later a group of African Americans landed, cleared undergrowth, and started to restore some of the smaller buildings in connection with their aim to create a rehabilitation center for drug addicts and convicts (von Pressentin Wright, 1983, p. 75).

Contemporary artists are eager to contribute to the new meanings that may be given to, or negotiated for, otherwise neglected and ruined structures and landscapes (Dickinson, 1999). For example, in the 1960s Dennis Oppenheim began using what he called "ravaged sites" – waste dumps, chemical plants, parking lots, freeway embankments – for his conceptual art projects. Also, Robert Smithson sought to give new meaning to a variety of marginal, derelict or otherwise worn-out industrial sites. For example, he turned a visit to the derelict industrial riverfront in nondescript Passaic, New Jersey into an exploration of the ever-present forces of entropic dissolution (Holt, 1979). Smithson also explored the aesthetic potential of ordinary and ruined landscapes in his many photographs of industrial sites in New Jersey and Germany, and in his proposals for the reclamation of brutalized landscapes such as the sites of open-pit mining (Los Angeles County Museum of Art, 1993).

In a similar vein, William Christenberry's work explores the aesthetic of worn, collapsing and vegetation-encrusted structures scattered throughout the rural American south. Bernd and Hilla Becher's taxonomic photographs of industrial facades, blast furnaces, coal tipples, and water towers establish new meaning and aesthetic value for these formerly utilitarian structures. In many cities, derelict industrial zones offer a refuge of large work spaces and cheap rents for artists. For example, in Philadelphia – a typical city in this regard – a new generation of artists (including Michael Frechette, Treacy Ziegler, Thomas Parker, Bob Nesbit, Jennifer Baker and many others) is forging a new "warehouse aesthetic" in which the ruined landscape is treated directly as subject matter, and the detritus of the industrial past incorporated into artworks of considerable originality (Dickinson, 1995). Also, the city is developing a rich stock of mural art. For example, the Mural Arts Program, a community-based mural painting organization supported by the city government (formed partly to domesticate and discipline an earlier generation of freelance graffiti artists) has completed over 1700 indoor and outdoor murals since 1984, and has a backlog of several thousand requests for additional neighborhood murals (Rice, 1999).

Industrial ruins are ominous reminders of lost civilizations of antiquity. As Vergara (1999) points out, while the early modernists described American cities as "unbridled, mad, frenetic, lusting for life", today, although still a leader in science and technology, the United States "now leads the world in the number, size and degradation of its abandoned structures" (p. 12). Indeed, for Van Wyck Brooks (1958) the "heaps of ashes, burned-out frames, seared enclosures, abandoned machinery, and all the tokens of a prodigal and long spent energy" betray the true impulse of American civilization, "a civilization that perpetually overreaches itself only to be obliged to surrender again and again to nature everything it has gained" (p. 92). Vergara's suggestion that the abandoned skyscraper ruins of downtown Detroit be turned into an American Acropolis anticipates one possible future for the shards of the industrial city (Vergara, 1995b, pp. 33–38). One commentator has imagined the death of Manhattan and its rebirth as a ruin: "Time and wind and water have ravaged the natural and man-made alike, and the city's familiar face has eroded in some places to the skeleton underneath. Manhattan strikes the observer the way ancient Athens or ancient Rome strikes us now – as fragmentary, enigmatic, a *memento mori* at once powerful and poignant" (Frank, 1997). Yet how much greater will these ruins of today be than those of antiquity!

CONCLUSION

We live in a New Futurism, an age where once again "speed and cybernetic disposability are advanced as the order of the day" (Frampton, 1997, p. 28).

As cycles of economic activity accelerate and intensify, societies are organized increasingly around abstract systems of knowledge and information. This point suggests that in the future, meaning will be found less in the privacy of psychological space, in the constitution of the individual subject, than in the realm of public, cultural space (Krauss, 1977, p. 240). As art historian Jack Burnham (1968) presciently observed: "We are now in transition from an object-oriented to a systems-oriented culture", a culture less concerned to invent solid artifacts with fixed, static qualities than with the "matter-energy-information exchanges" that organize relationships between people and their environment (p. 30). Thus the real, lasting, palpable, finite, static, and inert world of objects is replaced by an ephemeral, provisional, perishable, mobile, and disposable world dedicated to abstract information and communication. Therefore exchange value prevails over use value, the portable computer reduces the need for office and factory, and the Internet may make many of the transportation industries obsolete. In this future, aesthetics will be miniaturized and privatized, no longer centered on the architecture of grand structures but on the micro-architecture of the chip, the design of computer program icons, the decoration of the body.

Ours is the last culture that will produce gigantic structures able to run the course from utility to ruin. Indeed, the industrial ruins of today, produced in abundance in the entropic "factory" of contemporary change, are likely the last existing class of structures whose massive scale, specialized function, and solid construction make them suitable candidates for survival in the landscape and thus for evolution into the monuments of tomorrow. As Frampton (1997, p. 25) observes, perhaps only in the durability of such structures can architecture continue to "stand against the fungibility of things and the mortality of the species".

In the future, our era's power stations, steel mills, floating exploration platforms, oil refineries, cracking towers, tank farms, automobile factories, freeway interchanges, and shopping malls will appear as fantastic and as worthy of preservation as do step pyramids, Greek temples, and Gothic cathedrals today.

NOTES

1. For a discussion of how American architecture serves as "Europe's unconscious", see Cohen (1995).
2. Rem Koolhaas (1994) later took Le Corbusier to task, arguing that the "campaign of denigration" he waged against New York's skyscrapers in *The Radiant City* (1935) was possible only because "its strategist has never beheld the object of his aggression ... Le Corbusier's portrait of New York is an identikit: a purely speculative collage of its 'criminal' urbanistic features" (pp. 253–254).

3. For a contemporary appreciation of these "timeless cathedrals" of the Plains, see Mahar-Keplinger (1993).

4. In some instances, cultural memory sustaining a monument's meaning and existence may fade or disappear. For example, we can no longer recall the symbolic universe that gave rise to the stone figures on Easter Island. In other cases, a new regime may destroy the monument in an effort to expunge memory of the person or event so commemorated, as when the Nazis destroyed the memorial (designed by Mies van der Rohe) to Rosa Luxemburg and Karl Liebnitz. Again, the style in which a commemorative structure is rendered may fall out of favor. Consequently the intentional monument may be discarded or may disappear from the landscape.

5. Two structures in Philadelphia provide an interesting counterpoint. First, in June 1998, explosive charges failed to bring down a large part of the Jack Frost sugar refinery, a complex of eighteen interconnected buildings scheduled for demolition by implosion. The building collapsed unexpectedly after the wrecking crew spent most of a day trying to pull it down with cables. According to proletarian legend, the building's resistance was the result of its "Drummond steel" construction. See Vergara (1999, p. 40). Second, because of its location in the heart of Philadelphia, the spectacular 38-story fire ruin, One Meridian Plaza, was dismantled piece by piece; this process cost $25 million and took two years. See Kanaley (1998).

6. In 1993 the Pennsylvania Supreme Court ruled that the Historical Commission had jurisdiction only over building exteriors, not their interiors, thus severely impeding the task of historical preservation.

7. In an unintended irony, the entropic landscape of ruined and toppled buildings created as the set for Stanley Kubrick's 1987 film, *Full Metal Jacket*, occupies a site on the eastern perimeter of Docklands.

8. A comprehensive overview of commercial and housing developments on the Isle of Dogs is Cox (1995). See also al-Naib (1994).

9. Emergent urban and suburban landscapes in the United States are described variously in Blakely and Snyder (1997); Davis (1992, 1990); Lang (1995); Soja (1995); and Vergara (1995a). The psychogeography of upscale residential enclaves is the subject of several works of fiction by J. G. Ballard, including *Running Wild* (1988) and *Cocaine Nights* (1996).

10. This transformation also occurs in rural areas. For example, the Amish of Pennsylvania are commercially successful farmers despite their use of largely premodern technology. In the process they have created a distinctive landscape (in Lancaster County), which itself is now a major tourist attraction. Globalization and recession affect agriculture, leading – at least in Europe – to the conversion of farm buildings into tourist facilities. See Tagliabue (1998).

11. For a discussion of the politics of waterfront redevelopment projects, see Gordon (1997).

12. Steam engine rallies, highly popular events in the British countryside, as well as in the United States, are pure indulgences in the delights of the mechanical sublime. Displays of engines, many originally developed for agricultural applications, operate en masse for no other purpose than to show off "the poetry of mechanical motion" (Nye, 1994, p. 121). The accompanying cacophony is surely a realization of Luigi Russolo's "noise orchestra" dreams. (Old Tractors Never Die, 1998.)

13. As Fitch (1990) notes, where rebuilding occurs, as in the aftermath of World War II, either the "uncontrolled erosion of peacetime construction threatens to wreak as much

damage ... as did the mindless violence of war" or an effort is made to return the city to "something of its pre-war physiognomy" through a program of preservation, restoration, and reconstruction (p. 375). The latter policy has been pursued with some success in eastern European cities such as Warsaw. Lebbeus Woods, however, argues that the architectural reproduction of the past leads to social amnesia. He believes that restoration of cities destroyed by violence is a "reaffirmation of a past social order that ended in war ... (a) parody, worthy only of the admiration of tourists". It serves only to prop up "decrepit hierarchies, struggling to legitimate themselves ... through sentimentality and nostalgia" (Woods quoted in Roth, 1997, pp. 36–37). Hans Riemer (1946) photographed the ruins of Vienna as an antidote to man's "wonderful gift for forgetfulness ... as a continual reminder to ourselves and as a warning to future generations" (pp. v-vi).

14. Villa Savoye was declared an historical monument in 1964. See Jenger (1996, p. 53). Peter Blake (1996) thought it "a delightful ruin, perhaps much better left that way" (p. 64).

15. In J. G. Ballard's novel *High Rise* (1977), ruination of the monolithic apartment block – a sort of Le Corbusier's *Unité d'Habitation* on steroids – stems not so much from physical deterioration or poor construction as from the residents' reaction to the building's utopian aspirations, which cause them to descend into physical and psychic barbarism.

16. Incredibly, some "showers of bricks" result from failure to secure the facade to the building frame with metal ties during construction. Thus responsibility cannot be laid unequivocally at the doorstep of modernism. In other instances, failure is related to deterioration of the ties. See Kennedy (1998).

17. Mies van der Rohe captures the primacy of structure over facade in modern architecture: "Skyscrapers reveal their bold structural pattern during construction. Only then does the gigantic steel web seem impressive. When the outer walls are in place, the structural system which is the basis of all artistic design, is hidden by the chaos of meaningless and trivial forms" (Johnson, 1978, p. 187). Since gigantic structures of engineering such as bridges, dams and industrial installations are pure, unencumbered expressions of technology (they have no need of an enclosing outer skin or envelope), these modern structures are the "symbols of their epoch" which best express the "potential greatness of our times" (p. 191–192).

18. The paradox is also apparent in plans for restoring and modernizing ancient cities such as Fez, Morocco (Kimmelman 1998).

19. For an early discussion of how the "visual charms" and "appealing associations" of architectural decay might be incorporated into the "aesthetic repertoire of the modern planner", see Piper (1947).

ACKNOWLEDGMENTS

I would like to thank Camilo Vergara, Edward Tenner, Susan Roschke, and Jerilou Hammett for helpful comments and ideas. I am indebted to Karen Feinberg for her excellent editing of an earlier version and to John LeMasney, Office of Information Technology, Rider University for assistance in preparing the photographic illustrations. The writing of this paper has been generously supported by a Rider University Research Leave and Summer Research Fellowship. A version was presented at the Urban Lifestyles conference, Newcastle upon Tyne, England, and appears in the conference proceedings (Benson & Roe, 2000).

REFERENCES

Ballard, J. G. (1977). *High Rise*. London: Triad/Panther.
Ballard, J. G. (1988). *Running Wild*. New York: Farrar, Straus, Giroux.
Ballard, J. G. (1996). *Cocaine Nights*. London: Harper Collins.
Banham, R. (1967). *Theory and Design in the First Machine Age*. New York: Praeger.
Banham, R. (1986). *A Concrete Atlantis*. Cambridge, MA: MIT Press.
Barber, B. (1998). Corporate Citizens. *Metropolis*, (October), 169.
Barry, E. (1999). Down–and Out–in the Valley. *Metropolis*, (May), 69–73.
Belluck, P. (1998). Razing the Slums to Rescue the Residents. *New York Times*, (September 6), A-1, 26–27.
Benson, J., & Roe, M. (Eds) (2000). *Urban Lifestyles: Spaces, Places, People*. Rotterdam, NL: Balkema Publishers.
Bird, J. (1993). Dystopia on the Thames. In: J. Bird, B. Curtis, T. Putman, G. Robertson & L. Tickner (Eds), *Mapping the Futures: Local Cultures, Global Changes* (pp. 120–135). London: Routledge.
Blake, P. (1996). *The Master Builders*. New York: Norton.
Blakely, E., & Synder, M. (1997). *Fortress America: Gated Communities in the United States*. Washington D.C.: Brookings Institution.
Bogner, B. (1997). What Goes Up, Must Come Down. *Harvard Design Magazine*, (Fall), 33–43.
Brodie, A., Croom J., & Davis, J. O. (1999). *Behind Doors: The Hidden Architecture of England's Prisons*. Swindon, U.K.: English Heritage.
Brooke, J. (1997). Sleeping Below Plains, Missiles Stay on the Alert. *New York Times*, (December 15), A-14.
Brooks, V. W. (1958). *America's Coming of Age*. New York: Doubleday.
Buck-Morss, S. (1991). *The Dialectics of Seeing: Walter Benjamin and the Arcades Project*. Cambridge, MA: MIT Press.
Burnham, J. (1968). *Beyond Modern Sculpture*. New York: Braziller.
Cawthon, R. (2000). A Massive Rebuilding of Housing in Chicago. *Philadelphia Inquirer* (February 13), A-1, 6.
Central Philadelphia Development Corporation (1997). *Center City Digest*, (Summer).
Cohen, J-L. (1995). *Scenes of the World to Come: European Architecture and the American Challenge, 1893–1960*. Montreal: Canadian Center for Architecture.
Cox, A. (1995). *Docklands in the Making: The Redevelopment of the Isle of Dogs*. London: Athlone Press.
Curtis, W. (1987). *Modern Architecture Since 1900*. Englewood Cliffs, NJ: Prentice Hall.
Davis, M. (1998). Old Buss Co. Site Will Be Recycled into a Golf Course. *Philadelphia Inquirer* (August 11), A-1, 18.
Davis, M. (1990). *City of Quartz*. London: Vintage Books.
Davis, M. (1992). *Urban Control: The Ecology of Fear*. Westfield, NJ: Open Magazine Pamphlet Series.

Dickinson, J. (1995). Entropic Zones: The Art of Ruins and the Contemporary City. Catalog essay for the exhibition *Entropic Zones: Buildings and Structures of the Contemporary City*. Lawrenceville, NJ: Rider University Gallery.
Dickinson, J. (1996). Entropic Zones: Buildings and Structures of the Contemporary City. *Capitalism, Nature, Socialism, 7*(3), (September), 81–95.
Dickinson, J. (1999a). Journey Into Space: Interpretations of Landscape in Contemporary Art. In: D. E. Nye (Ed.), *Technologies of Landscape: From Reaping to Recycling* (pp. 40–66). Amherst, MA: University of Massachusetts Press.
Dickinson, J. (1999b). Entropic Zones. *Designer/builder*, (November), 19–27.
Dickinson, J. (2000). A Cheerless Blank: Philadelphia's Eastern State Penitentiary and the Architecture of Solitary Confinement. *Designer/builder*, (November), 17–27.
Dunlap, D. (1998). Grand Central, Reborn as a Mall. *New York Times*, (August 2), B-33, 38.
Fitch, J. (1990). *Historic Preservation: Curatorial Management of the Built World*. Charlottesville and London: University Press of Virginia.
Ford, E. (1997). The Theory and Practice of Impermanence. *Harvard Design Magazine*, (Fall), 12–18.
Forster, K. (1982). Monument/Memory and the Mortality of Architecture. *Oppositions, 25*(Fall), 2–19.
Frampton, K. (1997). Intimations of Durability. *Harvard Design Magazine*, (Fall), 22–28.
Frank, M. (1997). New York Skyscrapers as Ancient Temples. *New York Times*, (July 4), Weekend section, 26.
Garn, A. (1999). *Bethlehem Steel*. New York: Princeton Architectural Press.
Glancy, J. (1997). Ooh, You are Brutal. *The Guardian*, (London) (July 21), Arts section, 12–13.
Goin, P. (1991). *Nuclear Landscapes*. Baltimore: Johns Hopkins University Press.
Goin, P., & Raymond, E. (1999). Recycled Landscapes: Mining's Legacies in the Mesabi Iron Range. In: D. E. Nye (Ed.), *Landscape and Technology: From Reaping to Recycling* (pp. 267–283). Amherst, MA: University of Massachusetts Press.
Goldberg, C. (1997). Costly Energy Failure Presents New Challenge. *New York Times*, (March 9), A-9.
Gordon, L. (1997). Managing the Changing Political Environment in Urban Waterfront Redevelopment. *Urban Studies, 34*(1), 61–83.
Halbfinger, D. (1998). Fires Are a Blow to Bronx As It Tries to Lure Business. *New York Times*, (August 24), B-3.
Harbison, R. (1991). *The Built, the Unbuilt and the Unbuildable: In Pursuit of Architectural Meaning*. Cambridge, MA: MIT Press.
Hardingham, S. (1996a). *London: A Guide to Recent Architecture*. London: Ellipsis.
Hardingham, S. (1996b). *England: A Guide to Recent Architecture*. London: Ellipsis.
Holt, N. (Ed.) (1979). *The Writings of Robert Smithson*. New York: New York University Press.
Imrie, R. (1997). National Economic Policy in the United Kingdom. In: M. Pacione (Ed.), *Britain's Cities: Geographies of Division in Urban Britain* (pp. 88–107). London: Routledge.
Infield, T. (1997). In Pennsylvania, Industrial Giants Are Now History. *Philadelphia Inquirer*, (March 16), E-1, 3.
Iverson, M. (1993). *Alois Riegl: Art History and Theory*. Cambridge, MA: MIT Press.
Jackson, J. B. (1980). *The Necessity of Ruins*. Amherst, MA: University of Massachusetts Press.
Jacobs, J. (1961). *The Death and Life of Great American Cities*. New York: Random House.
Jenger, J. (1996). *Le Corbusier: Architect, Painter, Poet*. New York: Abrams.
Johnson, D. (1999). A Suburbiascape Grows in Inner-City Chicago. *New York Times*, (October 20), A-18.

Johnson, P. (1978). *Mies van der Rohe*. New York: Museum of Modern Art.
Johnson, T. (1996). Razing Philadelphia. *Philadelphia Inquirer*, (October 13), Magazine section, 14–31.
Johnston, N. (1994). *Eastern State Penitentiary: Crucible of Good Intentions*. Philadelphia: Philadelphia Museum of Art.
Johnston, N. (2000). *Forms of Constraint: A History of Prison Architecture*. Urbana and Chicago, IL: University of Illinois Press.
Kanaley, R. (1998). At Last, Meridian is Coming Down. *Philadelphia Inquirer*, (April 26), A-1, 16.
Katz, A. (1998). Dismantling the Motor City. *Metropolis*, (June), 33.
Kennedy, M. (1997). Old Workhouse to Stand Again. *The Guardian*, (London) (July 24).
Kennedy, R. (1998). In Midtown, Yet Another Shower of Bricks. *New York Times*, (August 11), B-1, 3.
Kimmelman, M. (1998). Preserving a City's Soul. *New York Times*, (August 18), E-1, 3.
Koolhaas, R. (1986). Response to Questionnaire. *Zone 1/2*, 448.
Koolhaas, R. (1994). *Delirious New York*. New York: Monacelli Press.
Kostof, S. (1982). His Majesty the Pick: The Aesthetics of Demolition. *Design Quarterly*, (118–119), 33–41.
Krauss, R. (1977). *Passages in Modern Sculpture*. Cambridge, MA: MIT Press.
Le Corbusier. ([1926] 1986). *Towards a New Architecture*. Mineola, NY: Dover Books.
Loewen, J. W. (1999). *Lies Across America: What Our Historic Sites Get Wrong*. New York: New Press.
Lowenthal, D. (1985). *The Past Is a Foreign Country*. Cambridge, U.K.: Cambridge University Press.
Mahar-Keplinger, L. (1993). *Grain Elevators*. New York: Princeton Architectural Press.
Marx, K. (n.d.) *Capital*, Vol. I. Moscow: Progress Publishers.
McCoy, C., & Sataline, S. (1996). Who's Burning Philadelphia? *Philadelphia Inquirer*, (October 13), A-1, 24.
McNeil, D. (1988). For Rising Crime Rate, High-Rise Jail to Match. *New York Times*, (February 28), A-4.
Mendelsohn, E. ([1926] 1993). *Erich Mendelsohn's "Amerika"*. New York: Dover Publications.
Missile Base Living on the Kansas Prairie (1999). *Designer/builder*, (May), 5–7.
Morrison, K. (1999). *The Workhouse: A Study of Poor Law Buildings in England*. Swindon, U.K.: English Heritage.
Mulvihill, G. (1998). In Camden, Waterfront Dream Builds. *Philadelphia Inquirer*, (August 9), B-1, 4.
al-Naib, S. K. (1994). *London Docklands: Past Present and Future*. London: Thames and Hudson.
Newman, A. (1997). Fill 'er Up? We Do. *New York Times*, (June 15), Sec. 13–1.
Nicholas, P. (1998). PHA Is Urged to Downsize Housing. *Philadelphia Inquirer*, (March 29), B-1, 2.
Nye, D. E. (1994). *American Technological Sublime*. Cambridge, MA: MIT Press.
Office of Housing and Community Development (1997). *Neighborhood Transformations: The Implementation of Philadelphia's Community Development Policy*. Philadelphia: City of Philadelphia, Office of Housing and Community Development.
Old Tractors Never Die: They Go to the Threshing Bee (1998). *New York Times*, (August 31), A-16.
Park, R. E., Burgess, E., & McKenzie, R. (1925). *The City*. Chicago: University of Chicago Press.
Parmley, S. (1997). A Crumbling Brewery Raises Ire in East Falls. *Philadelphia Inquirer*, (July 11), B-1, 5.

Pigott, R. (1997). Behind the Facade. *The Guardian*, (London) (July 30), Society section, 2–3.
Piper, J. (1947). Pleasing Decay. *Architectural Review*, (September), 85–96.
Preservation Alliance for Greater Philadelphia (1998). Preservation is Central to Center City Development. *Preservation Matters*, (September-October).
von Pressentin Wright, C. (1983). *Blue Guide to New York*. New York: Norton.
Rice, R. (1999). Art as Civic Biography: Philadelphia's Murals Project. *New Art Examiner*, (April), 18–23.
Riegl, A. ([1903] 1982). The Modern Cult of Monuments: Its Character and Its Origin. Translated by Kurt Forster. *Oppositions*, (Fall), 25, 21–78.
Riemer, H. (1946). *This Pearl Vienna*. Vienna: Jugend und Volk.
Roth, M. S. with Lyons, C., & Merewether, C. (1997). *Irresistible Decay: Reclaiming Ruins*. Los Angeles: Getty Research Institute.
Samuels, D. (1977). Bringing Down the House: An Explosion in Las Vegas Plays as Performance Art. *Harper's Magazine*, (July), 37–52.
Sheffield Development Corporation (n.d.). *Annual Report and Financial Statements, 1988–89*. Sheffield, U.K.: Sheffield Development Corporation.
Simmel, G. (1965). The Ruin. In: K. H. Wolff (Ed.), *Georg Simmel: Essays on Sociology, Philosophy and Aesthetics* (pp. 259–266). New York: Harper & Row.
SITE (1980). *SITE: Architecture as Art*. New York: St. Martin's Press.
Sobieszek, R. A. (1993). *Robert Smithson: Photo Works*. Los Angeles: Los Angeles County Museum of Art.
Soja, E. (1995). Postmodern Urbanism: The Six Restructurings of Los Angeles. In: S. Watson & K. Gibson (Eds), *Postmodern Cities and Spaces* (pp. 125–137). Oxford: Blackwell.
Speer, A. (1970). *Inside the Third Reich*. New York: Simon and Schuster.
Squiers, C. (1986). A Short History of Beirut in the 20th Century. *Zone 1/2*, 369–422.
Stamp, G. (1997). The Durability of Reputation. *Harvard Design Magazine*, (Fall), 55–57.
Steinglass, M. (2000). The Machine in the Garden. *Metropolis*, (October), 126–33, 166–167.
Tagliabue, J. (1998). Preserving a Heritage via Bed and Barn. *New York Times*, (August 13), Business section, 1, 4.
Tenner, E. (1997). Aging Gracefully. *Metropolis*, (September), 76–79.
Tindall, G. (1997). The Village That Hasn't Forgotten. *New York Times*, (January 5), Travel section, 8–9.
United States Department of Energy (1992). *Legend and Legacy: Fifty Years of Defense Production at the Hanford Site*. Washington D.C.: Department of Energy.
United States Department of Energy (1996). *Closing the Circle on Splitting the Atom*. Washington D.C.: United States Government Printing Office.
Vergara, C. (1995a). *The New American Ghetto*. New Brunswick, NJ: Rutgers University Press.
Vergara, C. (1995b). Downtown Detroit: American Acropolis or Vacant Land? *Metropolis*, (April), 33–38.
Vergara, C. (1997). Gorgeous Hulks. *Philadelphia Inquirer*, (August 2). Editorial section.
Vergara, C. (1999). *American Ruins*. New York: Monacelli Press.
Weyrich, N. (1998). Cul-de-Sacs Amid Decay: Can a Shrinking City Renew Itself by Learning from Suburbia? *Philadelphia Weekly*, (April 15), 11–13.
Wilford, J. N. (1998). Where Have all the Launch Pads Gone? *New York Times*, (October 13), F-5.
Woods, L. (1995). Everyday War. In: P. Lang (Ed.), *Mortal City*, (pp. 46–53). New York: Princeton Architectural Press.
Zink, C. (1998). Save the City's Iron-Clad Legacy. *Trenton Times*, (July 19), Commentary section, 1, 3.

Zink, C., & Hartman, D. W. (1992). *Spanning the Industrial Age: The John A. Roebling's Sons Company, Trenton, New Jersey 1848–1974*. Trenton, NJ: Trenton Roebling Community Development Corporation.

Zucker, P. (1968). *The Fascination of Decay*. Ridgewood, NJ: The Gregg Press.

Zukin, S. (1991). *Landscapes of Power: From Detroit to Disney World*. Berkeley, CA: University of California Press.

Zukin, S. (1995). *The Cultures of Cities*. Cambridge, MA: Blackwell.

CITY REDEVELOPMENT POLICIES AND THE CRIMINALIZATION OF HOMELESSNESS: A NARRATIVE CASE STUDY

Adalberto Aguirre, Jr. and Jonathan Brooks

INTRODUCTION

We see them pushing shopping carts filled with clothes, plastic bottles, newspapers, and discarded furniture. A dog is sometimes leashed to the shopping cart. We often see them going through trash dumpsters behind grocery stores. We see them at freeway exists holding up cardboard sings with a simple message – *Please Help! Need Food for my Family!* Sometimes we see children not too far away, watching and waiting. Who are these people? They are the homeless.

Homelessness is a pervasive feature in most urban areas in America.[1] Congress defines a *homeless person* in the Stewart B. McKinney Homeless Assistance Act (1988) as a person who lacks a fixed, regular, and adequate nighttime residence and is living either in a shelter, on the streets, or in public places not ordinarily used as living accommodations. Estimating the size of the homeless population in the U.S. or in any given city is problematic. The homeless are constantly on the move and they shy away from public contact. Most estimates of the homeless population place its size between 300,000 and 7,000,000, with at least 750,000 persons being homeless on any given night[2]

(Burt, 1995; Interagency Council on the Homeless, 1994; National Coalition for the Homeless, 1998). By conservative estimates, the homeless population includes between 150,000 and 300,000 children (Foscarinis & Scheibel, 1993; Stronge & Helm, 1991; Thompson, 1998). Single mothers and their children often find themselves homeless due to a lack of affordable housing or because they are fleeing domestic violence or an abusive household (Franzese, 1999; O'Flaherty, 1995; Mullins, 1994; Soine & Burg, 1995).[3]

The problem with most enumeration studies of the homeless population is that they attempt to only count the *literally homeless*, persons in shelters or on the streets. However, they fail to take into account the *hidden homeless*, persons squatting in buildings, living in automobiles, or other places, or the *proto-homeless*, persons living in garages or other non-conventional dwellings. Including these persons in estimates of the homeless population would noticeably increase the population's size. Many advocates for the homeless suspect that government estimates for the homeless population are low because of the "political implications involved in estimating the extent of homelessnes" (Yeich, 1994, pp. 5–6). One, however, must heed Kozol's (1988, p. 3) observation: "We would be wise to avoid the numbers game. Any search for the 'right number' carries the assumption that we may at last arrive at an acceptable number. There is no acceptable number whether the number is one million or four million or the administration's estimate of less than a million *there are too many homeless people in America*."

While the sociological literature is replete with studies of homelessness, most have focused on demographic characteristics: who and how many are homeless (Burt, 1992; Jencks, 1994; Wright, 1988); the causes of homelessness, ranging from personal factors – mental illness, alcohol and/or substance abuse (Barrow et al., 1999; Henderson, 1999), to structural factors – poverty, unemployment, and welfare and housing policy (Belcher & Singer, 1988; Drier & Applebaum, 1991; Elliot & Krivo, 1994; Timmer & Eitzen, 1992); ethnographic works that have given us a more subjective understanding of homelessness (Kozol, 1988; Liebow, 1993; Snow & Anderson, 1993). Few studies, however, have examined the *criminalization* of homelessness (Aulette & Aulette, 1987; Barak, 1991; Barak & Bohm, 1989; Goetz, 1992; Kress, 1994). In particular, few studies have examined the use of city redevelopment policies in criminalizing the homeless.[4]

The National Law Center on Homelessness and Poverty (1994, 1996, 1999) has documented an increasing trend among city governments in the use of redevelopment or revitalization policies to criminalize the homeless. For example, in an effort to make downtown areas more attractive to business investors and tourists, city governments will pass ordinances that restrict the use of public space by the homeless. These ordinances usually prohibit begging, sleeping or camping on public benches, sitting or lying down on public sidewalks, and being in public areas during certain hours (Ellickson, 1996; Foscarinis, 1996). To enforce the ordinances, police conduct sweeps of public areas usually inhabited by the homeless (Foscarinis, Cunningham-Bowers & Brown, 1999). As a result, police sweeps of public areas tend to capture a sizable number of homeless persons. In addition, by using zoning and building code ordinances to limit the creation of homeless shelters, city governments increase the chances that homeless persons will be caught in police sweeps of public areas.

Our purpose in this essay is to examine the relationship between downtown redevelopment and homeless policy. We will discuss how city redevelopment issues shaped homeless policy in the city of Riverside, California between the years 1986 and 1998. The narrative case study for the city of Riverside we present in this paper is drawn from articles in the city's major newspaper, *The Press Enterprise,* and interviews conducted with service providers to the homeless, homeless advocates, and homeless persons. We also attended several meetings held by the Riverside Downtown Partnership (an organization of downtown business owners, residents, and politicians) and the Riverside Homeless Coordinating Council (an organization of service providers, advocates, and politicians).[5] We argue in this paper that city redevelopment policies transformed public space into *contested terrain* in which the homeless became victims of business elite interests. The goal of the city's business elite was to conceal or isolate the homeless at any cost.[6]

THE POLITICAL ECONOMY CONTEXT OF HOMELESSNESS

The causes of homelessness are numerous, from personal problems to social structural forces. According to Nelson (1986, p. 48), an examination of the social structural forces associated with homelessness is necessary because the "problems of homelessness stem from something much deeper than individual affliction; homeless people are poor. There are simply not enough jobs or houses to go around for the country's lowest class, those who have fallen through the safety net." Social structural forces such as the non-availability of low-cost housing, gentrification and urban renewal, cutbacks in welfare, deindustrialization, and global economic restructuring have contributed to increases in the homeless population, especially over the last fifteen years (Barak, 1991; Marcuse, 1988; Timmer & Eitzen, 1992; Timmer, Eitzen & Tally, 1994; Yeich, 1994). For example, in a 1995 survey of twenty-nine cities in the U.S. conducted by the United States Conference of Mayors it was found that while 73% of the cities reported increases in requests for assisted housing from low income families, 71% of the cities reported closing down some types of assistance because the waiting lists were too long (Foscarinis, 1996). Advocates for the homeless argue that the lack of housing assistance programs for low income residents increases their chances of joining the homeless population.

Without question one cause of homelessness is a lack of affordable housing. As the need for affordable housing increased in the 1980s due to structural changes in a receding economy, the number of affordable housing units declined (Berger, 1996; Ladd, 1986; Lang, 1989; Turner, 1991). The non-availability of affordable housing may have contributed to the public's perception that homelessness was *cresting* in the 1980s (Ellickson, 1996).[7] Many of the housing programs designed to construct new low-cost housing or reconstruct old housing were eliminated. Much of the existing supply of public housing was subjected to demolition, sale, abandonment or conversion (Nelson, 1986).

As cities slowed down the construction of affordable housing for low income and homeless persons, they embarked on a new initiative during the late 1980s, to revitalize downtown areas in order to attract high income residents and tourists that could turn the downtown into a productive economy. However, in order to promote downtown revitalization among a city's business elite, cities found it necessary to expel unattractive elements, such as the homeless (National Center on Homelessness and Poverty, 1999). In a sense, the homeless became a threat to the political economy infrastructure of downtown revitalization.[8] Feagin and Parker (1990) have described city revitalization as an *urban renaissance* designed to bring affluent, and mostly white persons, back to the city. Interestingly, Marx (1990, p. 812) observed in *Capital* how the business elite pushed the removal of undesirables in order to promote the creation of capital: "Improvements of towns which accompany the increase in wealth, such as the demolition of badly built districts, the erection of palaces to house banks, warehouses, etc., the widening of streets for business traffic, for luxury carriages, for the introduction of tramways, obviously drive the poor away into even worse and more crowded corners."

The Politics of Public Space

In their attempts to revitalize downtown areas cities and the homeless have come into conflict over the use of public space. Henri Lefebvre, a French philosopher and urbanist, identifies two dimensions of space in everyday life, *social space* and *abstract space* (Gottdiener, 1994). *Social space* refers to how people use space as a place to live and interact with others. Homeless people tend to think of public space this way, space in terms of its use-value. Marx (1990, p. 126) writes in *Capital* that the "usefulness of a thing makes it a use-value . . . Use-values are only realized in use or in consumption." One can make the general observation that people use or consume space in their day-to-day activities. The homeless, however, having nowhere else to go, unless they find shelter, must use *public space* as a place to live and engage in their everyday activities. For the homeless, public space becomes their social space.

Abstract space refers to its "abstract qualities of dimension – size, width, area, location – and profit" (Gottdiener, 1994, p. 127). The elite – politicians, business owners, developers, and investors – think about space in these abstract terms, particularly in terms of its exchange-value. *Exchange-value* refers to the "quantitative relation, the proportion, in which use-values of one kind exchange for use-values of another kind" (Marx, 1990, p. 126). *Public space*, in this sense then, is a commodity because "any given piece of real estate has both a use value and an exchange value" (Logan & Molotch, 1987, p. 1). In addition, because public space is a commodity "individuals and groups differ on which aspect (use or exchange) is most crucial to their own lives" (Logan & Molotch, p. 2). Cities and the homeless will thus come into conflict over public space because they will differ over the use or exchange value of space. That is, for the homeless, public space is their social space because it has use-value. In contrast, for the business elite, public space is abstract space because of its exchange-value. As Logan and Molotch (1987, p. 3) have noted: "The ensuing conflict between those seeking gain from exchange values and those from use values is by no means a symmetrical one, for differently equipped contenders mobilize their individual, organizational, and class resources on behalf of place-related goals. The ability to manipulate place successfully, including altering the standing of one place to that of another, is linked to an individual's location in the stratification system generally."

With the cards stacked against them, and sitting at the very bottom of the stratification system, the homeless have little hope of being able to manipulate public space in their favor when facing powerful elite interests, e.g. politicians and business owners. Public space thus becomes an arena of struggle and the "politics of place creates a dynamic of exclusion whereby the homeless are uprooted from each consecutive living space (however inadequate) they manage to secure in the city" (Wolch & Dear, 1993, p. 200). In an attempt to win the battle with the homeless over public space, cities have resorted to criminalizing behaviors associated with the homeless in order to drive them out of revitalized areas. Foscarinis (1996, p. 30) has observed that some cities "state expressly that their intention is to drive their homeless residents out of the city" by criminalizing behaviors associated with the homeless.[9]

Neutralizing the Homeless

Many cities across the U.S. are more interested in the problems caused by homelessness than in the problems that cause homelessness (National Law Center on Homelessness and Poverty, 1994). The business elite in cities has shown that it is more concerned with meeting its need to revitalize the downtown areas in order to attract tourists, customers, and business investment to the downtown area than meeting the needs of its homeless population. As Marcuse (1988, p. 91) has noted: "The shock of homelessness would not be so great if the homeless were only in ghettos or slum areas remote from downtown. But homelessness has settled down in the middle of the central business district, threatening to drive away business and the tourist dollar".

City governments are primarily interested in economic growth such that "economic growth has become the defining issue of urban politics as cities are competing with each other to attract investment of mobile capital" (Goetz, 1992, p. 540). To attract mobile capital, cities invest in the *second circuit of capital* or *real estate* (Gottdiener, 1994). The homeless who are seen as a threat to business must be eliminated from the downtown areas in order for investment to occur because homelessness has a depressing impact on the exchange value of land (Goetz, 1992). Furthermore, services for the homeless, such as shelters or feeding programs, are seen as counterproductive as well because they allegedly might "attract" or "anchor" a large homeless population (Feagin & Feagin, 1990).

Cities employ several neutralizing techniques with the homeless in order not to legitimate their presence. One neutralizing technique is to blame the victim. Blaming the victim treats the homeless persons as undeserving (Gans, 1992; Wright, 1988). Blaming the victim serves a number of important functions for the "better-off population" (Gans, 1994). According to Yeich (1994, p. 1), "Attempts to pathologize homelessness have offered policy makers a convenient justification for inaction." It "justifies political ideologies and interests that oppose solutions, and thus increases the likelihood that nothing will be done" about homelessness (Gans, 1992, p. 465). Accordingly, Feagin and Feagin (1990, p. 296) argue that the homeless are "viewed not as people suffering because of societal conditions, such as unemployment and the lack of affordable housing, but rather stigmatized as highly deviant and as 'social refuse' to be discarded". Blaming the victim thus turns attention away from the social structural forces that give rise to homelessness and instead focuses on the homeless themselves as the source of the problem (Aulette & Aulette, 1987). As such, homeless persons are blamed for their own victimization because "That's how they want to live".

Another neutralizing technique is to label the homeless as hard-core criminals. Research studies, however, challenge the depiction of homeless persons as criminals. For example, Snow, Baker, and Anderson (1989) in their examination of the link between homelessness and criminality in Austin, Texas found that most arrests of homeless men were for non-violent, victimless offenses. Furthermore, they noted that rather than being a threat to public safety "at worst, they appear to be public nuisances who 'contaminate' public space and who are frequent but small-scale property offenders" (p. 546). Aulette and Aulette (1987, p. 247–248) argue that the crimes committed by the homeless "boil down to living on the street. That is, the activities of everyday life, eating, sleeping, carrying around one's belongings, looking for work, just being outside are 'crimes'." Other researchers have noted that the homeless are more often the victims of crime rather than the perpetrators (Barak, 1991; Barak & Bohm, 1989).

Criminalizing Homelessness

If these neutralizing techniques are not successful then criminalizing homelessness is a strategy employed by city governments to expel the homeless. In a survey of the 50 largest cities in the U.S. conducted by the National Law Center on Homelessness and Poverty (1996), it was found that 38% of the cities had initiated crackdowns on homeless people in the past several years, 54% had engaged in recent police "sweeps" of homeless people, and 77% had passed ordinances prohibiting or restricting begging. In San Francisco, for example, police took photographs of homeless persons they identified as "habitual drinkers" and distributed the photos to liquor stores (Foscarinis, Cunningham-Bowers & Brown, 1999). Liquor store owners were warned by the police that if they sold alcohol to the persons in the photos then the store owners would be cited for violating an "old" law prohibiting the sale of alcohol to habitual drinkers. Even liberal-minded cities such as Santa Monica, Berkeley, Santa Barbara, and West Hollywood have changed their policies toward the homeless and adopted more aggressive strategies (National Law Center on Homelessness and Poverty, 1994). Berkeley, Santa Barbara, and West Hollywood use their trespassing laws to keep downtown sidewalks clear of homeless persons. Santa Monica has adopted an anti-camping ordinance that bans sleeping in public parks although there is nowhere else for homeless persons to sleep.[10]

A NARRATIVE CASE STUDY

Like most cities, Riverside has a noticeable number of homeless individuals and families. Riverside has a population of close to 300,000 people. Estimates of the homeless population's size range between one thousand and three thousand persons. The city features hot and dry summers, and usually mild winters, often with rain from November through March. There are only two full time shelters operating within the city that are open to the homeless population, and one temporary shelter, the National Guard armory, that is open during the winter. The National Guard armory emergency shelter was started by former California Governor George Deukmejian in 1989 via the Emergency Cold Weather Shelter Program (ECWSP), and it can accommodate up to 165 persons on any given night. However, if the National Guard is conducting exercises then the armory is closed to the homeless; resulting in no alternate cold weather site. From November 1 to December 1, and from March 15 to March 31, the armory is open if the temperature falls below 40 degrees or if it is raining and below 50 degrees. Homeless persons using the shelter must be in by 10:00 p.m. and out by 6:00 a.m. The two year-round shelters, the Riverside Men's Shelter, with fifty beds, and the Genesis Shelter, with twenty beds (open to single women and women with children) have long waiting lists. Clearly even at the lower figure of 1,000 homeless persons within the city limits, there is not enough adequate shelter.

Riverside has had a sizable homeless population dating back to the 1970s. David, a former deputy chief of police, identified changes in Riverside's homeless population since the 1970s:

> In the early '70s most of those that were out on the street chose to do so. I would say that there were two different groups, the "truly homeless" and the "street people". Street people weren't necessarily homeless. They were the type who didn't work just hung out on the streets and drank. Back then the numbers were pretty small and the population was pretty stable, the same regulars out there every day. Then, in the early 1980s the economy changes, there were changes in government funding and we began to see more children out on the streets. We used to refer families to the Holiday Inn when there was more funding, we would put them up for the night. There were also big increases in the numbers of mentally ill because of deinstitutionalization.

Similarly, Steven, an advocate for the homeless and pastor of a downtown church noted that the most visible homeless persons in the city were the "single, adult homeless person ... disheveled perhaps, perhaps substance abusing, perhaps panhandling."

For a time, particularly during the latter part of the 1980s, the homeless were tolerated in downtown parks and streets. The city and downtown businesses, however, began to worry that the downtown area would become a mini skid-row where tourists and shoppers would fear to go; thus detracting from the city's vision of itself as a *world class city*. The city's mayor, Ron Loveridge, was one of the first city officials to voice the city's growing frustration with the homeless: "You can't let the streets of downtown [Riverside] become a campground for the homeless. It puts an almost impossible burden on those trying to entice people to come downtown and shop" (Morgan, 1994, p. B1). The city manager, John Holmes, echoed the mayor's observation: "We cannot allow this area to become a squatters camp" and ordered the police chief to "take whatever action you can to alleviate this deteriorating situation" (Morgan, 1994, p. B1).

The city's downtown residents also voiced concern at the growing numbers of homeless people in their midst. "They make a mess, and people don't feel safe when they are around," noted a downtown resident and the chairman of the city council's Downtown Neighborhood Advisory Committee. Another downtown resident stated that, "There is a sense of unrest just knowing that you're living that close to people who are not leading normal lives" (Morgan, 1994, p. B1). These statements are an example of a neutralizing techniques, blaming the victim; the homeless are portrayed as being dirty, unpleasant and criminal.

Attitudes and policy toward the homeless in the city began to change in the late 1980s reflecting what Kress (1994) has termed "homeless fatigue syndrome." In this case, downtown residents and city officials grew tired of seeing the homeless every day. Compassion turned to contempt as people grew tired of seeing the homeless on the streets. David, the former deputy chief of police, observed:

> A lot of people began to sleep along the downtown mall and in the parking structures, so we stepped up patrols and around from about 6:00 p.m. to 7:00 a.m. we would patrol the mall and parking structures to get people out. I hate to say it but we would harass them. There was also increased beggings, particularly at ATM's so the business owners would hire private security. I feel we were put in the middle of this whole mess. A turning point occurred in the late '80s when Chief Richardson led a sweep of the river bottom and threw out all of the homeless people out that were camping down there.

The reason given by Chief Richardson for the sweep was that he was clearing the river bottom of sexual deviants.

Particularly disturbing to Riverside's business elite was that as the homeless migrated from a temporary shelter in Fairmount Park, on the periphery of the downtown's limits, to White Park, in the downtown area and across from city hall, the homeless would pass through their posh neighborhoods. David lived in one of these neighborhoods, along with the mayor. David observed:

> In the late 1980s the armory opened. They would stay there during the night and during the day they would walk through my neighborhood, even going through the trash, on their way to White Park. We had a couple of judges living there, attorneys, City Councilman Loveridge, a school board member and others. Soon, property crimes, burglary and theft, began to go up in the area. We would even catch them sleeping on the porch overnight. Property values went down as well. That's when we started to become concerned.

The major concern with a visible homeless problem arose when public officials began to talk about revitalizing the deteriorating downtown area.

Downtown Redevelopment and the Politics of Public Space

Like many cities, Riverside has sought to revitalize its blighted downtown area in order to compete with other cities for investment and tourist dollars. The strength of a city for many is having an economically viable downtown area (McAuliffe, 1991). Talk about downtown redevelopment began around 1985 with the closing of the Mission Inn, a historic city landmark. In the mayor's words, he wanted Riverside to become a "world class city" and it could only do this by having a viable downtown. The mayor remarked, "You can't find a downtown in California that you could consider effective and vital that doesn't have a history of redevelopment" (Stanton, 1986, p. B1).

The historic Mission Inn, site of several movies, served as a "catalyst" for the city's downtown renaissance. Ed, a service provider for the homeless, noted that, "Riverside was a deteriorating, older downtown, there were opportunities with the purchase of the Mission Inn and with the changes and developments in the whole area of redevelopment." The director of the city's development agency called the Mission Inn "the guts of downtown ... the soul of it (Elmer, 1988, p. B1). Similarly, an attorney and former president of the Riverside Downtown Partnership said that the Inn was "critical, essential, all-encompassing important" to the downtown's revitalization" (Elmer, 1988, p. B1). Another member of the Riverside Downtown Partnership described it as a "tourism magnet for downtown Riverside" (Smith, 1985, p. B1). Promoted by the City Redevelopment Agency and the Riverside Downtown Partnership, the city's revitalization plan was estimated to cost 23 million dollars to improve the downtown area. The complete plan for revitalizing the downtown areas was estimated to cost a total of $300 million. The plan included a domed sports arena, remodeling dilapidated housing to attract young, professional couples, a shopping mall, a complex of luxury town homes, and a performing arts center (Elmer, 1986). The city's revitalization plan has yet to reach any operative dimensions.

Homelessness, particularly in the downtown area, became problematic when the city began to revitalize and expand the downtown business district around a reopened Mission Inn. As Steven observed, "the area surrounding the Mission Inn is a very sensitive area because of its importance to the redevelopment process." Because the homeless were deemed "unsightly and a detriment to business" the city council and mayor praised a project that sought to "eliminate nooks and places where the homeless sometimes sleep" (Arancibia, 1994, p. B1). For example, landscaping was deliberately designed to deter homeless people from residing downtown. Thorny bushes and pungent smelling ground cover were planted. The water fountain in front of the city's library was turned off, and later covered with dirt. The water fountain was a principal gathering place for the homeless. The city used "bum proof" benches consisting of metal bends that made sitting on them uncomfortable.

Comments by the city's downtown residents and business owners reflect the widely held belief that the presence of homeless persons lowers land values and detracts from business (Goetz, 1992; Wolch & Dear, 1993; Wright, 1997). One resident remarked, "It brings the neighborhood down" (Smith, 1991, p. B1). And a business owner lamented, "We lose business because people are afraid to come downtown" (Smith, 1991, p. B1). The persons we interviewed pointed out that the city's redevelopment plans excluded the homeless from downtown:

> That was being talked about a lot by the business community, that, in general, it dragged the overall health of the economy down because of an 'appearance factor' and the fact that people won't want to go downtown to do their shopping and their eating if they have to be confronted by people who are aggressive panhandlers. Maybe its just the problem at all, just seeing the homeless at all. It doesn't necessarily add to the image of Riverside as a "world class city" that has economic health and takes care of it's citizens. This is how the business community looked at it (David).

> If you're going to focus on the revitalization of the downtown area, if you want to get people to staying in hotels, and you want to build on tourism, you want to have as nice of an area as you possibly can. If you have large numbers of people who are unkempt and panhandling then it's going to detract from that. I understand the perspective of the city and private enterprise who want to make the city as nice as possible (redevelopment official).

> In general, the desire of any city is, and should be, its health on many different fronts. And without a healthy downtown that's not going to happen. I remember going downtown and seeing people sleeping in the mall and my perspective on it is that it's horrible, but I can see that its other impact – that it is not a pleasant thing – it's not necessarily conducive to making people feel comfortable in the downtown area. I think a lot of officials, people, wanted to help downtown. People wanted businesses and things to come in down there and that wasn't going to happen with that visible problem, literally on the front door of city hall (city council member).

When asked whether *homelessness* was the problem or that the homeless themselves were the problem, one interviewee (male, 43, pastor/advocate) responded, "I think the homeless were the problem, especially the visible homeless individuals that had been occupying downtown." Again these statements reflect a common belief among public officials and others that the homeless detract from the exchange-value of the land downtown. As Wright (1997, p. 89) has noted, "such redevelopments necessarily demand the removal of the very poor and homeless populations."

In addition to the homeless "problem" downtown, the city pressured the owners of two popular youth nightclubs to close because of the numbers of teenagers that congregated downtown. A popular weekly event, Downtown Wednesday Night, that featured a farmer's market was closed by the city council because too many teenagers were showing up. Then a city councilman, the mayor stated that "I don't care if they [the nightclubs] cater to Eagle Scouts or political science majors. You can't have a group of young people taking over like that . . . and expect the downtown to thrive" (Ogul, 1992, p. B1). Also removed from the downtown area was a homeless shelter for men that was purchased by the city for $300,000 and relocated to another area (Smith, 1990a, 1990b). Several downtown businesses were also forced out when rents began to rise with the opening of the Mission Inn. Those that did close were the result of what the city's redevelopment director termed a "natural recycling", forcing stores that catered to the poor – pawn shops, bargain stores and the like – to relocate somewhere else (Elmer, 1988). David explained that the main concern of the city was to, "Create an environment that does not attract the homeless. The city wanted to get rid of the magnets [such as feeding programs] and deal with the high visibility problem."

This theme, that Riverside would become a magnet and anchor for the homeless, is one that has been expressed by a number of elected officials and others. What is feared is that if Riverside does offer a number of shelters, services and feeding programs, it will attract homeless persons from neighboring areas. As Feagin and Feagin (1990, p. 299) note, "These officials seem to believe that if the homeless are made comfortable, homeless people will flock to the shelters to live at public expense". Steven called this view the "field of dreams" thesis, "if you build it they will come." Some of the persons we interviewed countered the claims of public officials:

> More services will attract more homeless. In reality, however, people come for the same reasons others do, economic opportunity and family. People don't come across the country for a temporary stay in a shelter. Usually the primary motive is a job or when you have family there (female, 47, shelter manager/activist).

> They hear about the booming economy here or just the climate. And they land here and maybe there isn't work, you can't get it right away. I don't think humane shelter services are going to attract a lot of people. I don't buy into that at all (male, 43, pastor/advocate).

> It's a very common argument. But to me it comes out ending up if the choice is between having something and having nothing and having the risk that there's going to be a magnet . . . I don't think its wise to say that nothing is going to exist because that's what we did for a good ten years and the problem did not get better by not having services. I've never heard of anyone saying, "Wow, we've come all the way from San Francisco because you guys are so great here. We heard of all the programs." I just don't think that happens that often (male, 35, service provider).

One of the services that the city wanted to shut down were the feeding programs held at White Park and that have since moved to a downtown church.[11] Reverend Lewis whose downtown church runs a feeding program every Wednesday evening and serves an average of 250 people each night, observed:

> It didn't matter ten years ago before downtown redevelopment. With that whole shift then I could understand why they would like to see us shut down. I know that there are people in the city that would like us not to do Project Food anymore. But we're not going to shut down!

A service provider for the homeless concurred:

> A lot of the churches had been feeding in the parks, long before I was an advocate, probably for fourteen or fifteen years with no problem from the city ... It was a very, very amicable relationship and there were no problems in the park. And then suddenly someone said that they needed the parks for the people, but who are the people? But it was really the Mission Inn (female, 47, shelter manager/activist).

Feeding programs downtown were undesirable to city officials because as Reverend Lewis describes, the area around the Mission Inn was a "touchy area". He recalled an encounter that he had with a public official:

> I had a visit from one of the city council members and the city council member expressed concern that having meals on Wednesday evening would have a negative impact on the Wednesday Night Market. I talked to other people, people who spend a lot of time downtown, who were never aware of any problems with guests of Project Food bothering people downtown. The problem was the teenagers.

Reverend Lewis also attempted to use his church as a shelter in December of 1996 when the armory was closed due to National Guard exercises. Public officials were again concerned with the impact sheltering would have on the downtown. Reverend Lewis explained, "I think that is the concern – having the church open as a shelter at night would have a negative impact on the appearance of downtown." Immediately, as the city council got word of this, Reverend Lewis recalled:

> Someone from city hall called up the Department of Community Action and said that "you can't do this ... we would require that the church have a conditional use permit." Also because the city contributed funds to the Department of Community Action and that perhaps the funds could be in jeopardy if this thing would be pursued. The Department of Community action had to back down with offering us supplies to shelter the homeless.

Reverend Lewis was frustrated because he thought sheltering the homeless "was a perfectly sensible plan, all the details worked out, and suddenly having city officials throwing a monkey wrench in the whole plan."

Despite pressure from the city, including the mayor, city manager and at least one council member, Reverend Lewis held his ground. To him it was an issue of religious freedom, a First Amendment issue. He stated that, "Shelter from the rain and cold, sanctuary from the elements, is not charity. It's simple justice" (Danelski, 1996). In order to keep the church from sheltering the homeless the city required that the church pay a $3,432 fee and wait three months for a hearing. The mayor remarked, "There are rules you have to follow. You do not just move into a place willy-nilly" (Morgan, 1996, p. B1). However, Reverend Lewis felt that "requiring a church – any church – to apply for an expensive conditional-use permit to make a building available as a temporary sanctuary of last resort for the homeless is an attempt to restrict the free exercise of religion." He stated that, "The precedence seems to be that churches are exempt from zoning regulations when they are seeking to do things that are part of their religious practice." Maggie Pacheco, the city's housing, neighborhoods, and community development manager disagreed stating, "Churches are no different than a business in terms of regulation" (Arrington, 1998a, p. B1).

Although a lawyer was contacted who felt that the case represented a clear First Amendment violation the case never went to court. While city officials hoped to find a shelter site outside of the city limits the city ended up allocating a temporary shelter, away from the downtown, in a nearby park (Rooney, 1996a, p. B1). According to Reverend Lewis, "When the city officials opened up a building at Fairmount Park that forestalled any legal confrontation with the city." Two weeks later another building in the same park was allocated for use as a temporary shelter on nights when the National Guard Armory was closed (Rooney, 1996b, p. B1).

Problems associated with the homeless downtown led to the formation of the Riverside Homeless Coordinating Council (RHCC) in 1993. Comprised of public officials, service providers, homeless advocates, members of the religious community, and concerned citizens, the RHCC was formed as a response to the city's concern with the, "Unfavorable impact upon White Park and the surrounding neighborhoods and businesses by the provision of meals for the homeless by a large number of churches and individuals." The RHCC's goals were to preserve White Park and improve the environment of the downtown area and to reduce or eliminate the use of public spaces not intended for providing services to the homeless. As one respondent noted, the forming of the RHCC was, "A political reactionary response to the redevelopment, to the opening of the Mission Inn. And they needed all of the homeless people out of the parks."

The RHCC meets monthly at the REACH (Riverside Efforts to Actively Counter Homelessness) Homeless Services Intake Center to discuss issues pertaining to homeless policy. REACH is located in a primarily industrial area and it is where homeless individuals can go to get food, haircuts, showers, laundry, dental and optical services, pick up mail, store their valuables, and get help finding work. REACH cannot, however, provide shelter because of zoning regulations. While service providers and the homeless wanted the site to be located downtown, opposition by downtown business owners prevented that from happening. They were particularly worried about the site attracting "undesirables". A member of the RHCC explained that "the political reality is a barrier to that" (Pitchford, 1995, p. B1). And one respondent noted, "They scrambled looking for a site. None of the service providers thought it was a good idea to have it over there [in an industrial area], but they had to get the people out of the parks."

The city even waived the conditional use permit fee of $3,773 in order to get homeless feeding programs out of the downtown area and into an industrial one (Pitchford & Arancibia, 1994). The REACH center was scheduled to open in 1994 but was halted due to lack of funds and community opposition. In 1995 the mayor of Riverside remarked, "There is no good site but it is the best one we could identify" (Pitchford, 1995, p. B1). Similarly, Chris Deurr, president of the Hunter Park Chamber of Congress in Riverside's Eastside said, "We support the idea, but we don't think that this is exactly the right spot for it" (Pitchford, 1995, p. B1). Ed stated "It seems like it would be easier to erect a skyscraper from scratch and take less time than to put up a few portable buildings for the homeless." The center finally opened in April of 1997 despite opposition from a council member in whose ward the center is located (Pitchford, 1997).

A City's Solution to the Homeless Problem

In order to implement its downtown revitalization plan, the city of Riverside approved ordinances in 1994 that attacked the homeless and the behaviors commonly associated with homelessness.[12] This was done in order to control what city redevelopment officials termed "anti-social behaviors" in the downtown pedestrian mall. The new ordinances outlawed such behaviors as panhandling, trespassing, and lying down or sitting on walkways or other public areas that were not specifically designed for that purpose (Pitchford, 1994). Some types of behavior were excluded from the ordinances, such as drinking a cup of coffee or eating a cup of yogurt while standing on a public sidewalk outside the store shop. In a sense, you could loiter if you were contributing to the local business economy by being a consumer. The new ordinances also allowed the empowered police personnel to remove shopping carts, and their contents, belonging to the homeless.

Not long after the ordinances took effect, the police moved on a downtown park in order to drive the homeless out (Arancibia, 1994). The city intensified its efforts to apprehend the homeless in parks and on downtown streets, especially those panhandling. According to one redevelopment official, "anti-social behaviors would not be tolerated and violators would be cited and appropriate actions would be taken." Ed, a service provider for the homeless, observed:

> In the case of Riverside and in a number of other cities the solution has been that the police had to have a role in it, that there was certain behavior that will not be tolerated in the community, sleeping on park benches, for example, or whatever the case may be.

One of the major problems identified by downtown business owners was panhandling by the homeless. According to downtown business owners, panhandling by the homeless was creating business losses because shoppers were afraid of going downtown. David explained his experience with homeless people begging for change:

> I'd go to the post office near my house when I was off duty and there would always be two or more people there panhandling and asking for money, usually the same people day after day. Regardless of the freedom of speech bullshit, I would pull out my badge and question them, harass them and tell them to get out of the area. If they were still around when I came back I would try to arrest them for something.

Another respondent talked about the concern with panhandling in the downtown mall:

> There was a café downtown near city hall where a lot of people, a lot of people doing business with the city courts, city government, county or whatever, who would be customers of that place. They drove up to go to park and there's two people up front, not of the most clean appearance and maybe mentally ill in a lot of cases and maybe somewhat aggressive in their panhandling. And that was tagged as the reason why the place was doing bad business.

At one time, the downtown pedestrian mall and nearby White Park, about a block from city hall, were havens for the homeless. As one respondent observed:

> I've always sort of taken the position that if you are homeless it doesn't mean that you don't want the same kinds of access to some sorts of amenities and things, some place where you can get access to food, go to the bathroom, a place where you can nap, or just watch people come and go. I think downtowns are natural places for people to gather, whether you are homeless or whether you have homes. However, the impact that this has in places like White Park that are gathering places are sometimes unintended and sometimes detrimental to a number of other uses that you would like to be happening at the park (female, 51, redevelopment department).

The city tolerated the homeless being there for a time but David explained how city policy began to change:

> The downtown area, what we call the "mile square", including White Park, the downtown mall, the bus depot, and the Vons shopping center became a magnet for the homeless. Homeless people from other areas began to migrate here as well. Basically we [the police] were called in because of this growing problem. We were supposed to put a large police presence in and chase them away because the homeless were seen as bad for business, even though we couldn't legally enforce it because it was unconstitutional.

He later added:

> We got a new city attorney, and closed the parks from sunset to sunrise. We cited anyone, homeless or not, who were hanging out in the park during those hours. They also made it illegal to park in certain areas, because some of the homeless had cars.

Thus, because the homeless were deemed unsightly and a threat to business the enactment of the anti-homeless ordinances and the closing of city parks from 11:00 p.m. until 5:00 a.m. restricted the presence of homeless persons and their use of public space. While not offering any alternative places for shelter, Chuck Beatty, a city councilmember whose ward includes the downtown, stated that "camping in our parks and mall must be eliminated" (Press Enterprise, 1994, p. A1).

A major force behind the anti-homeless ordinances was the Downtown Business Association, now called the Downtown Partnership. As a redevelopment department official stated, "the Downtown Business Association took a very strong position, and had what they called the anti-social or social issues committee . . . social issue was another word for how to deal with the homeless." The new city ordinances according to the official "were designed specifically to deal with the impact of the homeless downtown." There became a point when certain behaviors "would not be tolerated anymore" because of the "cost to our businesses and to our families that wanted to come downtown." Not surprisingly, the city ordinances forced homeless persons to move elsewhere in the city, away from the downtown to other areas including *marginal* spaces like abandoned buildings and fields. A service provider for the homeless explained:

> Because of the ordinances there has been a spreading out of the population there so they're not just in the downtown area any more. I've heard people from other areas where they didn't used to see people like that who now say that they see them all the time now.

Some of the homeless moved to the Santa Ana River bottom. An estimated three hundred homeless persons were thought to sleep in the river bottom since the police ordered them out of the parks (Morgan, 1997). One of the homeless women we spoke with at the National Guard armory said that there are easily "over a thousand people sleeping down there [Santa Ana river bottom]". Some of the homeless have dug tunnels underground, including one man who was reported to have tapped into the city's electricity lines.

The homeless who found shelter in the Santa Ana river bottom and in city parks away from the downtown soon found themselves forced out of those areas by periodic police sweeps. The police sweeps point out that what might be considered *marginal space* by the city can easily be redefined as *prime space* by those that have the power to do so (Snow & Anderson, 1993; Wright, 1997). One highly publicized police sweep happened on October 9, 1987. The month before, the Press Enterprise ran a story stating that homeless communities had "taken hold" in the Santa Ana River bottom (Clarke, 1987). A few weeks after the story's publication, a Riverside County Supervisor called for the police to remove the homeless from the river bottom (Gurza, 1987a). The supervisor stated that "it's a situation that we really cannot tolerate." The homeless were told by the police that they would be "chased out of the river bottom" if they did not leave voluntarily (Morgan, 1987). The homeless were routed from their campsites and seven homeless persons were arrested (Gurza, 1987b). The publicized police sweep sparked a protest by homeless activists. When asked why the homeless were routed from the river bottom, Sarah, an activist involved in the protest, recalled:

> From what I was told by the then city manager was that particularly, although there may have been several reasons, but that particular encampment was in full view of the very, very wealthy people who lived pretty much across from the river bed and they did not like, you know, that visualization. They did not like the view of the homeless, they could see their bonfires at night. And of course there was the crime factor and then the drug use.

However, she did not buy the official view that the homeless encampment harbored a number of criminals. As an activist who regularly went down to the river bottom to take food and supplies to the homeless she recollected:

> I knew all of the people. I was involved with that particular group at that time. It did not turn out that there was a high criminal population among those homeless people. And when they actually did the crime check, they did not find, except for one parolee, known criminals from the group that they ousted from the river bottom.

Portraying the homeless as criminals serves an important political purpose. Treating the homeless as a "problem population" diverts attention away from *homelessness* as a social problem and treats the homeless as the problem (Aulette & Aulette, 1987; Spitzer, 1975). As such, the homeless are treated as criminals, rather than victims of social forces, that must be dealt with by the criminal justice system. A homeless person arrested in the police sweep of the Santa Ana river bottom remarked, "all we want to say is we're tired of this persecution just because we're homeless. And we don't want to be called criminals" (Fulcher, 1987, p. B1). Another homeless person arrested in the sweep stated, "it's not a crime to be poor" (Espinoza, 1987, p. B1). A homeless activist observed about the homeless persons arrested in the sweep, "they just want to be located where it's safe and sanitary and not offensive" to public officials (Gurza, 1987b, p. B1). After the sweep, the Riverside City Council gave special recognition during one of its meetings to the police involved in the raid (Smith, 1987).

In 1997, another police sweep was conducted about a week before Thanksgiving. Several homeless individuals were arrested and cited for trespassing. Two children were found and placed in foster care. All others were ordered to pack up their belongings and move on. The encampments and shacks in which the homeless were living were bulldozed. After the sweep a homeless man who had been living in the riverbed for two and a half years remarked to a news reporter, "I never thought being homeless was a crime" (Stokley, 1997, p. B1).

Several of the homeless we interviewed commented about the mayor and seemed to be aware of city policy toward the homeless. The mayor of Riverside once remarked that there were no homeless persons in his city. The mayor expressed concern at the "chaos caused by the constant, uncontrolled feedings" of the homeless at White Park located near downtown. However, neither the service providers or homeless individuals we interviewed were aware of any problems with the feeding programs for the homeless. One homeless man, in response to the mayor's and city council's attitude toward the homeless stated that, "I think that the mayor and the city council members should try staying out on the streets for a week, with no money or anything and sleep in the shelters overnight. I bet they would have a different attitude about us then." A homeless female in her thirties remarked that, "They don't care about us, we might as well be invisible to them."

Attitudes such as the mayor's seem to be typical of the city officials and business owners in the revitalized downtown district. One councilmember remarked, "I've had it up to here with the homeless's rights. It's time for other people's rights to kick in." (Pitchford, 1994, p. B1). Similarly, a businessman who owns a downtown building, argued, "They (the homeless) are aggressive. Why shouldn't we be." Former Police Chief Linford "Sonny" Richardson seemed to summarize the current public opinion saying, "To hell with them" (Gurza, 1990b, p. B1). One reporter, commenting on the official attitude toward the homeless, summed it up by stating, "Riverside officials are more concerned about appeasing downtown residents and promoting economic growth than with meeting the needs of street people" (Gurza, 1990a).

DISCUSSION

While much has been written on homelessness as a social problem little research has looked at how cities respond to homelessness, particularly how their responses are related to urban redevelopment. The focus of this essay has been on downtown redevelopment issues and their use in repressing homeless persons. Public space in downtown Riverside has become what Wright (1997, p. 10) terms *polarized topographies*, "geographic areas with extreme examples of wealth and poverty closely interconnected." Although the homeless were tolerated in downtown Riverside for a time, intensified city efforts in the late 1980s and early 1990s to redevelop the downtown brought about increased hostility toward the homeless.

What space is used for and by whom it is used are points of contention. The homeless in Riverside who used "prime" or "pleasure" space as a place to eat, sleep and make a living – in short, doing what they need to do to survive – came into direct conflict with business people, politicians, and other citizens who each had their own definitions of what public space was to be used for. The "politics of place" eventually effected not only the homeless but also those who attempted to provide services to them, teens who used to go to concerts downtown and businesses that catered to poor people. Improvements in downtown Riverside forced the homeless to seek shelter elsewhere and deprived young people of recreational opportunities. A similar process of displacing the homeless has been observed in Austin (Snow & Anderson, 1993) and in San Jose and Chicago (Wright, 1997). As Snow and Anderson (1993, p. 104) note:

> The consequences of these processes of spatial transformation are twofold for the homeless: they are dislodged from areas traditionally appropriated by the down-and-out, particularly in central cities, and thereby the marginal space available to them is reduced; and they are therefore forced to spend more time in prime space, thus increasing their public visibility and the probability of complaints and eventual arrest.

Similarly, Wright (1997, p. 8) argues:

> The struggle over the definitions of and use of urban space may be clearly viewed in the city policies that actively disperse homeless street populations for being "out of place" and simultaneously attempt to contain them in institutional settings such as shelters, armories, prisons and rehabilitation centers.

The newspaper articles we reviewed and personal interviews we conducted suggest that Riverside promoted its downtown's revitalization by criminalizing the homeless. The city's response to the homeless was the outcome of several factors. The increased visibility of large numbers of homeless persons congregating in downtown areas was perceived as a threat by business owners to the exchange value of the land. Both patrons and business owners in downtown Riverside voiced concern over the impact that homelessness persons had on the downtown's economic viability. This led to action by the Riverside City Council. The city council approved ordinances criminalizing behaviors associated with the homeless in order to drive them out of the downtown areas. In the city of Riverside the criminalization of homelessness thus served a purpose – to drive the homeless away from the downtown areas.

The statements made by respondents suggest that there is an inextricable relationship between redevelopment and the city of Riverside's attempt to revitalize its downtown area. Riverside is not alone in its efforts to use redevelopment as a vehicle for removing the homeless from the city's limits. Similar responses to the homeless can be observed in other U.S. cities: Atlanta (Task Force for the Homeless, 1995), Chicago and San Jose (Wright, 1997), Los Angeles (Goetz, 1992), and in our nation's capitol (Kress, 1994). A respondent involved with the issue of homelessness in Riverside for over ten years observed, "there is a direct correlation between the city's redevelopment efforts and the anti-homeless ordinances passed . . . its pretty much cause and effect." Another respondent stated, "I do know that there is an increasing trend toward what some have called criminalization of homelessness. Ordinances are being passed which clearly are aimed at the homeless and wouldn't effect anyone else . . . this is largely because of downtown redevelopment."

The problem with criminalizing homelessness, however, is that it does nothing to solve the social problem of homelessness. It only tends to shift the problem around. According to the National Coalition for the Homeless (1997, p. 89):

> The flaws in the effort to criminalize homelessness are as numerous as they are obvious. Though no one should ever have to sleep in the park and beg for food, making these acts into criminal offenses does not, in any way, help the people who are forced to "commit" said offenses. These city ordinances (and similar state laws) are misguided because they seek to hide homeless people, not end homelessness. They are unjust because they seek to punish people for being poor. They are, in effect, persecution because people who are homeless do not have the option to rest, sleep, and set down their belongings in private. People who have no choice but to be homeless have no choice but to be public; to punish them for this heaps injustice on top of indignity . . . Are these cities asking people who are homeless to choose not to exist?

One of the respondents, Ed, admitted that criminalizing homelessness was "an illogical approach" and that "the laws are not applied evenly." He also stated, "eventually it becomes a self-defeating purpose . . . going in and busting up these groups and making people move around because they're going to show up somewhere else." Another service provider for the homeless remarked, "Unfortunately, the police are in the position of having to enforce these laws, but they don't have an answer of where they are supposed to go now" (Arancibia, 1994, p. B1). As a homeless man noted regarding the city's effort to drive the homeless away from downtown, "Out of sight downtown, out of mind is the aim for many" (Pitchford, 1995, p. B1).

Despite the city of Riverside's efforts to criminalize homelessness, alternatives to criminalizing the homeless exist.[13] A number of groups have formed at the grassroots level to challenge the city's anti-homeless policies. Frustrated with the city's unwillingness to help, these service providers and homeless advocates have attempted to respond to the problem in a variety of ways. For example, Steven stressed the need for "Doing organization among the churches, building an Interfaith Hospitality Network and advocating for a community-wide attempt to address the needs of both emergency and transitional shelter, especially for families." The Interfaith Hospitality Network (IHN) is a response to homelessness from the religious community, bridging a number of different faiths. Operating in some 49 cities across the country, the IHN offers shelter to homeless families in each church, up to fourteen individuals, for a week at a time on a rotating basis in a coalition of churches.

A group calling itself *Shelter Riverside*, a coalition of religious and community activists, frustrated with the city and county's inability to adequately address the needs of the homeless wrote a "homeless manifesto" and presented it to the county Board of Supervisors. The group was particularly concerned with the emergency shelter situation because starting March 31 1999 the National Guard armory would be closed to sheltering the homeless during inclement weather. The founder of the group remarked that "We've been relying on the National Guard armories for ten years and nothing's happened. We're just warehousing people. We do better at providing shelter for our animals than we do our people" (Petix, 1998a, p. B1). The Board of Supervisors welcomed the plan put forward by Shelter Riverside and expressed interest in developing a transitional housing and job training center on surplus government land located on a closed military base, March Air Force Base (Petix, 1998b). However, the redevelopment agency in charge of the surplus land postponed any action because it wanted to attract private business to invest in the surplus land. The agency felt that housing the homeless on the surplus military land would be a threat to business investors.

The use of land on closed military bases for homeless services was mandated by Title V of the Stewart B. McKinney Act. Service providers must apply to local redevelopment agencies to use the property at closed bases to assist homeless persons. Although the land and buildings are available, one respondent stated:

> The Joint Powers Authority's [who has jurisdiction at the military base] primary concern is with maximizing economic development opportunities that may be available at March Air Force Base. And that's generally inconsistent with providing homeless service ... because homeless services don't make money. So the homeless issue is not their highest priority.

Furthermore, the community surrounding the closed military has expressed its opposition to providing services and housing to the homeless on the surplus land.

While the city is in dire need of more permanent year round shelters for the homeless, one service provider stated:

> There are no plans to construct new shelters in the city of Riverside, its not even an option. It's not a problem if the political will is there and at this point the political will is not there ... It makes perfect sense to me to have people sleeping inside, rather than out on the streets sleeping in bushes, urinating in doorways, but apparently, some of our city officials do not see it that way ... That's why my plan includes the use of the Salvation Army as an emergency shelter. They want to do it. why on earth should they [public officials] stand in their way? It just mystifies me.

There are, however, empty buildings in the city that could be used as shelters. As one respondent noted:

> There are empty buildings that could be used but each of these empty buildings is in one or another city councilman's ward and they would fight tooth and nail not to have them in their ward ... NIMBYism [not-in-my-back-yard] is alive and well in Riverside.

Steven has presented a "homeless manifesto" of his own to the Riverside Homeless Coordinating Council entitled, "Hands Across the City." In it he calls upon "concerned citizens and public officials in Riverside to take the lead in developing our city's contribution" to a countywide continuum of care. He notes in the manifesto that "The quality of life in our city is measured not only by the opportunities it offers its successful and affluent citizens, but also by the care and assistance it provides for the homeless and poor in our midst."

People like Reverend Lewis and Mark are leading this grassroots movement. Mark, a service provider, who originally worked as a traveling sales and marketing representative for a major pharmaceutical company began to be exposed to homelessness during his travels across the United States. At first he was ambivalent toward the homeless, but his views rapidly changed, "I saw people on the street asking for money. I have real compassion for these people and felt something had to be done rather than just a free meal and a place to sleep for the night." His growing compassion for the homeless led him to start the *Concerned Family*, a shelter housing around twelve men, that provides food, clothing, medical care, crisis intervention, rehabilitation, and counseling. The shelter was funded by Mark and private donations.

The shelter started by Mark has been closed due to financial limitations. He is currently running a temporary, cold-weather shelter at the National Guard armory that houses over one-hundred men, women, and children on any given winter night. Mark is working toward opening a permanent shelter in the

Riverside area backed financially by several individuals and organizations. Mark has realized that the time for relying on the government is over.

> We can't rely on the government and city anymore for help. I have already had a thirty per-cent cut in funds to run the armory by the city. It's up to private individuals now to help deal with this cancer, because I am tired of the political games . . . of the shuffling of people around.

CONCLUDING REMARKS

Goetz (1992) argues that homelessness presents two distinct problems for local governments. The first problem is "the social problem of having people who are without shelter." (1992, p. 549) The second problem has to do with land values and the use of public space. The case study in this essay examined homeless policy in Riverside, California. By paying particular attention to the relationship between downtown revitalization and city ordinances toward the homeless, we argue that there is a direct relationship between the city's redevelopment efforts and the increasing criminalization of behaviors associated with homelessness. The city has been, and still is, actively trying to keep the homeless out of certain areas of the city, particularly the downtown area surrounding the Mission Inn. The city's objective has been achieved through the enactment of city ordinances that specifically target and criminalize behaviors associated with the homeless (Arrington, 1998b, 1998c). Other city ordinances have prevented service providers from operating their facilities in certain areas of the city. The observations we have made in this essay based on the case study are consistent with the findings of researchers in other cities (Aulette & Aulette, 1987; Kress, 1994; The National Law Center on Homelessness and Poverty, 1994; Wright, 1997).

While the primary focus in this essay has been on analyzing the relationship between downtown revitalization and repressive city ordinances toward the homeless, future research needs to be undertaken on how the homeless themselves have resisted these anti-homeless actions in cities such as Riverside. The homeless are not to be seen as passive agents who are always at the mercy of the "powers that be." They can and do actively and passively resist the actions taken by local elites.[14] However, as a Riverside lawyer stated, "The reality is that the homeless don't have as much political and financial influence as business interests in a downtown area that is revitalizing" (Arrington, 1998, p. B1). Advocacy groups such as Shelter Riverside and individuals such as Reverend Lewis and Mark may help to change that.

NOTES

1. The visibility of homeless persons in urban areas has led the general public to believe that homelessness is a relatively recent phenomenon. According to Smith (1996, p. 295): "Late in the 1970s, Americans began noticing more people sleeping in public places, wandering the streets with their possessions in shopping bags." Jencks (1994) has suggested that the term *homeless* had been used by sociologists as early as the 1950s in reference to migrant workers and single men living on skid-row. However, because they tended to sleep indoors, migrant workers and skid-row residents were not regarded as truly homeless. In contrast, today's homeless are *truly homeless* – most of them are unable to sleep indoors, instead they sleep in public places such as bus stations, parks, doorways, and abandoned buildings. An extensive historical examination of how American cities have treated the homeless is found in Simon (1992).

2. A national telephone survey conducted in 1990 found that between five and ten million Americans experienced some form of homelessness in the late 1980s (Interagency Council on the Homeless, 1994). The telephone survey was conducted again in 1994 and it found that twelve million Americans reported having experienced some form of homelessness (National Coalition for the Homeless, 1997).

3. The lack of reliable estimates for the size of the homeless population motivated the U.S. Bureau of the Census to use alternative approaches to counting the homeless in the 2000 Census. Census workers surveyed soup kitchens, homeless shelters, and city areas where the homeless gather – under bridges, in abandoned cars, empty industrial buildings, and rail boxcars (Cook, 2000; Peterson, 2000; Simakis, 2000). The effort to count the homeless was largely fueled by the states' interest in having a reliable measure of the homeless population's size for determining the allocation of federal funds for homeless projects (Broder, 2000; Loose, 2000; Schevitz, 2000).

4. For example, in an effort to promote downtown revitalization and tourism, the mayor of San Francisco, Willie Brown, urged city leaders to support the police's use of city ordinances in confiscating shopping carts away from homeless persons. By confiscating their shopping carts, homeless people would be less visible in the downtown areas, and they would be more likely to stay in peripheral, or hidden, areas away from tourists (Cohen, 1999).

5. A non-probability sampling technique was used to interview persons. The sample was drawn from persons at the meetings we attended. Interviews were open-ended and lasted about two hours. The interviews were tape-recorded and persons were free to talk about homeless issues. We have used fictitious names to identify respondents in the text.

6. The extent to which the homeless are isolated or concealed in society increases their chances of being exploited, especially by the business elite. For example, in February 1996 a New Jersey real estate developer was sentenced for using homeless men without protective clothing to remove asbestos from a commercial building (*United States v. Mohammed Mizani*, 1996). In Miami, two defendants were sentenced to five months in prison for using homeless persons to remove asbestos materials without the use of protective clothing (*United States v. Arthur Newman and Lawrence Rothman*, 1998). Finally, three men were sentenced to two year prison sentences for using homeless persons from Tennessee and Georgia to remove asbestos at a manufacturing facility in Marshfield, Wisconsin (*United States v. Frazier et al.*, 1999).

7. The perception that homelessness was *cresting* may have become even more noticeable in the 1990s. According to Foscarinis (1997) a survey of the nation's 50 largest cities found that San Francisco had fewer than 1,700 shelter beds for a population of 16,000 homeless persons; Atlanta had about 2,000 shelter beds available for a homeless population of 12,000; and Dallas had fewer than 1,730 shelter beds available for a population of 5,000 homeless persons. The non-availability of shelter beds forces homeless people to stay out on the streets, thus becoming more noticeable to the public.

8. The dynamics of the housing problem are a direct result of a free market ideology in which housing is supplied for profit like any other commodity. Lange (1989, p. 25) argues that: "The free housing market serves the class interests of those in power. The bulk of the population functions as passive consumers, a resource to be manipulated to further the profit making ends of the capitalist class. In such markets expensive new housing is constructed for the affluent and older housing is supposed to become available to the less affluent since it will drop in price as the increasing supply of new housing lessens its desirability. . . . In essence, housing is competed for like any other good, and the poor get the aging slums that are left over. In the free housing market, housing is viewed as a commodity to be sold to the highest bidder and the issues of equity and justice are largely ignored." Because housing is supplied on the basis of profit, the poorest, such as the homeless, who can afford to pay little if anything for housing are not likely to find any. Thus, it is not profitable to supply housing, or any other commodity, to the homeless because it is not profitable. Within the context of downtown revitalization, the homeless are excluded from a city's plans because they are not seen as economic contributors to the city's growth. It is for this reason that most downtown revitalization plans do not include construction of housing units or renovation of existing housing, such as abandoned hotels, for low income and homeless persons.

9. Cities use restrictions on the use of public space as a means of criminalizing the homeless. For example, some of the restrictions imposed by cities on public space are: prohibiting sleeping or dozing in a street, alley, park, or other public area; prohibiting the use of sleeping bags or occupying a temporary shelter; and prohibiting the use of public streets for living accommodations. An extensive review of court challenges to city ordinances that target the use of public space by homeless persons is found in Foscarinis (1996) and McConkey (1996).

10. Some advocates for the homeless argue that criminalizing homelessness is a poor use of a city's fiscal resources. Atlanta, for example, spends between $300,000 and $500,000 a year to incarcerate homeless persons for sleeping or begging in public (Foscarinis, 1996). For every 15 homeless people arrested in San Francisco, it costs the city 450 police working hours and $11,000 in costs. According to Foscarinis, Cunningham-Bowers, and Brown (1999), police time and resources used to arrest homeless persons is often more than the cost of providing the homeless with shelter. The cost in 1993 of holding a homeless person in jail was over $40 per day. In contrast, the cost of providing housing, food, transportation, and counseling services to a homeless person was about $31 a day.

11. White Park has become a symbol of how far the city was willing to go in its battle with the homeless. The city fenced off the park in 1997 in an attempt to drive away the homeless from the downtown. At the time, the city argued that the park's closing was necessary to upgrade it to a botanical garden where families could take a Sunday stroll among the rose bushes and other exotic plants. The park remains fenced and closed to everybody, not just the homeless.

12. According to Reid (1986), it was easy for society to attack the homeless in the 1980s because the court system, especially the U.S. Supreme Court, was unwilling to expand the fundamental rights under the Due Process or Equal Protection Clause of the 14th amendment to protect disadvantaged groups, such as the homeless. As Reid (1986, p. 116) notes, "instead of a desire to protect allegedly disadvantaged groups or to embrace the perceived need for recognition of certain interests as fundamental, the Supreme Court revealed a clear desire to leave any socioeconomic questions to the legislative chambers and the legislative process rather then resolving them through judicial decision."

13. According to the National Law Center on Homelessness and Poverty, alternatives to criminalizing the homeless would include: "adoption of laws that are designed to assist rather than harass homeless people; ensuring that law enforcement officials protect homeless people rather than single them out for law enforcement action; . . . and factor dialogue and outreach among city agencies, homeless people, advocates for homeless people, and members of the business community to find long-term solutions that address the underlying causes of homelessness" (Foscarinis, Cunningham-Bowers & Brown, 1999, p. 164).

14. For example, in *Johnson v. City of Dallas (1994)* homeless plaintiffs challenged the city of Dallas' enforcement of ordinances against them, especially prohibiting sleeping in public. In *Joyce v. City of San Francisco (1994)* homeless plaintiffs challenged San Francisco's use of city ordinances to remove them from public areas in the city. In perhaps the most notable case involving homeless plaintiffs, *Pottinger v. City of Miami (1992)*, they argued that Miami's use of city ordinances to arrest and harass them for engaging in basic everyday activities – sleeping and eating in public – was in violation of the eighth amendment. However, while homeless plaintiffs can challenge city ordinances in the courts, Daniels (1997) argues that homeless persons must overcome the court's perception of them as weak, dirty, lazy, addicts, and derelicts.

REFERENCES

Arancibia, J. (1994). Police Roust Homeless At White Park. *Press Enterprise* (Dec 8), B1.
Arrington, V. (1998a). Organization Lacks Responsible Parent. *Press Enterprise* (June 14), B1.
Arrington, V. (1998b). Homeless Shelter Options Under Discussion. *Press Enterprise* (June 15), B1.
Arrington, V. (1998c). No Welcome for the Homeless. *Press Enterprise* (June 16), B1.
Aulette, J., & Aulette, A. (1987). Police Harassment of the Homeless: The Political Purpose of the Criminalization of Homelessness. *Humanity and Society, 11*(2), 244–256.
Barak, G., & Bohm, R. H. (1989). The Crimes of the Homeless or the Crime of Homelessness? On the Dialectics of Criminalization, Decriminalization, and Victimization. *Contemporary-Crises, 13*, 275–278.
Barak, G. (1991). *Gimmie Shelter: A Social History of Homelessness in America.* New York: Praeger.
Barrow, S., Herman, D., Cordova, P., & Struening, E. (1999). Mortality Among Homeless Shelter Residents in New York City. *American Journal of Public Health, 89*(4), 529–534.
Belcher, J., & Singer, J. (1988). Homelessness: A Cost of Capitalism. *Social Policy, 18*, 44–88.
Berger, C. (1996). Beyond Homelessness: An Entitlement to Housing. *University of Miami Law Review, 45*, 315–334.
Broder, D. (2000). Leave the census alone, already! *Denver Post* (Apr 4), B9.

Burt, M. (1995). Critical Factors in Counting the Homeless. *American Journal of Orthopsychiatry*, 65, 334–339.

Clarke, L. (1987). Homeless communities take hold in Santa Ana river bottom. *Press Enterprise* (Sept 4), B1.

Cohen, W. (1999). No shelter for the homeless. *U.S. News & World Report* (Dec 20), 24.

Cook, R. (2000). Staff ensures homeless not left out. *Atlanta Journal and Constitution* (March 30), 1B.

Danelski, D. (1996). Righteous Indignation Over Homeless. *Press Enterprise* (Dec 10), B1.

Daniels, W. (1997). Derelicts: Recurring Misfortune, Economic Hard Times and Lifestyle Choices: Judicial Images of Homeless Litigants and Implications for Legal Advocates. *Buffalo Law Review*, 45, 687–737.

Drier, P., & Applebaum, R. (1991). American Nightmare: Homelessness. *Challenge* (March-April), 46–52.

Dwyer, D. (1988). Who Profits from the Homeless Plague: A Marxist Analysis. *Political Affairs* (Jan), 17–22.

Ellickson, R. (1996). Controlling Chronic Misconduct in City Spaces: Of Panhandlers, Skid Rows, and Public-Space Zoning. *Yale Law Journal*, 105, 1165–1248.

Elmer, V. (1988). Mission Inn May Waken Riverside's Downtown. *Press Enterprise* (Nov 20), B1.

Elmer, V. (1986). $300 Million Project for Downtown Outlined. *Press Enterprise* (July 17), B1.

Espinoza, L. (1987). Activist For Homeless Leads Protests Of Arrests. *Press Enterprise* (Oct 11), B1.

Feagin, J., & Parker, R. (1990). *Building American Cities: The Urban Real Estate Game*. Englewood Cliffs, N.J: Prentice-Hall.

Feagin, J., & Feagin, C. (1990). *Social Problems: A Critical Power-Conflict Perspective*. Englewood Cliffs, N.J: Prentice-Hall.

Foscarinis, M. (1996). Downward Spiral: Homelessness and Its Criminalization. *Yale Law and Policy Review*, 14, 1–77.

Foscarinis, M. (1997). Q: Is the rigorous enforcement of anti-nuisance laws a good idea?; NO: In the name of law and order, police are strong-arming the U.S. Constitution. *Insight on the News* (June 2), 24.

Foscarinis, M., Cunningham-Bowers, K., & Brown, K. (1999). Out of Sight – Out of Mind? The Continuing Trend Toward the Criminalization of Homelessness. *Georgetown Journal on Poverty Law and Policy*, 6, 145–165.

Foscarinis, M., & Scheibel, J. (1993). End Homelessness Now. *Georgetown Journal on Fighting Poverty*, 1, 44–46.

Franzese, P. (1999). Housing and Hope: The Crisis in Homelessness, Discrimination in Housing, and an Agenda for Landlord/Tenant Reform. *Seton Hall Law Review*, 29, 1461–1468.

Fulcher, R. (1987). Homeless Deny Link To Crimes. *Press Enterprise* (Oct 13), B1.

Gans, H. J. (1992). The War Against the Poor. *Dissent* (Fall), 461–465.

Gans, H. J. (1994). Positive Functions of the Undeserving Poor: Uses of the Underclass in America. *Politics and Society*, 22(3), 269–283.

Goetz, E. (1992). Land Use and Homeless Policy in Los Angeles. *International Journal of Urban and Regional Research*, 16(4), 540–554.

Gottdiener, M. (1994). *The New Urban Sociology*. New York: McGraw-Hill Inc.

Gurza, A. (1987a). Police Report on Homeless To Counter Activist. *Press Enterprise* (Oct 9), B1.

Gurza, A. (1987b). River Bottom Homeless Routed. *Press Enterprise* (Oct 10), B1.

Gurza, A. (1990a). Police Target White Park Squatters. *Press Enterprise* (July 20), B1.

Gurza, A. (1990b). Homeless Under Fire: Public Attitudes Move Away From Compassion. *Press Enterprise* (Aug 5), B1.

Henderson, C. (1999). Homelessness Has More Impact on TB Spread than HIV. *Tuberculosis & Airborne Disease Weekly* (April 26), 1081.

Interagency Council on the Homeless (1994). *Priority: Home! The Federal Plan to Break the Cycle of Homelessness.* Washington D.C.: U.S. Department of Housing and Urban Development.

Jencks, C. (1994). *The Homeless.* Cambridge, MA: Harvard University Press.

Kozol, J. (1988). *Rachel and Her Children: Homeless Families in America.* New York: Fawcett Columbine.

Kress, J. B. (1994). Homeless Fatigue Syndrome: The Backlash Against the Crime of Homelessness in the 1990s. *Social Justice, 21*(3), 85–108.

Ladd, C. (1986). A Right to Shelter for the Homeless in New York State. *New York University Law Review, 61,* 272–299.

Lang, M. H. (1989). *Homelessness Amidst Affluence: Structure and Paradox in the American Political Economy.* New York: Praeger.

Logan, J., & Molotch, H. (1987). *Urban Fortunes: The Political Economy of Place.* Berkeley: University of California Press.

Loose, C. (2000). Homing in on the Homeless. *Washington Post* (March 30), B1.

Marcuse, P. (1988). Neutralizing Homelessness. *Socialist Review, 18,* 69–96.

Marx, K. (1990) (original publication date 1867). *Capital: A Critique of Political Economy Volume 1.* London: Penguin Books.

McAuliffe (1991). Riverside's Downtown Economy Hangs On The Uncertainty Of The Mission Inn. *Press Enterprise* (Jan 13), B1.

McConkey, R. (1996). Camping Ordinances and the Homeless: Constitutional and Moral Issues Raised by Ordinances Prohibiting Sleeping in Public Areas. *Cumberland Law Review, 26,* 633–667.

Metzier, B. (1991). Officials Find, Demolish Deluxe Homeless Camp. *Press Enterprise* (June 20), B1.

Morgan, S. (1997). Homeless in River Bottom Await El Nino. *Press Enterprise* (Nov 13), B1.

Morgan, S. (1996). Riverside's Homeless Face Nights Out in the Cold. *Press Enterprise* (Dec 20), B1.

Morgan, S. (1994). Homeless People Overstay Their Welcome. *Press Enterprise* (Feb 5), B1.

Mullins, G. (1994). The Battered Woman and Homelessness. *Journal of Law and Policy, 3,* 237–255.

National Coalition for the Homeless (1997). The Criminalization of Homelessness: Waste and Injustice. *Safety Network* (Sept-Oct), 23–26.

National Coalition for the Homeless (1998). How Many People Experience Homelessness? http://nch.ari.net/wwwhome.html

National Law Center on Homelessness and Poverty (1994). *No Homeless People Allowed: A Report on Anti-Homeless Law, Litigation, and Alternatives in 49 United States Cities.* Washington D.C.: National Law Center on Homelessness and Poverty.

National Law Center on Homelessness and Poverty (1995). *No Room For The Inn.* Washington D.C.: National Law Center on Homelessness and Poverty.

National Law Center on Homelessness and Poverty (1996). *Mean Sweeps.* Washington D.C.: National Law Center on Homelessness and Poverty.

National Law Center on Homelessness and Poverty (1997). *Access Delayed, Access Denied.* Washington D.C.: National Law Center on Homelessness and Poverty.

National Law Center on Homelessness and Poverty (1999). *Out of Sight – Out of Mind?* Washington D.C.: National Law Center on Homelessness and Poverty.

Nelson, M. (1986). America's Homeless piled in the Streets. *Utne Reader* (Feb-March), 44–52.
O'Flaherty, B. (1995). An Economic Theory of Homelessness and Housing. *Journal of Housing Economics, 4*, 13–49.
Ogul, D. (1992). City Harasses Downtown Clubs Owner Says. *Press Enterprise* (July 22), B1.
Peterson, D. (2000). Census workers make greater effort to try to find homeless. *Star Tribune* (March 30), 1B.
Petix, M. (1998a). Strategy for Homeless Pushed. *Press Enterprise* (Jan 5), B1.
Petix, M. (1998b). Board Favors Plan to Help Homeless. *Press Enterprise* (Jan 14), B1.
Pitchford, P. (1994). Riverside Takes Legal Aim at Homeless. *Press Enterprise* (Oct 14), B1.
Pitchford, P. (1995). Eastside Homeless Center Approved. *Press Enterprise* (Jan 4), B1.
Pitchford, P. (1997). Homeless Services Center Opens. *Press Enterprise* (April 3), B1.
Pitchford, P., & Arancibia, J. (1994). Riverside City Council. *Press Enterprise* (May 5), B1.
Reid, I. (1986). Law, Politics and the Homeless. *West Virginia Law Review, 89*, 115–147.
Rooney, G. (1996a). Bringing Them Out of the Cold: Park House Open to the Homeless. *Press Enterprise* (Dec 6), B1.
Rooney, G. (1996b). 2nd Fairmount Building to Be Used. *Press Enterprise* (Dec 19), B1.
Schevitz, T. (2000). Counting on the homeless. *San Francisco Chronicle* (March 29), A19.
Simakis, A. (2000). Salvation Army is shutting out census takers: Soup kitchens are made off-limits. *Plain Dealer Reporter* (March 31), 1B.
Simon, H. (1992). Towns Without Pity: A Constitutional and Historical Analysis of Official Efforts to Drive Homeless Persons from American Cities. *Tulane Law Review, 66*, 631–682.
Smith, D. (1985). Downtown Businesses Express Optimism About Inn Renovation. *Press Enterprise* (June 2), B1.
Smith, D. (1987). City Council Stands By Police On Raid. *Press Enterprise* (Oct 14), B1.
Smith, J. (1996). Arresting the Homeless for Sleeping in Public: A Paradigm for Expanding the Robinson Doctrine. *Columbia Journal of Law and Social Problems, 29*, 293–335.
Smith, R. (1990a). Shelter Closing; Replacement Sought. *Press Enterprise* (July 7), B1.
Smith, D. (1990b). Homeless Shelter Debated. *Press* Enterprise (July 17), B1.
Smith, D. (1991). Shops' Neighbors Worry About Transients' Effect. *Press Enterprise* (Jan 4), B1.
Snow, D., Baker, S., & Anderson, L. (1989). Criminality and Homeless Men: An Empirical Assessment. *Social Problems, 36*(5), 532–547.
Snow, D., & Anderson, L. (1993). *Down On Their Luck: A Study of Homeless Street People.* Berkely: University of California Press.
Soine, L., & Burg, M. (1995). Combining Class Action Litigation and Social Science Research: A Case Study in Helping Homeless Mothers with Children. *American University Journal of Gender and the Law, 3*, 159–185.
Spitzer, S. (1975). Toward a Marxian Theory of Deviance. *Social Problems, 22*, 641–651.
Stanton, R. (1986). Redevelopment: Blighted Downtown Areas Being Reborn. *Press Enterprise* (Oct 5), B1.
Stokley, S. (1997). Police Order Homeless To Leave Riverbed. *Press Enterprise* (Nov 21), B1.
Stronge, J., & Helm, V. (1991). Legal Barriers to the Education of Homeless Children and Youth: Residency and Guardianship Issues. *Journal of Law and Education, 20*, 201–218.
Task Force for the Homeless (1995). The Criminalization of Poverty. Atlanta, Georgia: Task Force for the Homeless.
Thompson, D. (1998). Breaking the Cycle of Poverty: Models of Legal Advocacy to Implement the Educational Promise of the McKinney Act for Homeless Children and Youth. *Creighton Law Review, 31*, 1209–1249.

Timmer, D., & Stanley Eitzen, D. (1992). The Root Causes of Urban Homelessness in the United States. *Humanity and Society, 16,* 159–175.

Timmer, D., Eitzen, D., & Talley, K. (1994). *Paths to Homelessness: Extreme Poverty and the Urban Housing Crisis.* Boulder, CO: Westview Press.

Turner, M. (1991). Problems of Shelter: Homelessness and Affordable Housing in the United States. *Howard Law Journal, 34,* 71–73.

Wolch, J., & Dear, M. (1993). *Malign Neglect: Homelessness in an American City.* San Francisco: Jossey-Bass Publishers.

Wright, J. D. (1988). The Worthy and Unworthy Homeless. *Society* (July/August), 64–69.

Wright, T. (1997). *Out of Place: Homeless Mobilizations, Subcities, and Contested Landscapes.* Albany: SUNY Press.

Yeich, S. (1994). *The Politics of Ending Homelessness.* Lanham, MD: University Press of America.

INNOVATION, SPECULATION, AND URBAN DEVELOPMENT: THE NEW MEDIA MARKET BROKERS OF NEW YORK CITY

Michael Indergaard

INTRODUCTION

New media districts in major cities are becoming focal points for economic innovation and growth. The nebulous state of the new media sector (Scott, 1998; Pratt, 2000) makes it difficult to ascertain what the advent of these districts means for the larger city. The matter is complicated by the fact that new media firms attracted billions of dollars in speculative investment during the Internet stock market boom, injecting a tidal wave of wealth into local economies. Consequently, the new media sector in cities such as New York, Boston and San Francisco has generated large numbers of new jobs, stimulated demand for business services, and helped ignite astronomical rises in real estate prices. While the new media sector is acting as a supercharged engine, generating rapid growth in urban economies, it is also causing cities to absorb high levels of risk. Thus, the impact of new media districts on urban development depends upon the roles they play in the organization of speculation as well as innovation.

Recent accounts of Internet stock speculation in the business press cite a number of factors that shape investor perceptions. Shiller (2000) argues that the "new economy" thesis spread by prominent sources (e.g. the media, business schools, the Federal Reserve) has lead investors to indulge in psychological fantasy regarding the prospects of technology stocks. In a time of technological uncertainty, investors have been swayed by the notion that productivity-enhancing technology has brought the emergence of a new economy – one in which stocks can always be sold at a higher price. The "natural" effect is the same as that achieved in a "Ponzi" or pyramid scheme wherein a dishonest operator creates the illusion of a viable business by using money from new investors to pay off old. Similarly, Gross (2000) notes the idea that "new rules" are emerging has been promoted by leading new media figures and "spread by venture capitalists, Wall Street analysts and the financial media". Lipschultz (2000) contends that new media firms manipulate IPOs to gain publicity; they engineer spectacular first-day rises in stock valuations by releasing only a small portion of their stock to the market. Some analysts (Useem, 2000; Birger & Messina, 1998) charge that new media firms, venture capitalists, and investment banks that underwrite initial public offerings (IPOs) of stock are indeed operating Ponzi schemes.

The alleged role of new media entrepreneurs and venture capitalists in organizing speculation draws attention to a new kind of urban enclave – the new media district. Research have shown that innovative firms and venture capitalists are emeshed in local innovation systems such as Silicon Valley (Saxenian, 1994; Zook, 1999). It seems possible that the same social assets (Storper, 1997) that are used to support innovations – the networks and business visions constructed by entrepreneurs and their venture capitalists – have ended up abetting speculation.

I use the case of New York's Silicon Alley to examine the role of new media districts in organizing speculation and in diffusing its risks across the city. The problem of new media districts reveals gaps in how urban studies deal with the social organization of financial speculation. Leading approaches such as global systems models (Harvey, 1989; Sassen, 1991; Castells, 1996) and growth coalition models (Logan & Molotch, 1986; Zukin, 1991) do theorize the relation between cities and speculation, and the consequences for urban development. Yet, both approaches are limited in their ability to deal with the mechanisms and organizational forms involved in this kind of district. Moreover, the literature that focuses on innovation districts (Saxenian, 1994; Zook, 1999) does not consider how relations supporting innovation might come to support more dubious endeavors.

I explain how the ties that aid innovation can end up abetting speculation by using network models from economic sociology (Granovetter, 2000; Tillman & Indergaard, 1999; Burt, 1992) to supplement the account of relational assets offered in the innovation district literature. I argue that a new media district is a type of social terrain that gives rise to various kinds of market brokers, in this case new media entrepreneurs who use their stories to create markets and venture capitalists who weave new media firms into networks of risk capital. I propose that many entrepreneurs and their venture capitalists shifted their emphasis to positioning themselves vis a vis the stock market. Moreover, these networks of speculation extended into the rest of the urban economy as new media suppliers and other interests accepted new media stock as a sort of currency.

SPECULATION AND THE CITY

Some of the most prominent contributors to critical urban studies have analyzed the relationship between speculation and urban development. Their efforts largely fall into two approaches: (1) systems models that see urban development to be structured by global flows of capital and information and (2) growth coalition models that focus on how local real estate coalitions influence the flow of capital between and within cities. These differences in emphasis reflect a larger debate about whether the mechanisms that shape urban development are to be found in the logics of translocal systems or in the contingencies of local action.

Speculative Systems

Systems theorists assume that the vissitudes of urban development reflect the logic of the system, conceptualized as "capitalist" (Harvey, 1989), "global" (Sassen, 1991), or "informational" (Castells, 1996) in nature. In these accounts the logic that holds a system together also promotes speculative swings. The roles that major cities play in supporting speculation brings costs and benefits that are unevenly distributed within and between urban populations.

In his analysis of the urban process under a capitalist system Harvey argues that the circulation of capital, which is inherently speculative, contours urban development; he focuses on shifts in investment from the productive sector to the real estate sector. His treatment (1989) of recent economic shifts depicts cities as having new kinds of speculative entanglements. He claims that capitalists have sought to boost profit rates by creating financial mechanisms and image-making industries that accelerate the circulation of capital. Harvey thinks financial capital has gained unprecedented autonomy and drives a "casino economy" devoted to "financial speculation and fictious capital formation" (p. 332). Moreover, the financial system had become intertwined with cultural industries of cities: "On the back of this boom in business and financial services, a whole new Yuppie culture formed with its accoutrements of gentrification, close attention to symbolic capital, fashion, design, and quality of urban life" (p. 332). New York is cited as the center for making and processing fictious capital and Los Angeles as an "image production machine". New York also hosts a "cultural mass" engaged in producing "images, knowledge, and cultural and aesthetic forms" (p. 331).

Harvey's account implies that cities play basic roles in weaving together the financial and cultural fabric of contemporary society. However, the analysis is sociologically hollow in that he neglects to identify "mechanisms connecting structure and agency" (Katznelson, 1992, p. 140). We are left without an adequate account of the "dispositions and understandings" that contribute to the social organization of capitalism – what it is, for example, that "joins together the different fractions of finance capital, 'rentiers', and manufactureres into a coherent, unified agent" (p. 129). Agency becomes little more than a "reflection of structure" (p. 139) – a problem that other systems theorists tend to share.

Sassen (1991) argues that a limited number of major cities have the capacities to meet the functional requirements of a globalizing economic system, namely its command, control, and servicing needs. These "global cities" are spaces where the "work of globalization" gets done – much of it involving financial matters. The financial sector is the leading globalizing segment and the most important sector for the global city. Global cities are key sites for the production of financial "innovations" that meet the needs of restructuring and globalizing corporations. These innovations carry higher risks and as such, are more speculative in nature.

Sassen identifies a structural connection between innovation and speculation, but her neglect of the mechanisms involved makes it difficult to distinguish the two phenomena. In fact, recent work exploring financial manipulations (e.g. the savings and loan scandal) shows that many "entrepreneurs" committed deceptions and frauds (Pontell, Calavita & Tillman, 1997; Tillman, 1998). Market shifts, combined with regulatory gaps, creates openings for dishonest intermediaries (or "brokers") who manipulate relationships and perceptions through webs of deceit (Tillman & Indergaard, 1999). Sassen's account does contain evidence of such conditions: The expansion, reorganization, and fragmentation of financial markets lead to the expansion in the numbers and types of market actors and the emergence of new kinds of market "intermediaries" (1991, p. 84). But instead of examining how intermediaries *organize* to exploit these conditions Sassen comments that such conditions "require a vast infrastructure of highly specialized services" (p. 84).

Castells (1996) touts the "informational" nature of global systems. Information technology supports global flows of capital that serve as "meta" networks in the global economy. He uses currency trading to illustrate the speculative nature of these global flows: "computer programs and skillful financial analysts/computer wizards sitting at the global nodes of a selective telecommunications network play games, literally, with billions of dollars" (p. 436). He argues that all enterprises are becoming linked to this system. Capital is "realized, invested, and accumulated mainly in the sphere of circulation, that is as finance capital" (p. 471); in turn, global financial networks distribute capital in "all sectors".

Of special importance is the growing interdependence of the financial and technology sectors. Technology firms "depend on financial resources to go on with their endless drive toward innovation, productivity, and competitiveness" (p. 473) while financial capital "needs" knowledge and information generated and enhanced by information technology. Major cities have critical roles here as nodes and hubs of information exchange – "information-based, value production complexes, whose corporate headquarters and advanced financial firms can find both the suppliers and the highly skilled specialist labor they require" (p. 384).

Speculation has a number of consequences that are germane for our purposes: the increasing appeal of financial manipulation, dependence on capital movements shaped by "subjective perception and speculative turbulence", and the erosion of other cultural frameworks by a "quick bucks" ideology (p. 436). This hints at institutional and cultural processes that need to be explicated but Castells fails to examine which social segments develop particular kinds of perceptions or embrace risk cultures, and the circumstances and processes involved.

Local Growth Coalitions

The literature on growth coalitions or "machines" is one of the most important attempts to demonstrate how circumstances and actors within the city shape its development. The emphasis of growth coalition models is on organized efforts to manipulate market situations, particularly regarding real estate. The most influential statement is Logan and Molotch's (1987) account of coalitions that unite to turn the city into a "growth machine" that can maximize property values by drawing in outside investment. Although a variety of social actors are said to have an interest in property value, the core actors are members of the "rentier" class who seek to act as "structural enterpreneurs". In contrast to "active entrepreneurs" who use information from their networks to "speculate on the future of particular spots" (p. 30) structural enterpreneurs "speculate on their ability ... to alter the conditions that structure the market" (p. 30).

These rentiers and their allies in local business and government seek to intensify future land use in an area by creating a "good business climate", reducing regulation, and defusing public activism. These enterepreneurs engage in "organizational manipulations" (p. 31) and work to marshall an elite consensus so as to minimize the risks of real estate speculation. To this end they utilize relational resources that are locally embedded – a "knowledge of local markets and connections to community poltical and financial networks" (p. 40). With regard to effects growth coalitions tend to externalize the costs of development to other social segments or places and forestall the formation of alternative policies.

Zukin's revision (1991, 1995) argues that property interests and their allies mobilize development networks that connect with other sectors of the "symbolic economy" – a postindustrial ensemble of "cultural industries, business services, and real estate development" (1995, p. 270). This ensemble is the product of a new accommodation between cultural and business interests and a more general increase in "abstract financial speculations" (p. 11). The focus is organizing consumption activities within the city, an endeavor facilitated by devices such as museums and historical preservation, which help reframe devalued areas (e.g. industrial lofts) as desirable spaces for middle class consumers. While Zukin considers culture to be relatively autonomous, she argues that is is patterned by a structural dynamic: circuits of culture and capital interact in cyclical fashion.

Zukin's account stresses the power of representation, particularly the ability to "impose a vision on" or "frame" space (1997, p. 226). Respresentations involving art are powerful because they serve as indicators of "cultural vitality" (1995, p. 288). The mere presence of artists can alter the social perception of a place. Thus, various parties with an interest in a city's symbolic economy – real estate firms, non-profit institutions, city government, and artists – participate in cultural development coalitions that implement consumption themes across assorted spaces. But Zukin considers culture to be a problematic foundation for a city's economy – a temporary fix. The speculative nature of these cultural development strategies heightens the city's exposure during cyclical downturns.

In various respects the image of cities as sites for symbolic consumption fits well with new media districts – especially given the association of the Internet with creative fantasy. New media districts such as Silicon Alley also arose from the same artist enclaves that supported cultural development networks. And images of new media activity are being used to reframe and revalorize swathes of depressed real estate. However, new media districts are engaged in new forms of production. To the extent to which they are sites for organizing consumption, it is for markets that extend far beyond the city. Moverover, the main vehicle for financial speculation is the new media firm itself, in the form of its stock.

INNOVATION DISTRICTS

At first glance the literature on innovative enclaves (Storper, 1997; Scott, 1998; Saxenian, 1994) would seem of little help for analyzing speculation as it is based on the "rosy assumption that ties in general are of a virtuous nature" (Amin, 2000, p. 122). Accounts of innovative districts focus on the stock of "trust" and other "relational assets" (Storper, 1997; Pratt, 2000) they provide, which help economic actors to work together under conditions of uncertainty.

Recent treatments of innovation districts (Hall, 1998; Scott, 1998; Saxenian, 1994) feature "relational assets" such as rules of action, customs, conventions, and ties that facilitate collaboration. This work draws heavily on network models (Granovetter, 1985; Tilly & Tilly, 1998) that depict markets as consisting of webs of social ties and meanings. In the case of enclave economies (such as industrial districts and high tech regions) it is said that these relational and cultural webs are based on proximity and face-to-face interaction (Storper, 1997; Amin, 2000; Pratt, 2000). The literature suggests there is a dual nature to such districts: they seek ties with local supply markets that will aid their efforts to gain credibiliity with a broader circle of potential partners, customers, and investors. To this end community organizations and various kinds of district networks provide firms with access to specialized workers, suppliers, and financing (Saxenien, 1994). Venture capitalists are portrayed as having a central role here, particularly as they serve as "systems integrators bringing otherwise independent firms together to join forces in making some new product or pursuing some new process development" (Harrison, 1994, p. 109).

Recent studies of new media districts (Pratt, 2000; Zook, 1999; Scott, 1998) also stress the role of face-to-face interaction and local ties in the making of the new industries. Scott (1998) claims multimedia districts in San Francisco and Los Angeles are each constructed around a "local social order". Based on a study of San Francisco and New York Zook (1999) notes "a new firm's ability to build relationships with customers, suppliers, and strategic partners is influenced by the ability to build local networks" (p. 3). Others (Indergaard & McInerney, 1998; Pratt, 2000) suggest these districts are arenas where firms use their networks and visions to influence the making of new markets.

Zook (1999) contends that the organization of venture capital for new media firms is similar to that of Silicon Valley computer firms in that it takes the form of "regional systems" that have their own "conventions and institutions" (p. 1). Furthermore, he claims that most venture capitalists create relationships through "daily involvement in the local venture capital and entrepreneurial community" (p. 3). It is generally thought that venture capital is supported by insider expertise and ties that greatly reduce the odds of what are otherwise high risk ventures (Florida & Kenney, 1988; Saxenian, 1994; Zook, 1999). However, some observers from other fields (Florida & Kenney, 1990; Harrison, 1994) associate venture capital with the speculative "casino economy" of the last twenty years. Following Florida and Kenney (1990) Harrison remarks:

> ... the explosive growth of high-tech U.S. start-ups in the 1980s was fueled mainly by the venture capital industry, as part and parcel of the general speculative financial boom that also gave us the savings and loan crisis. ... small firm start-ups themselves became commodities, to be bought and sold in order to generate capital gains (Harrison 1994, p. 112).

Instead of seeing venture capital networks to be either categorically virtuous or invidious I argue that their nature is variable, contingent on other factors. Moreover, these are situations where trust itself can misfire. The trust that an innovative district produces through social ties and face-to-face interaction is a key resource for actors struggling to make commercial applications of a new technology, However, manipulating relations to gain unwarranted trust from others is the hallmark of less scrupulous entrepreneurs.

THE SOCIAL TERRAIN OF MARKET BROKERS

If we take a closer look at network theories we can see that the same kinds of social conditions that create opportunities for entrepreneurs to gain credibility – as market brokers – also provide opportunities for less responsible, if not predatory entrepreneurs. This allows us to link the role of new media districts in constructing new businesses and organizing risk capital with their role in organizing speculation. I propose that new media entrepreneurs and venture capitalists are, in essence, brokers who use their networks and stories to gain favorable positions in nascent Internet markets. Recently, many new media entrepreneurs and financiers have reacted to the robust market for technology stocks by shifting the focus of their brokering efforts to the financial markets.

Network theorists propose that the ability of market actors to identify opportunities or minimize risks, largely depends on the capacity of their network to provide trustworthy market intelligence. Such ties are problematic during market transformations, for example, when existing markets are disrupted or when new ones are being created. A lack of ties between different clusters of market actors or between different sectors – what network theorists term a "structural hole" (Burt, 1992; Granovetter, 2000) – constrains most market actors; however, this condition provides opportunities for entrepreneurs who can position themselves as "brokers" bridging a structural hole (Burt, 1992; Granovetter, 2000). For example, entrepreneurs with this kind of positioning are better able than other enterprises to connect with (and sell to) new market niches. Grabher and Stark (1996) propose that it is the presence of myriad structural holes and market brokers that make some regional economies (e.g. Italian industrial districts) so innovative. Some authors argue that the presence of structural holes implies that gaps in shared understandings also exist (Emirbayer & Goodwin, 1994; DiMaggio, 1994). Thus, brokers who want to create new ties also may provide new accounts of the situation (Indergaard, 1999; Tillman & Indergaard, 1999).

Importantly, parties requiring a broker's mediations are at a disadvantage, in terms of positioning and information. For a broker to retrain trust over time requires that they come through "when you need it" (Burt, 1993, p. 72). A brokering relationship based on trust would correspond to the *virtuous* networks depicted in the analysis of innovative production systems (e.g. Grabher & Stark, 1996; Zook, 1999; Granovetter, 2000). But less trustworthy brokers exploit transactions that involving promises of future services or benefits (Tillman & Indergaard, 1999). Ponzi schemes are classic examples. Each wave of investors is led to believe that they will profit from a subsequent wave of investment when in reality there will come a point where new investment falls off. This scenario of *invidious* networks might also apply to the kind of speculative circuits depicted in the analysis of cultural consumption ensembles (Zukin, 1991, 1995).

I propose that the problem of creating Internet industries fits well with a particular structural hole scenario sketched by Granovetter (2000): connecting what have traditionally been unconnected "spheres of exchange" containing goods which have not previously been "commensable against one another" (Granovetter, 2000, p. 11). The problems involved in linking such sectors are cultural as well as relational. "Goods or services not commensable cannot be exchanged in part because people do not understand how to think about such an exchange, and often as well because such an exchange is considered to be highly inappropriate" (Granovetter, 2000, p. 11). Deregulation (e.g. the Telecommunications Act of 1996) and technological innovations have, in part, enabled efforts to link myriad sectors (e.g. media, computers, communications) previously kept separate by state regulations and independent delivery systems. However, creating new firms and markets that bridge these sectors also requires the creation of new conventions, understandings, and sensibilities among businesses and customers.

Thus, a challege for new media entrepreneurs is to tell stories of markets, products, and business models that are credible to customers, collaborators, and investors. The latter also often look to a firm's venture capital backing when assessing its credibility; venture capitalists are perceived to be "smart money" possessing insider expertise and connections that allow they to assess a firm's prospects in its niche. However, in the new media case many entrepreneurs and their venture capitalists turned their focus to positioning firms for initial public offerings (IPOs) of stock. When investment banks and venture capitalists lose an interest in or the ability to assess risk, a more speculative scenario results. Many financiers use their expertise and contacts to set-up an expanding web of speculative investments; a combination of general market momentum and the limited supply of Internet stock, is likely to lead to large profits for the early wave of investors (e.g. institutional investors) and high fees for the investment banks, while later waves of investors are unlikely to recoup their investment. This is what has led a number of observers (Useem, 2000; Birger & Messina, 1998) to claim that many internet stocks resemble pyramid schemes.

THE NETWORKS AND CONVENTIONS OF AN URBAN ENSEMBLE

Innovation districts result from both an unintended mixing of diverse cultural, business, and institutional elements (Hall, 1998) and the deliberate creation of relationships, conventions, and institutions (Storper, 1997). Silicon Alley's development involved various entities who created, drew on, and modified its rich stock of networks and conventions. The district is an institutional ensemble that interacts, in complex ways, with the rest of the city and with larger systems.

Spaces and Artists in Transition

For much of the 1990s the economic recovery that reduced unemployment in most of the U.S. eluded New York City. The onset of national recession in 1989 struck doubly hard in New York because it followed close on the heels of the 1987 stock market crash that reduced the Wall Street workforce by some 30,000 jobs. The debacle of Wall Street – often presented as New York's economic engine – sapped confidence and left large stretches of mothballed office space, especially in the financial district. And downsizing at Wall Street had a multiplier effect: many service firms in Manhattan had benefited from the prolific spending of high paid Wall Streeters. After the recession struck unemployment soared over 10%; by the nadir in 1992, the city had lost some 360,000 jobs (Kanter & Messina, 1999; Malanga, 1999). As of 1994 vacancies in commercial real estate still averaged between 15 and 20% in much of lower Manhattan (Ward, 2000). Unemployment stubbornly remained stuck in the vicinity of 8%.

What has transpired over the last five years is a novel variation on the cyclical ebb and flow of capital and art long evident in New York (Zukin, 1982). This episode not only altered the image of marginal spaces but also resulted in the rise of a new kind of cultural production. Silicon Alley originally referred to a corridor in Lower Manhattan that followed Broadway south from the Flatiron District through Greenwich Village and into SoHo – the haunts of young creative types attracted to New York University and old factory lofts (Zukin, 1982). In 1995 1,350 firms involved with the new media employed 18,300 (Lohr, 1996). By 1997 the totals had grown to 2,128 firms and 48,828 jobs (Block, 1997). Between 1997 and 1999 the number of firms and jobs doubled. The enclave now extends north into parts of midtown, further south on Broadway and into the Financial District; firms also have begun to settle other boroughs.

The roots of Silicon Alley lay in experimentation with online services, CD-ROMs, and the worldwide web by creative types in Lower Manhattan. This new strata of "technobohemians" (Hall, 1998) played key roles in the enclave's organic development, that is, in creating a distinctive web of social ties and cultural meanings. The district's original culture, a sort of cyber counterculture, motivated creative types to experiment on the Internet before the advent of commercial visions. The early cultural images of Silicon Alley were of "the rebel" or "the underground". This culture has roots in the artistic ethos of Lower Manhattan and utopian ideals and libertarian sensibilities associated with computer hackers and hobbyists. New York University was the epicenter for this synthesis. It had opened an Interactive Telecommunications Program in 1979. The program, organized around the belief that the computer could facilitate free communication and expression, contributed to the creation of a cadre of computer-literate young people and an idealistic cyber ethic. There was a belief that anyone could become an online publisher and a disdain for media giants.

Graduates of the NYU program were prominent in building Silicon Alley's culture and its firms. Many entrepreneurs started out with a mission to change society or remake culture. However, the cyber ethic began to evolve once it was harnessed to entrepreneurialism. This is evident in an assessment by a @NY editor of the cultural changes new media brought to New York City:

> There's a new culture, a new society, a new lifestyle growing up here around Internet media. First and foremost, its entrepreneurial ... denizens of New York's new digital society are

... constantly looking for the new – new ways of doing things, expressing themselves ... new markets to conquer (Chervokas, 1997b, p. 2).

The district's web pioneers were largely hobbyists who created web sites for friends and small firms. They became adept at putting together rather motley "web teams" to serve market niches or do customized work. An explosion in corporate interest funded more elaborate arrangements, for example, online versions of magazines or marketing sites that ran hundreds of pages long (Gabriel, 1996). But many small firms continued to use a much more streamlined approach to host their own web sites. They produced "webzines" (e.g. online magazines) that were as popular as corporate sites (or more so) but more sustainable cost-wise.

The enclave's commercial turn produced distinctive institutions. Three Silicon Alley industry newsletters now circulate: *Silicon Alley Reporter*, *AlleyCat News*, and *@NY*. A venture capitalist notes (interview, July 1998) they "are part of the fabric" and "give people a common language". They have had a major role in establishing the credibility of firms. A top *Silicon Alley Reporter* executive comments (interview, July 1998) that "we wrote about the firms when they were small. This made them real, validated in the eyes of others". The credibility of firms was linked to that of the district itself. The executive notes,

the people we put on the cover wanted to be on the cover of *Red Herring* or *Upside* or *Wired*. But those publications hate New York, disdain it. They are mostly about San Francisco. . . . before we existed Silicon Alley didn't get much attention.

Professional and trade groups are also important supports for community-building. The New York New Media Association (NYNMA), which formed in the fall of 1994 with 8 members now claims 7,000 members (firm owners, consultants, graphic artists, musicians, accountants and lawyers) from some 2,500 firms (Pricewaterhouse Coopers, 2000). Other groups are The World Wide Web Consortium of Artists, the New Media Forum and Web Grrls (a trade association formed for women employed in the new media). The cyber culture is also practiced in area nightclubs and the lofts of local notables, which serve as sites for gatherings such as "Cybersuds", a monthly NYNMA affair. The party draws large crowds that include performance artists, underground musicians, fashion industry notables, techies and corporate types. In a scene that mixes spectacle and smoozing, entrepreneurs vie for the attention of corporate executives and venture capitalists (Gabriel, 1995).

As the new media community has developed the district has become pocketed with spaces for building relations and exchanging ideas. The *Silicon Alley Reporter* executive remarks (interview, July 1998),

I don't know anyone who is successful who doesn't have an amazing network. There is starting to be an overlap in networks with media and industries such as publishing, lawyers,

Wall Street. Big companies that participate in Silicon Alley do better instead of trying to do it all by themselves.

Of special importance are relationships that provide access to information or capital. A co-owner of one small start-up notes (interview, July 1998) that his company is especially dependent on its connections to local financial networks:

> Being in New York is very important – We'll never leave. New York City has a network of potential investors. This is my playground. We'll always have our headquarters here. A company like this always has to be looking for money – new rounds of investment.

Financiers depend on their networks for trustworthy information about promising ventures. Venture capitalists seek ties with other well-connected entities and individuals. The *Silicon Alley Reporter* executive is a good example. He remarks,

> I have personal relationships with every CEO in Silicon Alley. They ask me for contacts, advice. Its very social. I helped lots of firms get funded. A lot of financiers in New York call me for my opinions on firms (interview, July 1998).

Institutional Entrepreneurs

Several kinds of interests have played entrepreneurial roles in setting up support institutions for Silicon Alley. These "institutional entrepreneurs" (Castells, 1996) include venture capitalsts as well real estate interests. The real estate interests fit the "structural entrepeneur" model (Logan & Molotch, 1987) in that they are organizing to reduce their risks by using the new media to shape the framework for real estate investments. Interestingly, some Silicon Alley venture capitalists seem to be playing comparable roles vis a vis the new media and financial markets. Assorted networks and coalitions supported by real estate interests and venture capitalists have helped reframe the enclave, translating the sense of cultural vitality into images of economic vitality.

The Venture Capitalists

Early on venture capitalists made themselves into network hubs in Silicon Alley by helping organize strategic institutions. A cohort of indigenous venture capitalists formed as financiers left established venture capital firms in order to become financial brokers for Silicon Alley. Three members of the NYNMA board of directors have such backgrounds as do the partners in the district's best known venture capital firm – Flatiron Partners. Two of the NYNMA directors played key roles in organizing the association; one of the two, a Wall Street ex-patriate, was the person who coined the name "Silicon Alley" in 1993. He recalls (interview, July 1998) being motivated by a rivalry with Silicon Valley:

> I spent much of my time going to Silicon Valley. From the beginning there was a sense that what was being done here was different from what was being done in San Francisco. But the only ones that were setting themselves up as leaders of the revolution were these assholes in Silicon Valley.

After leaving Wall Street he set up a "cybersalon" as a means to bring together his old Wall Street friends and his new acquaintances in the new media. He and another venture capitalist had an idea "to start a larger, public organization".

> We recruited companies as sponsors for $100,000. I got the State to provide $50,000. Some of it went to pay for the new media study; some went to support a venture capitalist conference. ... No corporate memberships were allowed. It was all individuals. We wanted it to be more like a community. ... Our main activity was to throw a party, which made CyberSalon into a public event. There were only 15 people at first. It grew to as many as two to three thousand (interview, July 1998).

One of the venture capitalists who founded Flatiron Partners in 1996 notes that he and his partner left established venture capital firms to take advantage of the promising situation in Silicon Alley – an organized new media community was forming but there was a tremendous gap as far as venture capital was concerned:

> Since 1993 there had been an active new media community centered around the New York New Media Association. This seeded the market. ... But it never had the venture capital to help firms grow to the point of becoming leading firms nationally. ... By 1996 there were thousands of firms looking for venture capital and guidance (interview, July 1998).

He reports that that when Flatiron Partners began, "we had 1,200 business plans sent to us in the first three or four months". Initially, he held firm to what were then the standard guidelines for venture capital (inteview, July 1998):

> A venture capitalist needs to put capital to work, to do deals that are in the two to three-million-dollar range. New firms couldn't use that much. They needed seed capital – up to $500,000 tops. ... We spoke on the circuit to educate them on what we are doing. We couldn't do $100,000 deals.

Modest amounts of seed capital did begin to flow into Silicon Alley as "networks of angels formed" – angels being well-to-do individuals who invest smaller amounts of capital. In fact, the Flatiron partner noted that he forms ties with angel investors to find out which firms are promising candidates for large investments.

> What I look for is whether a firms can attract the attention of a decent executive who will invest $15,000 as an angel investor. For example, there was StockObjects, which sell pieces

of computer code on objects, such as film clips. I saw a guy from Corbis, the Bill Gates firm that bought the Bettleheim Archives, and sent him to meet with them. He is now on their board. He gives them credibility.

Flatiron Partners, in cooperation with NYNMA, has helped bring in other groups to invest in leading Silicon Alley firms. For example, the New York Investment Fund, a consortium of 75 New York businesses that originally targeted industrial firms, joined Flatiron Partners and other venture capitalists to invest in firms such as StarMedia and theStreet.com. Furthermore, the New York Investment Fund helped provide financing for a "cyberlab" established by Ericsson, the Swedish telecommunications giant. The lab, modeled on an Ericsson lab in Silicon Valley, is stocked with equipment and spaces that can be rented at low rates; Ericsson engineers and programmers are available to assist with up to 15 projects at a time. Firms seeking entre have to develop plans complementing Ericsson's strategy to integrate the Internet and telecommunications (Chen, 1997a). The New York Investment Fund's network of venture capitalists, executives, and media experts help identify and "sponsor" Silicon Alley firms for Cyberlab projects.

The standing of venture capitalists with local government and the media gives them considerable influence in policy debates. One leading venture capitalist who is involved in the Democratic Party comments (interview, July 1998):

> The public sector should be an omsbudsman. They should be identifying problems, opportunities. I just met with the director of the Economic Development Corporation. Local industry is being strangled by a lack of labor. All the local universities should get together. Our competitor is a firm like Goldman Sachs. The new media can't pay $150,000 for computer science grads. We can offer equity but there is not culture of equity for pay here. ... As firms go public with this problem government could support a conference to get the universities together and then get out of the way. The universities probably could not set it up. ... You need strategic vision.

Not long after these comments were made the NYNMA and the city sponsored a public forum for local universities to discuss the Silicon Alley labor problem.

The Real Estate Coalition

Several powerful groups were exploring ways of boosting Manhattan's new media industry as early as 1995. One group, composed of media giants Time Warner, McGraw-Hill and Forbes, began discussing ways to start a midtown facility for new media companies – a hub that would strategically focus resources, interactions and attention on the new sector. They envisioned a facility offering subsidized office space, equipment and research facilities for 150 new firms (Pulley, 1995).

It was a mini real estate machine (Logan & Molotch, 1987) from the financial district, not the media giants, that brought to life the New York Information Technology Center. The Alliance for Downtown New York, a business improvement district in the Wall Street area, persuaded a developer to "wire" an office building left vacant by the bankruptcy of Drexel Burnham Lambert. The city government, itself committed to revitalizing the Financial District's real estate market, agreed to provide a tax-credit plan to lower real estate costs for the center. Forty-one million dollars were spent outfitting the building with satellite hook-ups, high speed video conferencing, fiber optics for local area networks and high speed Internet connections. In a move that helped establish the image of Silicon Alley locally (and to a degree, nationally) the developer and his allies began a PR campaign that generated heavy coverage in the *New York Times*, as well as in more locally-oriented magazines and newspapers (Macht, 1997).

By late 1997 the 30+ story technology center was full – tenants include small firms, corporate new media units, and various services. This success lead the Alliance for Downtown and the city to extend the wired space concept with a "Plug 'N' Go program". Thirteen more financial district buildings have created pre-wired office spaces with low rents and short leases. They offer flexible wired space so as to attract small firms that do not know whether they will grow, fail or be acquired (Block, 1997). The Plug 'N' Go buildings now have over 250 tenants.

The Industry Association

The NYNMA is another actor that has helped bring Silicon Alley into the public consciousness particularly by sponsoring surveys of the local new media industry. NYNMA's president stated that the surveys were intended to provide data on the new media's significance so as to "gain the industry some respect from the city's economic development policymakers" (Goff, 1995, p. 22). The surveys have documented the impressive growth in firms and workers, leading the local media and city hall to regularly acclaim Silicon Alley as a cutting-edge in the city's economy.

The NYNMA has begun to organize its own services to help small firms access critical resources, especially capital and workers. It has introduced a Silicon Valley style "angel investors" program. A group of fifty investors will put money into promising start-ups in return for a stake in the firm. Investors include venture capitalists, investment bankers, executives from communications firms and entrepreneurs. Applicants have business plans evaluated and receive mentoring, if not cash (Chen, 1997b). NYNMA has also created internship programs that allow high school and college students to work in new media firms.

One critical contribution of NYNMA is the creation of forums where firms and support organizations can come together to discuss issues and agendas. NYNMA sponsors monthly meetings by special interest groups (e.g. marketing, business development) and along with entities such as the *Silicon Alley Reporter*, sponsors larger gatherings such as an annual forum on "The State of the New York New Media" and topical panels such as a recent event named "Stock Trader Nation".

Such opportunites for exchanging views are also vital for the creation of novel business forms and industry conventions. A *Silicon Alley Reporter* editor (interview, July 1998) notes, "When you are doing things that haven't been done before, you need a sounding board. You are making up the rules as you go". Similarly, a partner at Flatiron Partners comments, "We are making it up as we go and making it up isn' a bad idea now" (Wilson, 1999). And one of the venture capitalists who helped found the NYNMA (interview, July 1998) remarks, "orginally, we were collectively thinking through business approaches and developing an ideological agenda ... thousands have struggled to figure out how to make money at this. It takes a town to do this".

The 'Idea' of Silicon Alley

Such discussions have produced shared understandings about the nature of Internet business and the comparative advantages of New York's as a new media center. Originally, two beliefs were central to district sensibilities: (1) the Internet economy is really a media and communications business, not a technology business per se; and (2) making content is how Internet firms add value. In the words of an *@NY* editor (Chervokas, 1999, p. 1) the "idea of Silicon Alley" is the notion that the main opportunity for Internet business will be "developing and distributing new modes of expression that use Internet technology – and then taking advantage of the audiences created in the new medium". This idea has evolved as most firms found that they have to develop special applications and/or market positioning. Now the view is that the special role of Silicon Alley is making "creative business applications" – an endeavor that often involves content. Its status as a convention was evident in an exchange at a 1999 NYNMA panel. The CEO of Agency.com, a prominent web design firm, proposed that,

> The concentration of people who creates ideas is very high here. We are reaching a point where we don't have 'tech envy'. What's left to figure out now? The way people use

technology. To assess this is something we can do. New York is the place where people are paying attention to this (Suh, 1999).

A senior editor at *Newsweek* made a similar observation:

> I enter Silicon Valley and I feel like I am in a technology bubble. New York is the media capital of the world. It is inevitable that it will be a center for the new media. It is also a creative center for the country and the world. That creativity will end up on the Internet. New York will apply the technology tools created elsewhere (Levy, 1999).

A partner at Flatiron Partners then completed the Silicon Alley "party line":

> For people interested in technology New York is not the place for them. Silicon Valley is. ... People here are more implementers. ... But something interestings is happening. Technology is becoming a commodity. Services is where the value is. That's very good for New York. It's building services on top of technology. That is where the value is (Wilson, 1999).

VISION THING

Innovation districts develop sets of conventions about how firms can exploit new markets for innovative products (Storper, 1997; Pratt, forthcoming). The Silicon Alley case suggests that such sensibilities about the nature of Internet business and the comparative advantage of a district do not represent a formula for firms to follow as much as they serve as a general framework that firms, individually and collectively, can draw upon as they seek to develop credible visions.

As the attention has been drawn to high-profile firms vying for national recognition and stock market riches it is commonly assumed that few Internet firms are profitable. In fact, the majority of New York-area new media firms are small subcontractors or service firms that earn modest profits and over half of them engage in content-related activities (Pricewaterhouse Coopers, 2000, p. 19). At the same time, New York's new media sector has been shaken up by a combination of new pressures and opportunities that have led many firms to alter their business models. In particular, the would-be national contenders are exploring paths of business development whereby the can gain credibility as something more than a subcontractor doing project work or an edgy online magazine.

From Alternative Content to Niches

After an initial period of glory, the lot of the start-ups following the idealistic, alternative media model steadily worsened due to shifts in corporate strategies and their own bottom-line pressures. Some firms enjoyed a measure of success by targeting market niches (Johnson 1997). For example, Virtual Melanin hosted a web site ("Cafe los Negroes") for blacks and Latinos (Bunn, 1997). One of the most noteworthy examples was Psuedo, an Internet TV site, which survived into the new millenium as an unrepentant purveyor of "edgy", urban content. Psuedo became something of an "American Bandstand" for the alternative music set (e.g. techno-rave). A Psuedo manager recalls (interview, July 1998), "Psuedo starting by running a chat area on Prodigy. We had this huge loft, had big parties that brought together artists, creative types. We had this idea of mirroring in real space what goes on in a chat room". Although many observers have long panned the business acumen of Psuedo its vision about "the future of Internet TV" brought in millions of dollars of investment. Pseudo proposed that TV would become more fragmented as the Internet affected the cable TV market in a way that was similar to the effect cable had on network TV in the 1980s. The manager argued that by providing "authentic, credible, quality programing ... and original content" Psuedo would appeal to advertisers that want to hit "niches that are not well-represented in other mediums". These advertisers included soft drink, clothing, and auto firms interested in the 15 to 25 year-old age bracket.

Advertising Networks

In contrast to Psuedo, most of the firms seeking national reputations have moved away from the original alternative media model. Many firms are playing strategic roles in tracking, channeling, or brokering online activity (Caruso, 1997; Rothstein, 1997). One of the best examples is the area of online advertising services an area, an editor of @NY observes, in which Silicon Alley "has been wrestling with ways of inventing new paradigms" (Chervokas, 1997a, p. 2). Silicon Alley firms faced a formidable rival here – the Internet "portal" – firms such as search engines (e.g. YaHoo), online services (e.g. AOL) and browsers (e.g. Microsoft) that receive high advertising fees because they are popular starting points for netsurfers. Some Silicon Alley firms have pioneered a new Internet form – the advertising network – which may link thousands of web sites. Acting as network representatives firms such as DoubleClick offer to package combinations of sites to advertisers based on the latter's market strategy (Messina, 1998). DoubleClick also touts a tracking software that makes online ads more effective. The software delivers ads based on how often a user has seen an ad, user clicking habits, and how long the user views certain areas of a web site (Tedeschi, 2000).

E-Commerce

Other firms have explored the viability of E-commerce, between retailers and consumers or between businesses. For example, one web design shop known as Sixth Gear renamed itself Electronic Sales Systems, remaking itself into a facilitator of online transactions for industrial suppliers. One co-owner notes (interview, July 1998) that the firm calls itself an "internet-based sales channel":

> We don't do project work anymore ... Now we are focusing on industrial distributors. ... They say, 'don't put us in a catalog side by side with our competitors and don't get between us and our clients'. We had to figure out how to do this. We decided to use the 'phone company' model. We don't get involved in the conversation, we just 'provide a channel'.

Another part of his pitch to investors claims that the firm is well-placed at the "intersections of converging firms" – a good position for exploiting various kinds of change or "convergence" associated with the rise of the Internet. He reports:

> I tell investors that there are several types of convergence: first, technologies such as commerce and communication; second, infrastructure, which is becoming affordable and pervasive; three, market acceptance – consumers and business are accepting the Interent as a way to do business; and four, the financial industry – for example, venture capital is fueling this fire. This all creates a window of opportunity.

Community Networks

Another major path Silicon Alley firms are exploring is the community network model – a version of the portal model. Online community networks are networks of sites organized according to some theme. The commercial aim is to bring together large numbers of users which allows the firm to receive premium rates for ads and for links to online retailers. The trick is establishing a network of web sites and a community theme that are credible. For example, iVillage is the leading "woman's" online community nationally. Its founder argues that the iVillage model is compelling because it is organized to meet the needs and sensibilities of women – particularly, when it comes to their role as consumers: "For women, buying products is a big part of what they have to do" (Gould, 1998, p. 24). iVillage's network of sites contains a plethora of opportunities for gaining "product information" and for actually shopping online; moreover, the founder claims the network structure fits the more integrated nature of women:

> Pulling together different sites and categories under 'women' turns out to be really positive. Women traditionally do not put up strong fences between the different parts of their lives.

Innovation, Speculation, and Urban Development

> If you watch a group of women at work, they will go from [talking about] what kind of stockings they're wearing to talking about their company's strategy, to their investment portfolio in about ten minutes. ... So when you create a women's network, you create a place where a woman can take all the things that are on her plate and deal with them all in that one spot (p. 52).

iVillage claims that it ability to bring together women will make it very popular with other businesses: "We will end up with a very important concentration of women ... important aggregations of very pure demographics that are very valuable to advertisers and commerce players [online retailers]" (Gould, 1998, p. 56).

StarMedia is a Spanish and Portuguese language network that claims to have invented the category, "Latin American" portal. The founder notes there have been a lot of companies set up soley to support IPOs but argues that StarMedia has,

> a bigger mission, the idea of unifying Latin people around the world. The idea of creating this central plaza, this place where people can find each other, can communicate, can share ideas is a very powerful concept (Calacanis, 1999, p. 130).

He notes his first challenge was conveying this concept to venture capitalists.

> It was very difficult to find the nexus between people who understand Latin America and people who understood the Internet. People who understood Latin America were totally trapped in the old paradigm of industrial companies ... people on the Internet have this idea of Latin America as this very vast desert of poor people ... And it was only with Flatiron [Partners] who understood the Internet very well, and Chase Capital Parners who understood the Latin America market very well, that we were able to say to them, look this is how big the opportunity is (p. 128).

Web Services

One of the most important and telling paths of business development is the evolution of web design shops. Often called "agencies" – a model derived from the advertising industry – these firms typically did contract work for corporations such as designing splashy web sites, developing banner ads and planning online ad campaigns (Chervokas, 1999; Warner, 1999). A number are remaking themselves into comprehensive corporate services or consultants. Most of these are distancing themselves from the legacy of hip subcontractors who think they are the only ones who "get it". The CEO of Agency.com advises (Ritzel, 1999, p. 2),

> We were evangelists in the beginning, telling everyone exactly what is is they need ... Its time to stop telling clients that they are clueless. ... In the Silicon Alley services sector we need to identify what is is we do otherwise we'll only [end up] a division of different businesses.

A RareMedium executive remarks (interview, July 1998) that his firm, which started as "five guys in a room", is now "building long term relations with Fortune 500 companies". He stresses its alignment with corporate sensibilities:

> RareMedium started with CD-ROM content and migrated to multimedia and Internet content. Now we are looking for 'full service' relationships, not one-project clients ... We have leveraged the oportunity of corporate outsourcing ... Our model is to penetrate large clients, prove ourselves and get new brands, areas ... as Corporate America steps up its budget, the bar will be raised. They won't tolerate you unless you act like a business ... You are going to have to morph into a business with a global strategy.

In contrast, a third firm, Razorfish, continues to embrace the "edgy" content image but seeks to distinguish itself from the passe agency model. One of its co-founders proclaims, "We make the thing, not the advertising about the thing" (Goldberg, 1998, p. 32). He proposes that the firm can provide a broad array of services to clients whose product can be manufactured and delivered digitally:

> We focus on four main sectors: media/entertainment, tech/telecom, financial services and cultural institutions. These kind of businesses can fully integrate digital technology into every core element of the organization. ... What we do is digital change management. Our goal is to move our clients from the traditional marketplace mentality to a marketspace mentality, the idea of integrating a company's operations, finances and marketing in a seamless digital environment (p. 32).

SILICON ALLEY'S BULL RUN

By 1998 the role of venture capitalists in certifying the credibility of new media start-ups was altered by another financial mechanism: IPOs. IPOs by a handful of leading Internet firms (e.g. Yahoo, AOL, and Amazon.com) and the extraordinary rise in their stock values propelled the Internet to prominence as a material and symbolic force: The market value of Amazon.com increased 638% (to $18.9 billion), Yahoo rose 452% (to $11.7 billion), and America Online increased 294% (to $40.9 billion) (Gimein, 1998, p. 40).

Global systems theorists (Sassen, 1991; Castells, 1996) might point to such an episode as showing how the fortunes of industries and cities depend on global flows of capital. However, my examination of Silicon Alley up to this point has revealed the role of new media industry-builders and real estate interests in positioning new media firms vis a vis financial interests and markets. It is more accurate to state that the context for Silicon Alley entrepreneurs was changed by shifts in the circulation of capital, as evidenced by the robust stock market of the United States. Importantly, their responses to this situation and motives are varied. Moreover, I will show the role of networks of entrepreneurs and venture capitalists in connecting Internet stock to the bull market for stocks.

Several Silicon Alley firms led the way in connecting the district to the Internet gold rush. The first success was that of Earthweb, an online service for "techies. " Earthweb was initially listed at $14 a share for the IPO, a valuation of $110 million. Its value peaked at $89 a share, a market capitalization that exceeded $400 million (Chervokas, 1998). Soon thereafter theglobe.com had the most spectacular opening price in Wall Street history. Listed at $9 a share it opened at $87 a share as a "queue" of surplus of buyers formed; by the time the average share had changed hands five times the stock had peaked at $97 (Gimein, 1998, p. 40).

One important consequence of these initial successes was that it greatly increased the credibility of Silicon Alley – even in the eyes of Silicon Valley interests (Evans, 1999). Dozens of Silicon Alley firms began to shift their priorities from getting established in some segment of the Internet market to positioning themselves vis a vis the financial markets. It is important to remember that the great majority of the New York's new media firms did not participate. And there were multiple motives for those that did pursue IPOs. One common motive was to use a successful IPO to gain validations and publicity so that a firm would stand out over rivals. Another was to gain resources that could be used to build the business. Many firms believed they had to raise large sums of money so that they could acquire other firms – either to gain new capacities or to eliminate competitors. And owners and venture capitalists alike thought that successful IPOs would also provide firms with the kind of incentive that was needed to attract or retain highly skilled workers – stock options. And some were indeed hoping to enrich themselves. For example, one firm co-owner recalls,

> One of our partners [another firm] offered to buy us on a strategic basis but we weren't interested. You don't make money on a strategic buy – you make it on financial buys. That was the kind of money we were interested in (interview, July 1998).

Examining the positions of new media entrepreneurs and venture capitalists in the IPO process reveals their strategic roles in organizing what has been called the "irrational exhuberance" (Shiller, 2000) of the recent stock market. Figure 1 helps illuminate the actors and dynamics involved by outlining the steps taken in the Earthweb IPO. Venture capitalists not only get a firm to the launch stage but are also likely to have ties to the investment banks that underwrite the IPO. The investment chain depends on serial "vouching" for the potential of a firm.

The venture capitalist give the firm credibility with pre-IPO investors and then with investment banks. The banks, in combination with the venture capitalists, give a firm credibility with prospective stock buyers. Buyers are themselves stratified between institutional investors for whom stock orders are

Step One: Informal Schmoozing
Make contacts among the investment banking community. When an investment bank underwrites a stock, it is essentially buying the stock from the company and reselling it to portfolio managers, who then sell it on the market on the first day of trading. Earthweb met investment bankers through personal relationships and through its venture capitalist.

Step Two: Bankers' Beauty Contest
Banks make formal presentations explaining how they would underwrite the offering if they were selected. Earthweb choose four banks with J. P. Morgan selected as the lead underwriter.

Step Three: Due Diligence
The underwriters study the company's business record and prospects in its market segment. Then the company files with the SEC its declaration of its intent to go public along with a copy of the prospectus.

Step Four: Quiet Period
After the filing company employees are prohibited from making public statements about the company until 25 days after the stock begins trading. The underwriters set the opening price for the stock offering based on a comparable group of companies with similar business models; Earthweb's stock was priced in the range of $12 to $14 per share. The number of shares to be sold is set to optimize demand (i.e. create a supply of stock that is less than the level of the expected demand).

Step Five: The Road Show
Company management goes on a whirlwind tour to get institutional investors interested in buying the stock. Earthweb management met with 65 fund managers over the course of two week tour of 13 cities.

Step Six: Taking Orders
The underwriters set the final price of the offering. Institutional investors then place their orders with the underwriting banks.

Step Seven: The Stock is Traded
In the first day of public trading the price of many Internet stocks shoots up several hundred% as stock changes hands as many as five times. Institutional investors can reap immediate profits from selling; company management must wait 180 days before selling its stock.

Source: adapted from Ball (1999).

Fig. 1. Steps in Earthweb's Initial Public Offering (IPO).

reserved at a set opening price and subsequent waves of less-connected buyers who may pay four or five times as much during the course of the first day of trading. One can also see that the involvement of new media entrepreneurs, venture capitalists, and various other financial interests was "rational", at least in terms of short-term gains. There is great uncertainty about the viability of individual firms and even whole segments in the nascent Internet industry. However, the way the process is organized the risks are largely bore by a mass of less-informed investors who join the investment chain in the later stages. Those who join early in the process on account of their priviledged position in the investment network (e.g. venture capitalists, investment banks, institutional investors) are less exposed to risk and are the ones most likely to receive huge returns.

The returns were indeed huge in the spring of 1999 during a protracted run of successful Silicon Alley IPOs (there were 34 IPOs in all) that elevated the standing of the enclave and of firms such as DoubleClick, Razorfish, StarMedia Network, and iVillage. It also brought a tidal wave of paper wealth. At the start of 2000 the combined market value of 29 Silicon Alley firms was $29.5 billion (Watson, 2000, p. 1). It seemed that the endings of several business stories would be quite happy as evidenced by the value of personal stakes that founders held in leading firms: $749 million for the founder of DoubleClick; $180 million for the founder of iVillage; $191 million for the founder of StarMedia Network; and $169 million each for the two co-founders of Razorfish (Kanter & Messina, 1999).

NEW YORK'S "NEW ECONOMY"

The new media, bolstered by high stock valuations has fueled explosive growth in the local economy, especially in the real estate sector; it has also induced a significant amount of restructuring in the business service firms. However, Silicon Alley's rapid extension has strained the markets for labor and real estate and is leading to the displacement of other kinds of firms. And given the role of speculation the sustainability of the growth achieved is problematic.

The local media claims that the new media is replacing Wall Street as the engine of the New York economy (Kanter & Messina, 1999) and represents the cutting edge of a "new economy" that is responsible for the city's comeback (Malanga, 1999). Several years of robust job creation (80,000 new jobs a year) have finally brought the employment totals back to the pre-1989 recession level. A PricewaterhouseCoopers study (2000) confirms the magnitude of new media job growth. The number of new media workers in the city grew from 55,973 in 1997 to 138,258 in 1999 (p. 29), an increase of 82,285 (147%) (see Table 1). A hiring boom at firms enjoying rich IPOs was one major factor. For example, between 1998 and 1990 the workforce at DoubleClick increased from 200 to 1,200; Agency.com went from 220 to 1,000, and Razorfish increased from 350 to 1,000 (Malaga, 1999, p. 52). Another factor is the large number of firms the district is drawing to the New York area; an estimated 20% of the

Table 1. New Media Employment in New York City and New York Area, by Employment Status, 1995, 1997, and 1999.

Employment Status	1995	1997	1999	Change 1997–1999	Percent Change
New York City					
Freelance	6,150	12,680	20,773	+8,093	+63.8
Part time	2,800	11,280	12,820	+1,540	+13.7
Full time	18,350	32,013	104,665	+72,652	+226.9
Total	27,300	55,973	138,258	+82,285	+147.0
New York Area[a]					
Freelance	17,000	24,478	38,659	+14,181	+57.9
Part time	8,600	22,526	25,363	+2,837	+12.6
Full time	45,900	58,767	185,617	+126,850	+215.9
Total	71,500	105,771	249,639	+143,868	+136.0

[a] The PricewaterhouseCoopers study defines the New York Area as including New York City, Long Island, the lower Hudson Valley of New York State, northeastern New Jersey and southwestern Connecticut.

Source: PricewaterhouseCoopers (2000, 29).

new media companies or divisions operating in the New York region had moved or expanded to the region from another location (PricewaterhouseCoopers, 2000, p. 37).

The Diffusion of Growth and Risk

Since mid-decade there have been many reports in the local business press of established service firms that were opening up new media units and of the rise of a whole ensemble of specialized business and technical services geared to serve Silicon Alley: technical trainers, language translation services, software duplication, recruiters and consultants (Messina, 1996a). It appears that the IPO run has boosted growth in these sectors and accelerated the restructuring of sevice firms. Local media reports claim that various sectors are creating "tens of thousands of jobs" in a "scramble to serve Silicon Alley" (Kanter & Messina, 1999, p. 30). This includes many business and professional services that were "once dependent on Wall Street" – "advertising firms, lawyers, accountants, architects, contractors, printers and caterers" (p. 30).

Changing their business focus is not the only risk these firms are taking on behalf of Silicon Alley. Because many start-ups were short on cash, many service firms agreed to "flexibility" on compensation: they defer fees, accept flat rates or even accept an equity stake. In fact, many firms have actively sought such arrangements as investment opportunities – a practice that is common in Silicon Valley where service firms build portfolios of equity in other firms. For example, when a construction firm rennovated the offices of eFit.com, a physical fitness web site, it accepted 75% of its $120,000 fee in eFit.com equity (Kanter & Messina, 1999, p. 36). An executive search firm, which previously used to ask for a third of the first year's salary that an executive receives in a new job, now asks for a third of their first year's stock options; the firm has reportedly make 20 such deals (Baer, 2000). Similar tactics became common in the interior design and construction sector where dozens of firms (architects, contractors, project coordinators, furniture dealers, and lighting designers) have reorganized, or started up, to rennovate offices for new media firms. The head of one architectural firms termed its reorganizations "an investment in a new wave of entrepreneurs" (Croghan, 2000, p. 58). These firms have tailored their efforts to become part of the system supporting new media financing. A flooring supplier created a project management firm that handles design, construction, and and equipment installation for new media firms; it also helps provide financing for the work if needed. The founder remarks, that his firm exploits the need for speed, which is essential because new media firms "run through a lot of money fast. The sooner they get set up, the sooner they can raise more money" (p. 58).

Originally, the new media, like other Manhattan industries (Zukin, 1995), took advantage of underemployed artists and musicians willing to do creative work for relatively low pay. As corporations became key new media clients, Manhattan's corporate services added new kinds of talent to the mix, such as computerliterate lawyers and accountants (Temes, 1995). When Wall Street and corporations began to use large salaries to woo technical workers, many new media firms turned to stock options, a Silicon Valley tactic, to recruit or retain "talent". The IPOs greatly increased the gap between executives and lower level workers – on paper. A stock benefit plan administrator estimated that more than 1,000 millionaires have been created among the founders and executives of Silicon Alley firms; another 2,000 has amassed between $500,000 and $1 million (Kanter & Messina, 1999).

Recycling Real Estate

One of the most important signs of the accelerating importance of the new media to the city economy has been their physical expansion into new spaces. They are filling spaces left vacant by the financial industry as well as pushing into the old industrial districts, many of which have fallen into marginality. However, the new media firms are not only chasing out the ghosts of the industrial past but also large numbers of contemporary manufacturers.

At the beginning of the new millennium the new media was making a pronounced contribution to a boom in the Manhattan real estate market that had been revived by growth in the city's financial and service sectors. The entry of a mass of new media start-ups into the real estate market was spearheaded by the IPO stars who were willing and able to pay premium rents for relatively large spaces. For example, About.com signed on for 170,000 square feet at $50 a square foot, while DoubleClick leased 400,000 square feet (Bernard, 2000, p. 14); in January of 2000 a report estimated that some 100 new media firms were searching for offices in the range of 60,000 to 100,000 square feet (Jakobson, 2000).

During the first half of 2000 the new media and other technology businesses leased 5.2 million square feet, displacing the financial sector at 3 million square feet as the number leaser of commercial space in New York City. In the second quarter the new media and technology firms accounted for nearly 31% of the total space rented in the city in comparison to about 17% for the financial sector (Chiu, 2000, p. 1, 54). Circumstantial evidence of the new media's impact on the revitalized real estate market can be found in submarket data (see Table 2).

Table 2. Change in Commercial Real Estate Market, Selected Areas in Manhattan, 1994 to 1999.

	Vacancy Average		Rent Rate (per sq. foot)	
	Dec. 1994	Dec. 1999	Dec. 1994	Dec. 1999
Garment District	13.3%	4.3%	$23.46	$32.64
Midtown South				
– Chelsea	9.8%	4.3%	$17.40	$36.05
– Hudson Sq./SoHo	14.9%	5.6%	$14.48	$34.48
– Park Ave. South/ Flatiron	21.9%	2.6%	$23.95	$33.93
– Union Sq./East Village/NoHo	14.3%	7.6%	$17.60	$32.93
– West Village	17.1%	0.2%	$11.95	$30.00
Downtown (includes financial district)	17.8%	6.8%	$31.32	$37.32

Source: Ward (2000, 42, 48, 50).

Vacancy rates for commercial real estate at the center of the original Silicon Alley enclave, the Flatiron District and SoHo Districts, had been 22% and 15%, respectively at the end of 1994. By the end of 1999 vacancy rates for the two districts had fallen to 2.6% and 5.6%, respectively. Furthermore, in the same period the average rent (per sq. foot) in the two areas had soared: from $24 to $34 in the Flatiron and from $14.50 to $34.50 in SoHo. The effect of the expansion of the new media firms into new territories can be seen in the cases of Chelsea and Downtown (the financial district). Vacancy rates for Chelsea and Downtown have plummeted while rents have skyrocketed during the same period.

As commercial spaces in the original center of Silicon Alley became rented up, new media firms led the way into new territories that included not only Downtown, but also increasingly, parts of Midtown such as Chelsea and the Garment district. Chelsea is now deemed the "epicenter" of new media-related activity. Local media reports credit new media firms with having "reclaimed buildings and neighborhoods forsaken by the traditional business community" (Kanter & Messina, 1999, p. 30).

However, in Chelsea, the Garment district, and other submarkets there are sizable clusters of firms occupying industrial space that are now being displaced by the new media firms – the latter backed up by big dreams and oftentimes, big money. A notable example in Chelsea is the massive Starrett-Lehigh Building, a 2.3 million sq. foot former railroad terminus. New media firms, paying nearly $30 a sq. foot in rent, are displacing mattress firms and warehousers who had been paying rents in single digits (Aron, 2000b, p. 26). Displacement is taking place even in the Garment District where zoning low preserves 50% of all side street space for garment-related businesses. The new media is rapidly penetrating the remainder of the side street space as well the unprotected avenues (Flamm, 2000, p. 24). The process of commercial gentrification is also pushing out other firms in areas favored by the new media. For example, Trinity Real Estate, the real estate arm of historic Trinity Church, has stopped renewing the leases of dozens of printers in what are now considered prime new media spots such as the Hudson Square neighborhood by SoHo (Kanter & Messina, 1999, p. 38). The printers had paid as little as $7 a sq. foot in rent; Trinity is now asking new media and telecom firms for as much as $38 (Aron, 2000b, p. 26). More recently, the new media has begun to push into Chinatown, the Flower District, and old industrial areas in Queens and Brooklyn.

At least some landlords were aware that the new base, consisting of large numbers of small start-ups with uncertain prospects, carried more risk than their traditional corporate tenants. As demand exploded after a new flood of venture capital was injected into the ranks of start-ups, landlords became more selective – at least in deciding which new media firm to lease to. Some, in the manner of venture capitalists, seek to assess the credibility of would-be tenants. They are "asking to see business plans, ask about a company's backers, the size and shape of the funding, its 'burn rate' and 'exit strategy'" (Anderson, 2000, p. 2). Start-up usually have to pay a rent "premium" of 10 to 20%, pay for one year's rent in advance, and "provide proof of strong venture capital backing" (Aron, 2000a, p. 28). Some landlords accepted equity in new media tenants to seal the deals (Aron, 2000a; Anderson, 2000).

BROKERING BETS

New media stock became a form of currency, used to purchase stakes in other firms and to compensate employees, landlords, and other service providers. Some firms even made charitable donations in stock. While many new "investors" sought to evaluate firms in the manner of venture capitalists, many of the latter relaxed their standards. The "smart" money became less so as assorted businesses rushed in to position themselves as intermediaries in new media financial networks.

The development of a venture capital infrastructure had long been a critical concern in Silicon Alley. In the wake of the string of lucrative IPOs, venture capital investments surged into the district. Whereas venture capital investment in Silicon Alley totaled only $49.5 million in 1995 (Chen, 1997c) it reached $536 million during the second quarter of 1999 – a jump of 320% from the 1998 second quarter total of $127 million (Li, 1999, p. 1). A dozen new venture capital funds had opened to serve Silicon Alley by the end of 1999 (Watson, 2000). This provided long-sought-after evidence of the district's credibility. At "Venture Downtown", a NYNMA event where venture capitalists pay $1000 to hear pitches by start-ups, a venture capitalist reflected:

> We didn't have the networks that existed in Silicon Valley. But now Silicon Alley is growing virally, like the Internet. The number of venture capital companies in New York has quadrupled in the past couple of years.... We've reached critical mass" (Tierney, 2000, p. 1).

The head of the NYNMA similarly exulted,

> At first new media entrepreneurs here didn't have the investment flywheel that existed on the West Coast. But thank God for lust and greed and competition ... We started getting venture capitalists in our own backyard" (Tierney, 2000, p. 1).

Not everyone was so sanguine. Early in 2000 a partner at Flatiron warned that an oversupply of venture capital was distorting the process of firm formation:

> I would not be surprised if there were 20 or 30 new Internet companies a day being formed in New York. That's scary. People are forming companies for the wrong reason. They aren't passionate about their ideas. ... there is so much capital in New York anybody can start a company (Hansell, 2000, p. 13).

Yet, Flatiron Partners continued to increase its own Silicon Alley investments as did its established competitiors. Bear Stearns and Wit Capital each raised new funds of $200 million while General Atlantic and Captial Z Partners started a $300 million fund. Silicon Valley funds are increasing their investment pools for Silicon Alley or are setting up affiliates there (Hansell, 2000).

Furthermore, there is a large number of "nontraditional players" who are now positioning themselves as Silicon Alley venture capitalists. This includes corporate giants like Time Warner and Anderson Consulting – the latter having earmarked $1 billion venture fund for web firms (Walsh, 2000c). It also involves several growing Silicon Alley web service firms, RareMedium and Concrete Media, who set up incubators to provide young start-up firms with office space and technical services in exchange for equity (Hansell, 2000). Other incubators include a local affliate of California's Idealab and i-Hatch, a venture founded by the ex-head of Bertlesmann's online Bookstore, the former head of CBS's internet unit, and a former Lazard Freres venture capitalist (Hansell, 2000).

Venture capitalists usually wait for start-ups to mature before they evaluate them; They typically come in the second or third rounds of financing. However, the process accelerated during the IPO craze. Venture capitalists felt compelled to get involved much earlier in a company's lifecycle because the stock market had been willing to embrace IPOs from young start-ups. A comment by an i-Hatch partner captured the new mood: "We want to find the guy as he is just looking for the cocktail napkin to write his idea on" (Hansell, 2000, p. 13).

AFTER THE GOLD RUSH

One might say that the long-anticipated bursting of the stock market bubble in April of 2000 "changed everything". However, this is not altogether true. It is the firms that broke from the pack in an effort to become national contenders (or simply rich in the stock market) that have been staggered by the end of the bull market. It is not clear how the fall of Silicon Alley's champions will affect the district's trajectory and the prospects of thousands of small new media firms in the city that do earn (modest) profits. In fact, the sector continues to expand in the city, although more slowly than before. Moreover, various forces of urban development that have drawn from the new media explosion remain in motion.

Exit the Standard Bearers

With the closing of the IPO "window" new waves of investment are no longer forthcoming or have been greatly reduced. Because most of Silicon Alley's would-be national contenders are cash-guzzelers that have yet to earn a profit many are desperately seeking new sources of operating capital – or profitable exits.

The crash of the NASDAQ technology index in April of 2000 and its continuing decline has hit Silicon Alley's public firms harder than most. The NASDAQ lost over a third of its value since its high in March of 2000. According to *At New York*, an online journal which tracks the "Alley Fund" – a composite of 46 Silicon Alley firms that are publically traded – their combined stock value fell from $1,519 at the beginning of the year to only $276 as of the end of October, 2000 – a drop of over 80% (*At New York* January 7, 2000; October 27, 2000).

The carnage experienced by some leading firms was remarkable. On its IPO day in March of 1999, iVillage staff had celebrated a peak price of $113 a share (Napoli & Hansell, 1999); by October of 2000 the value had plummeted to $2.09 a share. Other firms experienced drastic drops from January to October (2000): DoubleClick fell from $249 to $14.50; About.com from $90 to $26; Razorfish from $96 to $4.44; RareMedium from $80 to $5.03. Theglobe.com, at $0.78 a share lead a list of firms that are in danger of being "delisted" by the stock exchange for having values persist below $1 a share: Bluefly, Big Star Entertainment, GSO (aka Cybershop), Vitaminshoppe.com and Cornerstone Internet Solutions (Vargas, 2000). Of the firms that had gone out of business as of November 2000, the most reknown was Psuedo, which left 240 staffers jobless. Other substantial firms that disappeared included APBnews.com (140 jobs lost) and Urban Box Office (300 jobs lost). A number firms, which had been acquiring firms until recently, laid off workers in a bid to cut costs and/or impress investors. For example, StarMedia, which had acquired 10 firms, laid off 125 workers; About.com which had acquired 8 firms discharged 20 workers; Xceed which had acquired 3 firms, laid off 75.

Firms that have almost exhausted their capital are trying to get new venture capital or looking for a desirable "exit". The more desperate a firm is for capital, the less selective it can be. For example, InterWorld, one of the few Silicon Alley firms that develops software, was hit hard by the nosedive of dot-com firms it served. It sold a $20 million stake to Jackpot Enterprises, a Nevada gaming machine operator turned Internet investment firm; the latter is mulling a full merger (Walsh, 2000b). Conversely, About.com, with $133 million in cash, was acquired for $600 million by Primedia, an old media giant. Analysts speculate that a similar outcome awaits more viable firms, such as iVillage and StarMedia. But for other firms "no price appear to be too low" (Walsh, 2000a).

The total estimated layoffs for Silicon Alley stood at only 3,000 workers as of October 2000 (Eaton & Blair, 2000). Thus, while publically-traded firms are struggling and more layoffs are on the way, it is not clear what the stumble of these standard bearers means for the district as a whole. The future of Silicon Alley firms will come down to the positioning that they can achieve vis a vis restructuring markets and corporations. There is less talk about new media firms leading a revolution against the "old" economy; the debate is shifting to the question of how much leverage they will have vis-à-vis established corporations.

At this point it is useful to try to untangle the innovative and speculative dimensions of Silicon Alley's development. District insiders suggest that the pull of speculative riches distorted or undermined the efforts of firms and their institutional supporters to organize a viable position vis a vis the Internet industries. An @*New York* editor claims the "dull stock market chase" diverted Silicon Alley firms their real strength: "fearless creative thinking" (Chervokas, 1999, p. 2). The publisher of *Silicon Alley Reporter* remarks that firms need to "focus on the business models that generate revenue and that inspire the market that really matters . . . customers" instead of creating "business models that the stock market is rewarding" (Calacanis, 2000, p. 3). After the first Silicon Alley IPOs a venture capitalist warned of the "danger of a systemic problem from overvaluation of unsound businesses" and "young people with expectations out-of-whack". He warned that the easy money might "spoil a generation of entrepreneurs. . . . and go up the food chain . . . impact venture capitalists and other investors" (interview, July 1998). These words appear prophetic. More generally this case shows that the roles venture capital plays in innovation and speculation, while related, are distinguishable. Much is lost if an analysis assumes they are wedded by some system logic or that speculation itself is a constant state.

Advancing the Silicon Frontier

Despite the heightened uncertainty forces of development are still drawing off the new media's symbolic and material power to transform the face of the city. The new media has contributed to a sense on the part of many that a "new social reality" has arrived. Real estate speculators are investing in "chic areas" in the meatpacking district and flower district now associated with the "new economy" and "new media lifestyles" (Holusha, 2000b). In TriBeCa, formerly a "butter and eggs" district, warehouses that rented for $1 a sq. foot in the 1960s are being converted to condos that sell for $500 to $800 a sq. foot

(Dunlap, 2000). Some real estate firms are changing their maps, claiming that old place identities are fading while new "submarkets" are emerging. Most portentious is the expansion of the "business district" to new corners of Manhattan and even to other boroughs – creating a "new frontier" for providers of office space (Goff, 2000). Some real estate observers trumpet the emergence of a "seamless city" extending from 60th Street through Lower Manhattan (Holusha, 2000a).

Two major initiatives are using the imagery of the new media in order to extend the business district. The first is the Guiliani Administration's "Digital NYC" program, which seeks to create new media districts across the city through emulating the Financial District's Plug'n'Go program. Digital NYC encourages local development corporations, universities, building owners, and telecoms to set up wired enclaves outside Lower Manhattan. The city is offering to pay up to half of the marketing cost (up to $250,000 per district) and provisions would be made to lower property taxes, provide cheaper energy, and to reduce rental rates. Interest has been shown by 15 local entities in Chinatown, the Bronx, Brooklyn, Staten Island, and Long Island City (Queens) (Lentz, 1999).

It appears that the city is using this program and the aura of technology to create an umbrella for mini "growth machines" (Logan & Molotch, 1987) that would change the image and value of old industrial properties. In some cases new media firms are being packaged with some of the myriad telecom "hotels" that are re-cycling old industrial facilities throughout the area. For example, a telecom hotel is an anchor for a Digital NYC site at Brooklyn's massive Bush Terminal, a 12-building complex with 5 million sq. feet of space. In fact, this inititive is extending the zone of displacement for industrial firms. Manufacturers who recently settled in Bush Terminal after being displaced from Manhattan are being displaced again as the rent is raised from $5 a sq. foot to $20 (Boss, 2000).

The second initative was set in motion by a report issued in January 2000. It estimated that of 100 new media firms looking for large blocs of office space, only one out of four would find it in Manhattan. Armed with this report, New York Senator Charles Schumer called the shortage of office space (Jakobson, 2000, p. 70),

> a troubling storm cloud on the horizon that could cause New York City to stagnate or even decline ... If we don't address [the office space shortage] and mitigate it, another city – possibly a city even outside the United States – will surely replace us as the world capital of the ideas economy

To this end Schumer put together the "Group of 35", a rather extraordinary development coalition co-chaired by Secretary of the Treasury Robert Rubin that included not only the usual growth machine members (utilities, universities, real estate interestes) but also New York's Empire State Development Corporation, representatives from cosmopolitan entities (Time Warner, Goldman Sachs, CitiCorp, Chase Manhattan, and the Federal Reserve Bank) and CEOs from high-profile Silicon Alley firms (Agency.com, DoubleClick, and StarMedia). In setting up the coalition Senator Schumer offered the following strategy (Gordon, 2000, p. 3):

> To create the large blocs of space that are needed, we need a public-private partnership where the city puts together a large parcel of land – perhaps by condemnaton – provides the transportation infrastructure with help from Washington, zones the space for appropriate uses. ... , and then reclaims all or part of the expense by selling parcels of land to private developers who will build there with or even without signed up tenants.

The Group of 35 is considering three neigborhoods as possible high tech zones: the "far-West Side of Manhattan, Long Island City in Queens, and Downtown Brooklyn" (Lentz, 2000). While the latter two sites are also sites for Digital City initiatives, a West Side project would tread on turf that had been successfully defended by neighborhood groups against the development schemes of none less than Donald Trump. Furthermore, the football Jets would like a new stadium in the area as is a group that is preparing a New York City bid for the 2012 Olympics. Thus, there is an interesting possibility that the perceived new media imperative may be used to help a diverse group of development interests justify public condemnation of a contested space. The state's power to condemn land might allow this West Side project to bypass court-room challenges (Jakobson, 2000).

Interestingly, a number of real estate interests have noted that the new media firms are not really desirable tenants. Because they have had a high rate of failure and tend to lack revenue they have no credit rating. The head of the firm that is New York's second largest landlord remarks, "If you are a credit tenant landlords will work to get you in. We had lots of opportunities to lease to dot-coms and decided not to". As a result, the new media firms that bidded on his space, "probably thought it was full" (Jakobson, 2000, p. 146). Instead of accepting equity many Manhattan landlords and service providers are now presenting harsher terms; many landlords are demanding higher rents and up to three years rent in advance. Some are setting up "temporary" office facilities for new media firms so as to be able to regularly re-evaluate the riskiness of firms. Other landlords are simply turning away new media business. Furthermore, the massive telecom investments that are being packaged with the new media are themselves relativly riskly given the glut of firms competing to capture New York (McDonald, 2000).

Whereas Manhattan real estate interests are becoming wary their counterparts in other boroughs are willing to bet that the magic of technology can transform the image of their industrial properties. Thus, the development coalitions may be diffusing the risk of the new media (and telecoms) to the other boroughs.

SEAMLESS CITY, HYBRID NETWORKS

Diverse kinds of interests and networks have contributed to the formation of Silicon Alley, it ongoing development, its speculative entanglements, and a path of urban development that has resulted in the "seamless city", that is, the physical extension of the business district. The various theoretical approaches discussed in this paper seek to explain how networks link the development of cities with that of markets. Networks are explicit in innovation district models (Saxenian, 1994; Storper, 1997) and implicit in discussions of flows, circuits and coalitions by other urban theorists. A major debate concerns the position of cities vis a vis networks – whether cities are nodes in translocal systems (Harvey, 1989; Sassen, 1991; Castells, 1996) or arenas for forming networks (Logan & Molotch, 1987; Zukin, 1995; Saxenian, 1994; Storper, 1997). My study lends support for the latter position. The social terrain of cities makes them critical arenas for the social construction of various markets (not just those for real estate); this role, in turn, profoundly shapes the nature of their ongoing development.

My study revealed gaps in how urbanists explain the links between innovation and speculation. I argue that theories of urban development need to provide a broader account of the social and cultural processes involved in the dialectical development of cities and markets. Compared to systems theory the framework I derive from economic sociology better explains how a social terrain that supports innovation can also abet less virtuous maneuvers by speculators. Compared to growth coalition and innovative district theories my study better captures the diverse and oftentimes "hybrid" nature of the networks that are shaping new media districts and contemporary urban development; new media industry builders, while sometimes collaborators with real estate interests, have distinctive interests, agendas, and organizational forms. We also see that work on innovation districts should recognize the less virtuous networks they can harbor. Finally, urbanists need to explore the nature and significance of hybrid networks, and new forms of power associated with them – for example, the role venture capitalists play as structural entrepreneurs at the intersections of a district's various markets.

REFERENCES

Amin, A. (2000). The Economic Base of Contemporary Cities. In: G. Bridge & S. Watson (Eds), *A Companion to the City* (pp. 115–129). Malden, MA and Oxford, U.K.: Blackwell.

Anderson, D. (2000). The Race for Office Space. *TheStandard.Com* (Jan 11), 1–3.

Aron, L. J. (2000a). Web, Retail Surge Earns Full House for Midtown South. *Crain's New York Business* (January 17), 26, 48.

Aron, L. J. (2000b). Downtown Re-emerges as Office Mecca. *Crain's New York Business* (January 17), 28, 50.

At New York (2000a). Friday Silicon Alley Stocks. *At New York* (October 27). [http://www.atnewyork.com/stocks/article/0,1471,8491_497541,00html].

At New York (2000b). The Week: Silicon Alley Share Mixed as NASDAQ gets Hammered. *At New York* (January 7). [http://www.atnewyork.com/stocks/print/0,1471,8491_277641,00html].

Baer, M. (2000). The New Economy's Currency is Stock, Stock, Stock. *New York Times* (March 29), 12.

Bernard, S. (2000). Gotham Expansion Teams. *New York* (June 12), 13–14.

Birger, J., & Messina, J. (1998). Up in Smoke: Silicon Alley Companies have Burned Through $1 Billion Without Producing any Profits or Viable Businesses". *Crain's New York Business* (November 16), 1, 51–57.

Block, V. (1997). Rising Firm Tally Expands Geography of Silicon Alley. *Crain's New York Business* (November 24), 36, 37.

Bunn, A. (1997). Upstart Startups. *Village Voice* (November 11).

Burt, R. (1993). The Social Structure of Competition. In: R. Swedberg (Ed.), *Explorations in Economic Sociology* (pp. 65–103). New York: Russell Sage.

Burt, R. (1992). *Structural Holes*. Cambridge, MA: Harvard University Press.

Calacanis, J. M. (2000). The Internet 2.0 Manifesto. *Silicon Alley Dailey* (November 13), 1–5.

Calacanis, J. M. (1999). Star Gazing: SAR Talks Business with StarMedia's Million-Dollar Man. *Silicon Alley Reporter*, 3, 8, 126–136, 148–151.

Caruso, D. (1997). Digital Commerce: The Interactive Media Industry begins to Deconstruct its Self-made Myths. *New York Times* (April 7).

Castells, M. (1996). *The Rise of the Network Society*. Oxford: Blackwell.

Chen D. (1997a). Kravis-led Fund is Backing Internet Lab. *New York Times* (October 21).

Chen, D. (1997b). Now in New York: Angels that Rescue Companies. *New York Times*. (July 2).

Chen, D. (1997c). Venture Capital Discovers New York's New Technology. *New York Times* (February 5).

Chervokas, J. (1997b). New Media? How about No Media! *@NY* (September 1). [http://www.news-ny.com/view302.htm].

Chervokas, J. (1997b). How New York has Changed. *@NY* (September 5). [http://www.news-ny.com/view301.htm].

Chervokas, J. (1998). Looking Beyond the Madness: What IPO Fever Really Means. *@NY* (November 13), 1.

Chervokas, J. (1999). The Good Guys Won the Battle but Stand to Lose the War. *@NY* (February 19), 1–2.

Chervokas, J. (1999). Its Time to Turn New York into a High Tech Hothouse. *@NY* (March 26), 1–3.

Chiu, J. (2000). Space Grabbers. *Crain's New York Business* (August 7), 1, 54.

Croghan, L. (2000). Silicon Alley Style. *Crain's New York Business* (February 3), 3, 59.
DiMaggio, P. (1994). Culture and Economy. In: N. Smelser & R. Swedberg (Eds), *The Handbook of Economic Sociology* (pp. 27–57). Princeton, NJ: Princeton University Press.
Dunlap, D. (2000). For Once-Gritty TriBeCa, a Golden Glow. *New York Times* (July 30), Section 11, 1, 8.
Eaton, L., & Blair, J. (2000). Dot-Com Fever Followed by Dot-Com Chill. *New York Times* (October 27), Section B, 1, 10.
Emirbayer, M., & Goodwin, J. (1994). Network Analysis, Culture, and the Problems of Agency. *American Journal of Sociology*, 99, 1411–1455.
Evans, J. (1999). Alley to Valley. *TheStandard.Com*. March 8, 1–2.
Flam, M. (2000). A Booming Economy Remakes Manhattan: Renovations, an Internet Invasion, Glitzy Headquarter Boost Rents. *Crain's New York Business* (January 17), 24, 43, 44.
Florida, R., & Kenney, M. (1990). *The Breakthrough Illusion*. New York: Basicbooks.
Florida, R., & Kenney, M. (1988). Venture Capital, High Technology and Regional Development. *Regional Studies*, 22, 33–48.
Gabriel, T. (1995). Where Silicon Alley Artists go to Download. *New York Times* (October 8), 49, 52.
Gabriel, T. (1996). The Meteoric Rise of Web Site Designers. *New York Times* (February 12), Section D, 1, 7.
Gimein, M. (1998). Around the Globe, New Stock Market Mania. *The Industry Standard* (December 28), 40–42, 44.
Goff, L. (2000). Tenants' Tastes Remap Manhattan: New Borders Reflect Changing Hot Spots. *Crain's New York Business* (October 16), 63–64.
Goff, L. (1995). From Wall St. to Publishing, High Tech Powers Job Gains. *Crain's New York Business* (November 27), 22, 29.
Goldberg, M. (1999). Razorfish's Secret Recipe. *Silicon Alley Reporter* (Summer), 32–34, 84.
Gordon, C. (2000). You Take Manhattan. *At New York* (February 21), 1–3.
Gould, G. (1998). Alley Girl: iVillage's Candice Carpenter on the Power of the Woman's Brand and How it May Change Silicon Alley. *Silicon Alley Reporter* (September), 20–22, 24, 52, 56.
Grabher, G., & Stark, D. (1996). Organizing Diversity: Evolutionary Theory, Network Analysis and Postsocialism. *Regional Studies*, 31, 533–544.
Granovetter, M. (1985). Economic Action and Social Structure: The Problem of Embeddedness. *American Journal of Sociology*, 91, 481–510.
Granovetter, M. (2000). A Theoretical Agenda for Economic Sociology, revised version of a paper presented at the Second Annual Conference on Economic Sociology, University of Pennsylvania, March 4, Philadelphia.
Gross, D. (2000). Desparation of the Dot-coms. *New York Times* (April 7), 23.
Hall, P. (1998). *Cities in Civilization*. New York: Pantheon.
Hansell, S. (2000). Gold Rush in Silicon Alley. *New York Times* (February 7), Section C, 1.
Harmon, A. (1999). Stocks Drive a Rush to Riches in Manhattan's Silicon Alley. *New York Times* (May 31), Section A, 1, Section B, 6.
Harrison, B. (1994). *Lean and Mean*. New York: BasicBooks.
Harvey, D. (1989). *The Condition of Postmodernity*. Malden, MA and Oxford, U.K.: Blackwell.
Holusha, J. (2000a). An Upcycle Just Keeps Rolling. *New York Times* (September 24), Section 11, 1, 6.
Holusha, J. (2000b). A Developer Puts a Bet on the Meatpacking District. *New York Times* (July 16), 9.
Indergaard, M. (1999). Retrainers as Labor Market Brokers. *Social Problems*, 46, 67–87.

Indergaard, M., & McInerney, P. (1998). Making Silicon Alley, Remaking Cyberspace, a paper presented at the annual meetings of the American Sociological Association, San Francisco.
Jakobson, L. (2000). Back to the Garage: Sen Chuck Schumer and his Group of 35 Tackle the City's Space Crunch. *Silicon Alley Reporter, 4*, 7, 70, 148.
Johnson, S. (1997). Web Editor Argues: Believe the Hype. *@NY*. [http://www.news-ny.com/view221.htm].
Kanter, L., & Messina, J. (1999). Unexpected Riches Remake the City. *Crain's New York Business* (November 29), 28, 30, 32.
Katznelson, I. (1992). *Marxism and the City*. Oxford, U.K.: Oxford University Press.
Krugman, P. (2000). The Ponzi Paradigm. *New York Times* (March 12), 15.
Lentz, P. (2000). Pursuing West Side Glory. *Crain's New York Business* (September 18), 1, 52.
Lentz, P. (1999). Tech Tonics: NYC Fostering New Districts for Start-ups. *Crains New York Business* (October 11), 4.
Levy, S. (1999). Comments at The State of the New York Media, a panel sponsored by the New York New Media Association, February 4, Fashion Institute of Technology, New York City.
Li, K. (1999). Urge to Merge Pervades Silicon Alley. *TheStandard.com.* (August 16), 1.
Lipschultz, D. (2000). Simple Case of Supply and Demand. *Red Herring* (May), 426, 428, 430, 432, 434.
Logan, J., & Molotch, H. (1987). *Urban Fortunes*. Berkeley and Los Angeles: University of California Press.
Macht, J. (1997). It Takes a Cybervillage. *INC Technology* (November 18).
Malanga, S. (1999). New Economy Propels City to Record Year. *Crain's New York Business* (December 13), 1, 52.
McDonald, M. (2000). Developers Ring Down the Curtain on Telecom Projects. *Crain's New York Business* (November 13), p. 14.
Messina, J. (1996). Software Pitchmen Strong Believers in New York's Burgeoning Sector. *Crain's New York Business* (February 5).
Messina, J. (1998). On-line Networks Confront Survival of the Clickest. *Crain's New York Business* (January 20), 17–18.
Napoli, L., & Hansell, S. (1999). For Silicon Alley, a Time to Rise and Shine. *New York Times* (March 29), Secion C, 1, 4.
Pontell, H., Calavita, K., & Tillman, R. (1997). *Big Money Crime*. Berkeley and Los Angeles: University of California Press.
Pratt, A. (2000). New Media, the New Economy and New Spaces. *Geoforum, 31*, 425–436.
PricewaterhouseCoopers (2000). *3rd New York New Media Industry Survey: Opportunities and Challenges of New York's Emerging Cyber-Industry*. New York.
Pulley, B. (1995). New York Striving to become Technology's Creative Center. *New York Times* (February 13), Section A, 1; Secion B, 2.
Rinzel, M. (1999). Agency.com CEO Chan Suh Says Content Sites Will Reign in '99. *Silicon Alley Dailey* (February 25), 1–2.
Rothstein, E. (1997). Connections: Downloading Virtual Underwear and Rethinking the Grand Design of the Web. *New York Times* (March 17).
Sassen, S. (1991). *The Global City*. Princeton, NJ: Princeton University Press.
Saxenian, A. (1994). *Regional Advantage*. Cambridge, MA: Harvard University Press.
Shiller, R. (2000). *Irrational Exhuberance*. Princeton, NJ: Princeton University Press.
Scott, A. (1998). From Silicon Valley to Hollywood: Growth and Development of the Multimedia Industry in California. In: H. J. Braczyk, P. Cooke & M. Heidenreich (Eds), *Regional Innovation Systems* (pp. 136–162). London: UCL Press.

Storper, M. (1997). *The Regional World*. New York: Guilford Press.
Suh, C. (1999). Comments at "The State of the New York Media", a panel Sponsored by the New York New Media Association, February 4, Fashion Institute of Technology, New York City.
Tedeschi, B. (2000). Critics Lauch Legal Assault on Tracking of Web Users. *New York Times* (February 7), Section C, 1, 10.
Temes, J. (1995). Silicon Alley and Other Hot Spots: Companies Cluster in New City Districts, Finding Staff, Ideas – and Business. *Crain's New York Business* (June 12), 17, 31.
Tierney, J. (2000). In E-World, Capital is Where it's @. *New York Times* (May 3), Section B, 1.
Tillman, R. (1998). *Broken Promises*. Boston: Northeastern University Press.
Tillman, R., & Indergaard, M. (1999). Field of Schemes: Health Insurance Fraud in the Small Business Sector. *Social Problems, 46*, 572–590.
Tilly, C., & Tilly, C. (1998). *Work Under Capitalism*. Boulder, CO: Westview.
Useem, J. (2000). New Ethics or No Ethics? *Fortune* (March 20), 83–86.
Vargas, A. (2000). Falling Dot.coms Confront Delisting. *Crain's New York Business* (July 10), 3, 31.
Walsh, M. (2000a). Old, New Media Rush to E-Merge. *Crain's New York Business* (November 6), 3, 57.
Walsh, M. (2000b). World Closes in on Net Firm. *Crain's New York Business* (October 23), 1, 57.
Walsh, M. (2000c). Venture Capital Investors Crowd Internet Gateway. *Crain's New York Business* (January 17), 63, 66.
Ward, U. (2000). Special report: Commercial Real Estate. *Crain's New York Business* (January 17), 42, 48, 50.
Warner, B. (1999). Silicon Alley's Cool Fish. *TheStandard.Com*. (February 15), 1–6.
Watson, T. (2000). Billions Flowed into Silicon Alley, as Internet Industry Grew Up in 1999. *At New York* (January 6), 1–4. [http://www.atnewyork.com/views/print/0,1471,8481_274901,00.html]
Wilson, F. (1999). Comments at "The State of the New York Media", a panel Sponsored by the New York New Media Association, February 4, Fashion Institute of Technology, New York City.
Zook, M. (1999). Regional Systems of Financing: The Impact of Venture Capital on the Emerging Intenet Content and Commerce Industry in the United States. A paper presented at a conference, Global Networks, Innovation and Development Strategy: The Informational Region as a Development Strategy, University of California at Santa Cruz, November 11–13.
Zukin, S. (1982). *Loft Living*. New Brunswick, NJ: Rutgers University Pess.
Zukin, S. (1991). *Landscapes of Power*. Berkeley: University of California Press.
Zukin, S. (1995). *The Cultures of Cities*. Oxford: Blackwell.
Zukin, S. (1997). Cultural Strategies of Economic Development and the Hegemony of Vision. In: A. Merrifield & E. Swyngedouw (Eds), *The Urbanization of Injustice* (pp. 223–243). New York: New York Unversity Press.

REGIME STRUCTURE AND THE POLITICS OF ISSUE DEFINITION: URBAN REDEVELOPMENT IN PITTSBURGH, PAST AND PRESENT

Gregory J. Crowley

INTRODUCTION

This paper compares political processes of redevelopment in Pittsburgh during two phases of urban Renaissance with the purpose of demonstrating the connections between the structural principles of urban regimes and the collective processes of setting development agendas, defining alternative solutions, and making public decisions. The paper is intended to contribute to recent scholarly attempts at reconstructing the growth machine thesis (Cox, 1999; Jonas & Wilson, 1999; Short, 1999; Wilson, 1996) and urban regime politics (Feldman, 1996; Goodwin, & Painter, 1996; Jessop, 1996; Lauria, 1996, 1999), the two prevailing perspectives on local development politics in the past two decades. I share with these appreciative critics the view that most research informed by the growth machine and urban regime approaches has paid insufficient attention to the discursive strategies that stakeholders in local development use to allocate the attention of the public and to generate consensus for their development agendas.

The costs to urban scholarship in failing to analyze processes of problem definition and consensus formation have been considerable and the new generation of political economists has only just begun to address some of the problems. Feldman (1996) has criticized the rational choice conception of "interests" in regime theory, arguing that interests of groups are not given for any situation or social role but are "socially defined, contradictory, strategic, spatial, and not objectively knowable" (p. 47). Jessop (1996), Wilson (1996), Cox (1999), and Short (1999) have begun to address the failure of earlier studies of local growth coalitions to investigate how land-based elites impute interests to groups in the urban community as a means of legitimizing their development agendas to the public. For example, Cox demonstrates how growth elites utilize "territorial ideologies", which "postulate unities of interest at the local level", and "ideologies of local community" that mobilize the "we feeling", in order to draw the public's attention to the often uneven racial, class, and other social consequences of growth politics while enhancing forms of identity between groups defined on the basis of locality (Cox, 1999, p. 21–22). Similarly, Short uses the term urban representation to refer to the carefully chosen ways in which urban boosters write and speak about their cities in order to promote them as sites for investment. "Regimes of representation" are "whole sets of ideas, words, concepts and practices ... in which particular forms of urban representation take on more specific meaning" (Short 1999, p. 38).

In this paper I address two other problems in the growth machine and regime approaches that have not been addressed systematically in recent critical literature. First, recent critics have made little effort to analyze how specific urban development controversies are constituted and their outcomes influenced by the discursive features of political interaction such as the creative use of symbols and rhetoric to define problems or to redefine problems and thereby expand conflict. The recent literature has also ignored other important discursive aspects of political interaction such as the efforts by combatants to control the production and flow of information or the staging and timing of events. Yet if urban political economists are to broaden their understanding of how specific "policy images" or problem definitions come to dominate given policy debates or sets of debates, it is necessary to show how these discursive objects are constituted (or transformed) in social interaction.[1] Second, at the heart of regime theory is the problem of how to represent the structural features of local governance (Elkin, 1987). Though ironically, both regime theorists and their recent critics have given short shrift to the question of how structural features of regimes shape the collective definition of local policy problems and solutions.

In what follows, I utilize empirical data from a longitudinal case study of economic development politics in Pittsburgh to demonstrate the correspondence between key changes in the structural features of Pittsburgh's regime over the past fifty years and shifts in the forms of interaction available to organized stakeholders who strive to influence the formation of local development policy. Thus I move beyond approaches that view strategies of problem definition and consensus formation in isolation from either broader institutional structures or contingent processes of political interaction.

After a brief discussion of how this paper defines the logical connection between regime structure and key aspects of the process of collective definition, I compare variations in how such processes are constituted under different structural conditions, using longitudinal data from different phases of Renaissance[2] in Pittsburgh to construct a case study. The paper closes on a normative note by arguing that as the composition of Pittsburgh's regime has evolved to become more pluralistic, less centralized, and less unified over the past fifty years, organizations outside the traditional growth elite, have become increasingly successful at using strategies of issue redefinition to generate controversies and expand them into public arenas (Baumgartner, 1987). The gradual restructuring of the processes of collective definition has forced elites – who still largely set the local development agenda – to yield some control over the consideration of policy alternatives to nonelite groups. Thus while Renaissance III (1990–present) is a significantly more pluralistic process than was Renaissance I (1945–1970), nonelites enjoy only limited influence over policy design due to the mechanism of their access, conflict expansion.

FOUR ELEMENTS OF STRATEGIC INTERACTION

In this section I outline four categories useful for analyzing how political actors attempt to define an agenda, link the agenda to a specific policy alternative, and construct a public consensus around their favored solution.[3] The categories include: (1) the use of rhetorical techniques to frame a condition as a problem, thereby allocating public attention (Gusfield, 1981; Rochefort & Cobb, 1993, 4; D. Stone, 1997); (2) the expansion of conflicts through the redefinition of issues to broaden participation (Schattsneider 1960; Baumgartner 1987); (3) the management of the production and flow of information in and out of different social contexts (Goffman. 1959; Hall, 1972); and (4) the structuring of social and material contexts in which problems and solutions are linked and then presented as alternatives to authoritative decision makers (Callon, 1985; Latour, 1991; Hajer, 1995–1996). Next I discuss these elements in detail.

Political actors use sophisticated *rhetoric* to construct relatively clear and simple images of problems from disparate and complex conditions. Social constructionists since Kitsuse and Spector (1973) take as their unit for analyzing social problems the "claims making" or "definitional" activities that constitute public discourse about social problems. The constructionist emphasis on definitional activities reflects epistemological debates in philosophy and the social sciences about the role of language in social life. With their roots in symbolic interactionism, ethnomethodology, and social phenomenology, social constructionists believe that language functions not as a mirror of the external world but as a set of tools for constructing models and causal stories of it. Understanding the constructionist position is relevant for how we view the policy process because it connects questions of how political actors and policy planners interpret social conditions with questions concerning how those actors exercise power. Fischer and Forester (1993) state the problem eloquently: "If analysts' ways of representing reality are necessarily selective, they seem as necessarily bound up with relations of power, agenda setting, inclusion and exclusion, selective attention, and neglect (p. 1).

The discursive aspects of political processes have important if often overlooked implications for how we understand public policy (March & Olsen, 1984). Rational decision-making models of politics predict behavior of legislators, voters, or political leaders in terms of rational calculation of means for maximizing the fulfillment of preferences assumed to be given and stable. This model, however, ignores how the definition of people's preferences depends upon how choices are presented to them, when, and by whom (Quattrone & Tversky, 1988; D. Stone, 1997). Political actors must constantly monitor the evolution of their policy debates in order to determine strategically how, when, and by whom choices should be packaged and presented to relevant publics (Cobb & Elder, 1972).

Research on growth machines and urban regimes of the past two decades has established that business interests and land-based elites in cities dominate the agenda for local development. According to Elkin (1987) the division of labor between market and state is the most basic feature of the American political economy. States control legal and public powers while productive assets are concentrated in the private sector. Regime theory emphasizes the limitations upon local authority in the United States and the need for city officials to form governing coalitions with local business and other institutional actors who possess crucial resources necessary for effective governance, as electoral success of politicians (Elkin, 1987), revenue generating capacities of government (Peterson, 1981), and policy planners' capacities to achieve important results (Stone, 1989) all depend in important ways upon assets controlled by growth elites. Central to the regime perspective is the notion that the specific set of informal arrangements defining a local regime shapes the city's economic development policy in important ways (Stone et al., 1991).

But elite power is not total and challengers to the growth machine can often influence which alternative solutions to elite agendas will be considered. Efforts at countermobilization against elite agendas often take the form of *conflict expansion through redefinition of elite problem solutions* (Schattsneider, 1960; Baumgartner, 1987). When challengers redefine problems they can generate controversies that shift the arena of decision from private meetings and exclusive elite civic associations into the public realm of the media, legislative debates, and community-based organizations, where nonelite problem definitions often have an advantage in gaining public sympathy. Public controversies force elected officials to consider the legitimacy of their power and electoral implications of their choices. Expanders of controversy will attempt to demonstrate that an issue, such as government consideration of tax abatement programs to lure investment to the city, has broad – usually negative – implications for the community. This helps them build a coalition of opponents. Growth elites, on the other hand, will tend to define growth issues more narrowly, usually in terms of the competitive pressures on city officials to employ a variety of policy tools to help the region attract global investment.

Growth elites and opponents alike will strive to *manage the production and flow of information* required to project a favored definition of the situation to relevant publics (Cobb & Elder, 1972). Goffman (1959) used the term arts of impression management in referring to attributes of teamwork required for a group to manage the impressions it wishes to project to an audience. Hall (1972) applied similar ideas to the practices of politicians. The key difference between rhetoric and impression management is that the latter is concerned less with the ways in which language allocates attention and persuades than with the strategic flow of information, including the timing and staging of the release of information. Successful partnerships between business elites and government officials often hinge upon their ability to keep crucial information about development projects out of the hands of would-be opponents at key moments in a debate. Thus controversies in local development often revolve around the struggle to determine if an economic impact study should be conducted, who should study the costs of a proposal to the public, or when a private company's contribution to public-private project should be announced. The successes of conflict expanders often depend upon their capacity to produce economic impact reports, for instance, that show discrepancies in the studies produced by the growth elite.

Finally, many political struggles are determined by the success of combatants in making favored solutions appear inevitable by linking them with vast economic and technical resources prior to formal consideration by policy makers. Beck uses the term *subpolitics* to refer to how "important choices are made before or after a development becomes an 'issue' for policymakers" (quoted in Hajer, 1995–1996, p. 27.) One advantage growth-oriented mayors typically have over their opponents is the capacity to form partnerships with business groups for working out technical, economic, or political solutions to their development proposals before other stakeholders or the general public are even made aware of the issue. Civic elites lobby state legislators to put tax referenda on ballots for raising local development revenues before county commissioners have a chance to give their own input. Mayors work out large-scale renewal agendas with private developers behind closed doors and only after financing and planning arrangements are complete is city council solicited for its approval. In examples like these mayors can, with apparent legitimacy, brand opponents as naysayers who "have no plan of their own". The problem is that such appearances of feasibility and practicality, which lend a sense of inevitability to an agenda, come at the expense of substantive public debate (Bachrach & Baratz, 1970).

Each of the four dimensions of political interaction discussed above – rhetoric and problem definition; problem redefinition and conflict expansion; management of the production and flow of information; and subpolitical processes for linking favored problems with favored solutions – is a key analytical component of the local policymaking process. It remains to specify how political interactions are conditioned by the structural principles of regimes. In the next two sections I discuss how variations in three dimensions of regime structure correspond to variations in the composition of political interactions across two major phases of redevelopment in Pittsburgh. The three dimensions are: (1) the extent of unity among participants in regime politics and the forms of coordination between public and private actors; (2) the distribution of capacities among organizations in the community to produce expert knowledge on topics related to local policy, and; (3) the extent of organization and incorporation into the regime of groups outside the growth elite.

PITTSBURGH'S CORPORATIST REGIME AND THE SOCIAL CONSTRUCTION OF RENAISSANCE I

Party Unity and Coordination of Public Institutions

Trust is generated in social relationships through repeated interactions that permit actors to see the deficiencies in mutual noncooperation (Hardin, 1982). In long-term relations members recognize that it is in their interest to subordinate short-term gains to the goal of achieving trust and predictability in the future. The condition of trust aptly describes relations among key actors in the public sector as well as those between the public and private sectors in the early postwar period in Pittsburgh. This unity was an important feature of Pittsburgh regime politics during the first Renaissance and it limited local political actors' interest in expanding policy conflicts through strategies of redefinition.

More than any other factor, it was the highly unified Democratic Party in Allegheny County that facilitated coordination of public actors for development and renewal during Renaissance I. The party owed much of its strength and identity to the leadership of David Lawrence, chairman of the party since 1920 (Stave, 1970). Lawrence built the Democratic machine, was instrumental in its ascent to power during the New Deal, and while Mayor of Pittsburgh from 1945–1956 utilized the party machine he had built to conduct the largest urban reconstruction project in America up to that time.

There were two key sources of party unity. First, Lawrence was both mayor of Pittsburgh and chairman of the county Democratic Party at the high point of Democratic control of politics in the Pittsburgh region. As chairman of the party's executive committee, Lawrence was responsible for selecting candidates among rank and file party regulars for positions on City Council and as board members for public agencies. The party gave its endorsement and strong support to the campaigns of those it wanted to see enter public office (Stave, 1970). Given that the Democrats enjoyed high rates of legitimacy after seizing power in the Roosevelt era, and had the loyalty of over 90% of the electorate (Jezierski, 1988), politicians found it difficult and costly to build a base of political support outside the party machine. Members of City Council and other elected officials thus had strong incentives to "go along to get along" with Lawrence's plans for radical smoke and flood control legislation and large-scale physical renewal that characterized the vaunted postwar Renaissance in the region. As Lawrence's biographer put it, "those who consistently opposed [Lawrence's] program were almost without exception dropped in a subsequent election" (Weber, 1988, p. 232).

A second source of unity was to be found in the structures and capacities of government itself. In 1911 a business-backed reform effort led to the establishment of a strong-mayor form of government and a nine member council elected at large, which freed Lawrence's hand-picked legislators from neighborhood and sectoral pressures that might have caused them to voice strong independent views on renewal projects affecting residents and businesses in their districts. Furthermore, a partially autonomous Urban Redevelopment Authority, of which Lawrence served as the first chairman in 1948, possessed the legal authority of eminent domain and garnered the development expertise necessary to efficiently conduct large-scale land-clearance and assembly for redevelopment. While the URA has functioned effectively as an extension of the mayor's office since its inception, its independence from electoral politics has served to shield the mayor's office from conflict associated with the extensive clearance and removal projects of Renaissance I (Jezierski, 1988).

Sources of Unity in the Private Sector

Beginning in the late 19th century industrial expansion in the Pittsburgh region was increasingly financed by a small number of downtown banks. With their growing concentration of investment capital and corporate ownership, Pittsburgh banks exercised immense control over economic and industrial growth throughout the region. Mellon Bank in particular became the largest source of capital in western Pennsylvania, with controlling interests in Gulf Oil, Koppers Company, ALCOA, Pittsburgh Consolidated Coal, and Pittsburgh Plate Glass, to name a few of the bank's largest interests. In fact, most companies in the region had significant ties to Mellon capital. Muller (1989, p. 194) estimates that in 1941 Pittsburgh banks held more than 50% of the region's banking capital.

Moreover, Mellon Bank's wealth was overwhelmingly concentrated in western Pennsylvania. With over $3 billion in assets throughout the region (Weber, 1988, p. 235), Mellon Bank was more dependent than any other business in the area upon the continuation of a strong and sustainable regional economy. Recognizing the need to sustain the profitability of his investments, Mellon Bank chairman Richard King Mellon established an extensive network of interlocking directors that gave his bank enormous direct influence over executives of firms throughout the region. The bank's 28 directors held 239 seats on the boards of 185 different companies in western Pennsylvania.[4]

The centrality of Mellon Bank within the financial community, and the symbolism of R. K. Mellon as the civic statesman of the regional community, constituted a very powerful unifying force within the private sector. Executives, as well as public officials and industrial workers, strongly associated the collective interests of the region with the private interests of the bank. Equally important, through the bank's substantial debt and equity holdings in the region, and the extensive web of interlocking directorates, Mellon wielded effective mechanisms for mobilizing the business community around interests of the bank.

Institutions of Public-Private Cooperation and the Concentration of Expert Knowledge

The end of World War II brought a sharp decline in national demand for steel. Decentralization in the steel industry, rising costs of labor in the United States after the war, and the American steel industry's technological lag combined to paint a bleak future for the steel capital of the U.S. Moreover, like many industrial cities in decline at the time, Pittsburgh suffered from serious environmental problems, especially smoke pollution and river flooding. With the city's infrastructure in decay and its housing stock seriously inadequate, at the end of the war major corporate patrons began plans for relocation. It was under such conditions of imminent decline that civic elites and public officials identified the need to achieve cooperation across institutional and party lines if the city was to recover from its economic, social, and environmental ills.

With financial backing and support from R. K. Mellon, Pittsburgh's business elite founded the Allegheny Conference on Community Development in 1943 (Lubove, 1969). The Conference became the key business-backed civic organization in the region and, along with the Pittsburgh Regional Planning Association (PRPA) and Western Division of the Pennsylvania Economy League (PEL), dominated the planning and policy agenda of postwar renewal. By convening the region's most important economic players, the Conference was able to mobilize funding and technical support for most of the studies, reports, and renewal plans that informed Renaissance I. Given Mellon's control of most corporations represented in the Conference, unity of action within the executive committee was nearly guaranteed on all matters important to the renewal effort.

The Conference operated through standing committees – e.g. housing and neighborhood development, land use and zoning, legislation – which included selected public officials as well as members of the corporate community. The committee system institutionalized a relationship of cooperation between the public and private sectors that has characterized the Pittsburgh policy community ever since (Ferman, 1996). Research conducted by the Economy League and various standing committees led to recommendations for the creation of other key institutional players in the renewal effort, including the nonprofit ACTION-Housing – established in 1957 to plan and develop low and moderate income residential units – and a host of new public authorities, modeled on the URA, to insulate from party and public scrutiny the building of parking facilities, public auditoriums, tunnels, and sanitation facilities.

Coleman (1983) has described the Pittsburgh public-private partnership as a form of corporatism: the local state, the URA and other authorities, and the Allegheny Conference and its affiliates, comprised an interpersonal network of elite decision-makers organized around Lawrence and Mellon. Other interests existed – neighborhoods, women, ethnic groups, and labor - but were not organized and incorporated into Pittsburgh's policy-making institutions.

With this background of long-term reciprocal exchanges, basic to machine-style politics, members focused on the incremental yet consistent gains to be achieved by going along with Lawrence's radical redevelopment agenda. Lawrence's hand-picked legislators and administrative officials had little incentive to calculate short-term costs and benefits of each mayoral initiative. Protracted controversy was therefore quite rare, a condition not conducive to the opposition politics of issue redefinition. The civic and quasi-state institutions extended these relations of trust across the boundary of public and private power, making it possible for leadership to act quickly and efficiently in setting the agenda for development, considering alternatives, and pressuring and persuading the appropriate officials to make policy choices. Perhaps it is only a small exaggeration to claim, as one neighborhood activist has, that "seven phone calls are all that's needed" to begin a redevelopment plan (quoted in Jezierski, 1988, p. 176).

Elite Unity, Control Over the Production and Flow of Expert Knowledge, and the Social Construction of Renaissance I

In this section I examine how the three key characteristics of Pittsburgh's postwar regime discussed above – a highly unified public-private elite; their domination of the production of expert knowledge required to define and control the policy agenda; and the relative lack of organizational and technical capacities of alternative political actors to the elite public-private partnership – shaped and constrained the four dimensions of political interaction earlier in this paper.

As we have seen, a key feature of corporatist governance is the emphasis on peaceful, incremental bargaining and the corresponding absence of incentives for policymakers to generate public controversies about government actions through strategies of issue redefinition. This collective interest in regime stability was nowhere more obvious than in City Council, where Lawrence established a policy of simply informing members each week of the activities of civic leaders (Weber, 1988, p. 232). Serious opposition to the mayor was viewed by the machine as a violation of norms of consensus and would usually result in the member losing his spot on the party ticket in the next election. Absent any institutionalized incentives for members to generate conflict by articulating alternatives to the mayor's problem solutions and soliciting them for public consumption through the media, the role of the council in the redevelopment agenda became "primarily a facilitating one" (Stewman & Tarr, 1981, p. 161).

In the private sector, interdependence and trust were equally strong and incentives for expanding controversy through issue redefinition minimal. Mellon's controlling interests in the region's major corporations and his ample resources for offering selective incentives to corporations threatened by proposed environmental regulations helped minimize dissent and stimulate collective action. A dramatic example of this occurred in 1946, when the Conference sponsored legislation aimed at eliminating the exemption of railroads from stringent smoke laws. Pennsylvania Railroad responded by gearing up for a controversy in the state legislature. The company sent lawyers to the state capitol to present to lawmakers a policy picture very different from that promoted by the Conference. The railroad supported the western Pennsylvania delegation's commitment to progress on smoke control, the lawyers argued, but added that if reforms were implemented too quickly consumer prices in western Pennsylvania would reach unacceptably high levels due to higher freight costs resulting from the railroad's forced conversion to diesel and electric locomotives. The lawyers were redefining the Conference's issue of protecting residents of Pittsburgh from smoke pollution to protecting working class consumers from price inflation.

R. K. Mellon, himself a director of the railroad, intervened in the fledgling controversy, demanding the company change its position and subtly threatening to organize a boycott of the railroad if action were not forthcoming (Weber 1988, p. 242). The railroad quickly shifted its position, agreeing that comprehensive smoke control legislation should be pushed no matter what the short-term costs to consumers and to the railroad.

But perhaps the growth elite's most effective instrument for dominating post-war renewal efforts was their monopoly over the production and flow of expert knowledge about development issues. The Conference's enormous command over resources allowed it to utilize the Pennsylvania Economy League, its research arm, to produce studies, reports, and recommendations for government action. There simply was no other organization that could compete with the information generating capacities of the Conference. For example, all research and planning for Gateway Center – a set of new office buildings downtown financed by a New York Insurance company on property acquired and cleared by the URA – was conducted by the technical staffs of the Conference and its affiliates. As Stewman and Tarr (1986) point out, this condition resulted as much from a lack of technical and organizational capacities in the city government and among nonelite groups as from the superior capacities of the Conference and its network of research, planning, and leadership organizations.

> The city itself did not have either the technical staff or the funds to hire consultants. City representatives participated in planning sessions, but the conference formulated the essential program and submitted its plans to the city for approval and implementation. This arrangement required trust between private and public parties and agreement on common goals. The ultimate motivation was economic: to utilize public powers in coordination with private planning and investment to revitalize the economics of the city and the region. A more democratic process might have produced a different set of plans and goals, but there were few voices in the post-war period offering other options (p. 156–157).

The civic elite's monopoly over the production of expert knowledge functioned to produce "nondecisions" (Bachrach & Baratz, 1970), by keeping alternative definitions of development issues effectively off the public agenda. Protest to large-scale clearance and displacement schemes was at times effective in slowing down policy implementation or forcing policy makers to consider electoral fallout of controversial plans (Lubove, 1969, p. 110). For example, protests by upper-middle-class residents of the city's east end successfully forced Mayor Lawrence to reconsider supporting the Conference's plan to locate a proposed Civic Arena in their neighborhood (Lubove, 1969, Ch. 6). But protests were largely ad hoc events staged in the eleventh hour of major public initiatives. At best, they could hope to stop major redevelopment plans from being implemented; protesters did not have the combination of financial, technical, and organizational capacities to sustain a campaign of rational arguments in favor of a particular definition of the issue.

When one organization controls the production of information relevant to a policy field it also maintains discretion in controlling the context and timing of the release of information throughout the policy community. The Conference chose when to consult with voluntary and public agencies affected by their proposals and when to inform the press (Martin, 1952). The civic elite thereby avoided protracted public debate over how social problems were to be conceptualized and which policy alternatives would be considered. Instead, neither policy makers nor the electorate heard about issues on the government agenda until they were already linked to specific solutions through sets of technologies and modes of financing previously set in place. Through these *subpolitical* processes the Conference's problem solutions were made to appear inevitable rather than as one among many possible approaches each with its own benefits and costs. As I shall show below, the latter is the norm in Pittsburgh today: solutions to social problems proposed by the civic elite are relativized by problem solutions generated from networks of alternative actors in the region.

PLURALIST CORPORATISM AND THE SOCIAL CONSTRUCTION OF RENAISSANCE III

The contemporary Pittsburgh regime is similar in many ways to the postwar partnership discussed above. Today, the institutions of the public-private partnership still dominate agenda setting and consensus formation. Yet the regime, though highly centralized by contemporary standards (Jezierski, 1988; Perry, 1990), has changed in significant ways as well (Ferman, 1996). The policy dialogue about redevelopment is complicated by a growing number of institutional and ad hoc actors struggling to define specific policy problems and competing to promote alternative policy solutions. Coalitions are more likely to form around specific agendas rather than long-term affiliations, and the scope of participation tends to vary dramatically across the life of any issue. Finally, the interests and preferences of participants appear to be less stable than was the case in the post-war period, making policy debates more contentious and confusing. Next I briefly discuss the conditions under which the regime has become more *pluralistic* while maintaining important elements of its corporatist past. Then I summarize the results of a case study that represents how the corporatist and pluralistic aspects of Pittsburgh's regime structure the political processes of economic development policy making.

I focus on three key areas of change in the regime: (1) decentralization in the production of expert knowledge; (2) partial erosion of unity within the political and business elite, and; (3) expansion of organizational actors outside the growth elite, including community-based organizations, preservation groups, urban planners, and conservative policy think tanks.

The Decentralization of Expert Knowledge

After Renaissance I, David Lawrence and subsequent mayors continued to expand public and civic institutions for mobilizing governmental and private resources. The growth of the civic sector has resulted in a gradual shift of key resources away from the political machine, which has weakened mayoral influence over economic development. Significant institutional changes began during the administration of Pete Flaherty (1969–1977). A populist mayor, Flaherty championed the interests of middle-class neighborhoods and distanced himself from both the business community and the political machine. His first anti-machine action involved sharp staff cuts in city government, including the replacement of major department heads and supervisors connected to ward bosses (Sbragia, 1989). This depleted government of many party loyals and shrank patronage resources necessary for Democrats to control the electoral arena.

Further, in 1971 Flaherty institutionalized formal linkages between the growing neighborhood movement and city government when he set up the Community Planning Program in the City Planning Department. The program opened up the planning process by giving community-based organizations official access to city government and intergovernmental resources. Before Flaherty, virtually all federal money went to development projects in the central business district. With the establishment of the Community Planning Program and the arrival of Community Development Block Grant (CDBG) funds, nearly half of all federal dollars was channeled to the neighborhoods (Stewman & Tarr, 1981). The Planning Department was also given authority over the evaluation and allocation of CDBG funds. These administrative reforms had the effect of separating neighborhoods from ward organizations and channeling intergovernmental grants to neighborhoods according to formal, administrative criteria, thus giving CBOs autonomy from party politics (Ferman, 1996, p. 100).

Further expansion of the corporatist structure resulted from reforms initiated by Mayor Caliguiri (1977–1988). The formal ties between CBOs and government were strengthened when Caliguiri created the Department of Housing, which established cooperation between CBOs, private lenders, and the URA on programs to build affordable housing (Ahlbrandt, 1986). Toward the end of Caliguiri's term, the Pittsburgh Partnership for Neighborhood Development (PPND) was incorporated. The non-profit organization was established to channel resources of the business community, city government, and foundations toward the CBOs. The Community Technical Assistance Center, set up in 1980, furthered this effort by providing a range of low-cost resources to CBOs. These reforms increased significantly the flow of money, technical resources, and information into the neighborhoods. Caliguiri began the practice of involving neighborhoods in the actual formation of the capital budget. To this day, the planning department provides a draft of the capital budget and six-year policy plan to the neighborhoods for comments and input. The document includes information on city investments in neighborhoods and statistics on neighborhood progress.

With 75% of federal money going to neighborhoods under Caliguiri, and a new institutional framework incorporating a large constituency of planners, CBOs, and neighborhood development personnel into the regime (Ferman, 1996, p. 100), a new political force independent of the mayor was established. This group, which remains strong today, has potentially very different views from the mayor or City Council on specific policy issues, especially on issues that pit downtown interests against those of the neighborhoods.

Growth in civic institutions since the Lawrence administration has created information generating and organizational capacity among nonelites tied more closely to neighborhoods and nonprofit organizations than to the downtown growth machine. Capacities of these groups to generate expert knowledge have been enhanced further by new communications and information technology. Most CBOs, for example, have World Wide Web sites that can be used to spread information quickly and cheaply and to borrow information from a vast network of other web-based economic development services, policy think tanks, and nonprofits. Further regime decentralization resulted from the incorporation in 1995 of the Allegheny Institute, a conservative public policy research group devoted solely to regional issues. Financed by Richard Scaife, conservative millionaire and champion of deregulation of markets and privatization of public services, the Allegheny Institute has posed a significant challenge to the monopoly on policy research and issue definition formerly held by the Conference's Pennsylvania Economy League.

Pittsburgh's Economic Transformation and the Declining Unity in Civic Leadership

The economy of Pittsburgh and the surrounding region had been deindustrializing slowly for three decades when, in the early 1980s, the steel industry collapsed during a period of intensified economic globalization and corporate restructuring (Clark, 1989). While Pittsburgh has been recovering gradually from the shocks of deindustrialization by utilizing its major research universities to guide growth into advanced technology and specialized manufacturing, the old mill towns still have not fully recovered from the decline of steel and the resulting loss of over 100,000 manufacturing jobs (Sbragia, 1990).

The transformation of Pittsburgh's economy has had a direct impact on the structure and dynamics of the governing coalition. The steep decline in population following job loss and outmigration from the region has undermined the city's influence in the state legislature. Increasingly, governance requires leaders to assemble broad coalitions among policymakers throughout the surrounding counties.[5] Decline in manufacturing employment also resulted in a corresponding drop in union membership, which challenged further the dominance of the Democratic Party in the region. The elections of 1995 marked a watershed in party politics in the region when Republicans captured the Allegheny County board of commissioners for the first time in sixty years. The era of one-party rule had come to an end.

In a period of increasing globalization of financial flows, production, and trade, capital mobility poses another serious challenge to the stability of the traditional corporatist structure. Mellon Bank now holds a diversified portfolio of equity no longer dependent upon the western Pennsylvania region. The welfare of the bank and other regional champions such as U.S. Steel (now USX), ALCOA, and Pittsburgh Plate Glass is no longer tied to western Pennsylvania. A fundamental political consequence of capital mobility is the gradual deterioration in unity and trust among members of the business community that in the past had persisted through long-term reciprocal exchanges and interdependency. Firms are simply less committed to growth and renewal in the region and, due to expansion of their responsibilities at the international level, are less able to devote time and resources to civic affairs than was the case during Renaissance I. As one local CEO characterized the contrast between past and present:

> When you look at my predecessors two or three back, they would have been able to devote their time to Pittsburgh on an 80:20 ratio. Today, I can hardly afford to give Pittsburgh 20:80. Although I have an interest in Pittsburgh, I have too many things elsewhere, so that I personally can't give it that kind of attention (quoted in Ahlbrandt & Coleman, 1987, p. 19).

Next I utilize materials from a detailed case study of a recent development controversy in Pittsburgh to illustrate how emergent pluralist dynamics in the corporatist regime have influenced processes of political interaction underlying Renaissance III. The case illustrates that, while many more institutional and ad hoc actors now struggle to define policy problems than was the case fifty years ago, it would be a mistake to view regional decision making in terms of classical pluralist theory. The classical pluralist image of politics as an equilibrium-seeking policy system of competing groups (Waste, 1986) does not apply to either Pittsburgh's past or present governing arrangements. It would be equally inadequate, however, to apply a classical elitist model to Pittsburgh development politics. To be sure, the URA, the mayor's office, and the Conference executive committee are still highly influential in setting the development agenda. But now community-based organizations, taxpayer coalitions, conservative think tanks, reformist city planners, historic preservation groups and other organizations have greater capacities to participate in policy design by generating public controversies about the elite development agenda, thereby forcing issues to be considered outside elite organizations – in public debates, in the media, on the internet, and in the CBOs. Thus, expansion of conflict through issue redefinition has become the principal mechanism for challengers of the elite growth agenda to achieve access to government.

Moreover, because coalitions tend to form around specific issues rather than long-term affiliations, actors take positions on the basis of their perceptions of how a policy action will benefit them directly and in the short-term. This emphasis upon the instrumental side of politics as opposed to the earlier emphasis upon politics as the construction and maintenance of stable social relations intensifies participants' interests in the contents of specific policy agendas. What is the nature of the problem? What is the most effective solution? The answers to such questions become highly contested as more participants enter the debate, challenge assumptions made by others, and propose alternative solutions. The outcome of policy initiatives hinges increasingly upon participants' capacities and skills in expanding conflict through issue redefinition rather than in principled dialogue and debate.

Issue Definition and the Expansion of Controversy in Renaissance III

In the past decade, major state-led redevelopment efforts in Pittsburgh have focused on fortifying the city's role as economic and cultural epicenter of the region. A cardinal objective has been to transform downtown into a 24-hour hub by enhancing retail, entertainment, and residential options and by constructing or expanding regional destination facilities to attract more tourists and visitors to the city. Vigorous controversy has surrounded the planning of new development – a baseball park and football stadium, expansion and improvement of the Cultural District, comprehensive redevelopment of the retail core of downtown, and a tripling in the size of the convention center – resulting in enhanced public participation in these projects.

In July of 1997 the administration of Mayor Tom Murphy contracted with Urban Retail Properties, a Chicago developer, to study redevelopment options for seven blocks in the city's retail core.[6] At the time of the study tenants in the retail district were mainly small stores, owned and run by local residents, and serving employees and the few residents of downtown. For decades, merchants in the district had been losing business to regional malls. Public officials, civic leaders, shoppers, and business owners agreed unanimously that something had to be done to reclaim years of business the city had lost to other parts of the region.

The Mayor's Opening Move: Rhetoric of Neglect and Necessity of Big Solutions

When in the summer of 1998 the mayor began discussing publicly[7] the intent of his relationship with Urban Retail, he located the central cause of retail decline in Pittsburgh's growing shopping and entertainment competition from suburban malls, *a causal story*[8] that key regional actors interested in the issue did not dispute. The result of regional competition, the mayor claimed, was a retail district largely neglected and left to decay by struggling tenants and landowners. Unlike his definition of the problem, however, Murphy's proposed solution eventually became highly contested. The mayor insisted there was only one effective policy response: In order to lure suburban shoppers back into Pittsburgh it was necessary to demolish most of the existing buildings in the retail corridor and replace them with an "urban mall" master planned and built by a major national developer like Urban Retail. In theory, urban malls offer a mixture of retail, service, and entertainment options that distinguishes them from their suburban counterparts. Urban malls include prestigious national retailers like FAO Schwartz, Tiffany & Co., Victoria's Secret, and other "one-of-a-kind" stores that prefer urban locations. They also include, or are located in proximity to, upscale dining and entertainment venues such as Hard Rock Café and Planet Hollywood, options not usually found in suburban locations.

Minimizing Conflict by Controlling the Flow of Information

In a series of press conferences that took place prior to the city's formal release of its intended development plan in October 1999, the mayor held that Urban

Retail would be interested in acting as project developer of a new retail corridor, but only if it could own most of the land slated for the site, lease the space, manage the buildings, and market the project to potential investors. Urban Retail wanted full control or would prefer not to get involved at all. The Chicago developer could not act alone, however. During the spring of 1998 the city had already begun raising money to purchase and demolish properties in the redevelopment area. The key actor was the URA, whose eminent domain authority could be called upon to assemble individual properties and prepare them for resale to the developer.

In July 1998, major newspapers in the region announced a deal had been struck between the city and Urban Retail. The city, however, refused to release key details of the plan to the public, explaining that it would be irresponsible to do so until particulars of the proposal were determined. Though the city did make it clear that the plan called for destroying many of the existing buildings in the retail corridor and relocating the majority of the hundred or so existing tenants to make way for major national chain stores. The plan was estimated to cost $170 million, 25% of which would involve public expense, including $10 million already pledged by the state. The city, the URA, and Urban Retail had thus managed to design key elements of a redevelopment agenda, link the plan with a major national developer with a strong reputation for success, and lay out a prospective role for government, without as yet seeking input from the public or from the owners or tenants of the 86 parcels of land in the redevelopment zone.[9]

Opposition and Counterstrategy 1: Problem Redefinition and Conflict Expansion

In the fall of 1998, key stakeholders in the region, acting on bits of unofficial information about mayor's plan, began to speculate about its likely consequences for the public. The *Pittsburgh Tribune Review*, a popular newspaper among conservative critics of activist government, attempted to incite conflict through a *counterrhetoric of injustice*[10] in several stories that dramatized how merchants in the redevelopment area faced an utterly uncertain future and lived under the constant threat of displacement by eminent domain.

The elements of controversy had begun to fall into place. Historic preservation groups joined the newspaper in criticizing the city's project, but they redefined the mayor's statement of the problem in different terms. On their websites, in press releases, and in public meetings, preservationists mobilized a *counterrhetoric of loss*[11] stating that if the mayor's plan were implemented, historically significant architecture in the area would be forever destroyed and replaced with "cookie-cutter" style national chain stores. The Allegheny Institute joined the fray, using its website, the Op'Ed pages of the *Tribune Review*, and a series of public meetings, to denounce the mayoral group for their threat of eminent domain authority for the purposes of taking properties from one set of private owners and handing them over to another.

Mayor's Response to the Opposition: Subpolitics and the Struggle to Control the Flow of Information

Yet the mayor managed to avert, or at least delay, a full-blown conflict by denying his opponents access to critical information he controlled and they needed to make meaningful arguments against the city's plans. For nearly a year following the July 1998 announcement of a partnership between the mayor and Urban Retail, conservative critics, preservationists, and tenants and landowners in the retail district, received no answers from the administration to questions such as: Which buildings in the redevelopment area does the mayor's plan demolish? Which existing businesses would the mayor like to remove or relocate? What national businesses are proposed for the district? Is the mayor's plan a sound economic investment for the city? Are there alternative ways for revitalizing the retail corridor? Will downtown businesses, preservationists, critics of big government, and other interested citizens ever be allowed to propose alternatives to the mayor's project? Without answers to questions like these, the mayor's would-be critics could not critically evaluate his plan. They couldn't conduct economic impact reports or benefit-cost analyses of public investment, nor could they consider alternatives in a meaningful way.

The mayor, Urban Retail, and the URA continued successfully to legitimize their unwillingness to release information on the grounds that many details had yet to be worked out and Urban Retail had yet to sign contracts with vendors. In press conferences the group promised more information about their plan when it became available. Whenever he was asked why other interested groups had not been allowed to contribute to the plan, Murphy persistently responded that public input would be forthcoming, "but first we've got to have a plan to debate." Yet while would-be opponents waited for answers from the city, mayoral *subpolitics* continued. Urban Retail had completed enough details of their proposal to begin marketing the Pittsburgh location to prospective tenants. The Chicago retailer was showing the plan, entitled "Market Place at Fifth and Forbes", to national retailers at shopping center conventions around the country, while tenants in the retail district continued to wait for the city to contact them about the proposed future of the properties on which they depended. Thus the mayoral group continued to assemble crucial components of their plan, making it harder to criticize and harder to reverse. But still no public discussion had occurred.

Opposition and Counterstrategy 2: Finding the Right Moment to Expand Conflict

The mayor's strategy seemed to be working. His group was dominating the public agenda, keeping would be critics in abeyance while consolidating his redevelopment proposal. At least this was the case until further details of Urban Retail's plan finally leaked out in marketing materials circulated by the developer at a Las Vegas shopping center convention in May 1999.[12] Brochures aimed at marketing the development to prospective tenants showed the footprint of the plan, indicating exactly where the project would be situated downtown, and defining the location of four anchor stores. When news media in Pittsburgh began covering Urban Retail's "behind-the-scenes" marketing activities, the mayor's adversaries seized the opportunity to try once again to trigger a public controversy through strategies of issue redefinition. Employing a *counterrhetoric of insincerity*, the director of Market Square Association, a group of downtown business and property owners, began to speak our boldly against the "secretive" planning activities of the mayoral group. Soon all the major television and print news media began to document opposition to the mayor's "ludicrous" approach of "lining up new tenants [downtown] without letting current one's know what's happening."[13] Many merchants decided not to depend any longer on the mayor's discretion in releasing information about his agenda. Key members of the Market Square Association began issuing statements redefining the problem of downtown decline and displacing what they called the mayor's "top down" solution. Utilizing planning theory and technical knowledge borrowed from preservationists, merchants told a *causal story* of decline entirely at odds with the mayor's. Landowners and businesses in the retail district had for years failed to work collectively to pursue an incremental approach toward property improvement, they argued. This failure of collective action, they asserted, explained why no long-term plan had been created earlier to upgrade the quality of stores and businesses downtown. Moreover, the argument went, because the city and the URA were perpetually preoccupied with large-scale redevelopment projects, they had done little over the years to promote incremental regeneration of properties in the district.

Members of the Market Square Association and leading preservationists, began using metaphors of "rebirth" instead of "renewal" to define the objectives of development, and opposed their ideas of "grassroots participation" to "top down planning" they accused the mayor of following. City Council was the key decision-making body whose definitions of the situation opponents wished to shape. The mayor's plan required approval from the URA, the City Planning Commission, the Historic Review Commission, and City Council. The first three, staffed with mayoral appointees, had predictably approved the plan shortly after the deal was announced between Urban Retail and the city in October 1999. Members of City Council, however, are more susceptible to defining the problem in terms other than those set forth by the mayor. There are two structural conditions for this. First, reformers had managed to pass a referendum in 1987 changing council from a body elected at large to one based upon districts. The new structure motivates district representatives in council to evaluate local development policy in terms of its effects on their constituents rather than upon the city as a whole. As district constituents become aware of controversies affecting their neighborhoods, many are inclined to pressure representatives in council to consider arguments being made by conflict expanders who are cultivating public sympathy for their position. Second, because the Democratic Party is not nearly so unified as it had been in Lawrence's day, neither the mayor nor any other party leader can easily control the voting behavior of council members.

Thus the requirements for political legitimacy forced members of City Council to listen to the views and concerns of the mayor's opponents, who had begun to urge their representatives in council to hold special hearings in order to stimulate public discussion on the project. Preservationists began holding meetings with City Council to present their alternatives to the mayor's agenda. Taxpayer groups and conservative policy analysts organized public meetings in which they promoted yet a third definition of the problem of downtown decline: government interference in the real estate market. The Allegheny Institute argued that major retailers have been discouraged from investing downtown because of the heavy taxes levied by the city and the state, a problem resulting from fiscal policies that favor subsidies for big business underwritten by residents, small businesses, and merchants in the city.

Though their definitions of the problem differed, preservationists, the activist downtown merchants, and conservative critics did converge upon a common solution that helped bring them into an alliance against the mayor's plan. This network of groups pushed the claim forcefully to City Council that a large-scale, master-planned, government-led approach that makes liberal use of eminent domain authority and public subsidies was the wrong way to redevelop the downtown retail corridor.

Paradoxically, the mayor's earlier strategy of immobilizing key information ultimately fanned the flames of controversy rather than containing them. As it turned out, City Council never voted on the plan because the mayor never presented it to them. In the spring of 2000 the member of council representing downtown set up a planning collaborative to hear all the voices and gather information to evaluate the different plans under consideration. By the time City Council was prepared to vote in the fall, a key player in the city's plans had backed out. Nordstrom, an upscale Seattle retailer decided to shelve its plans to build a store in the Pittsburgh location. This news prompted Murphy to drop the renewal plan for which he had battled so long.

CONCLUSION

The dynamics of policy controversy in Pittsburgh described above are not unique to the case of retail redevelopment. Nearly every other major development project in the past decade has followed a similar logic: elite partnerships pursue a project in secrecy, withhold information from would-be-critics for as long as possible, then begin compromise after opponents generate and expand controversies through strategies of issue redefinition.[14]

It is evident in the above examination of connections between regime composition and the dynamics of political interaction during two periods of Renaissance in Pittsburgh that conflict expansion through issue redefinition is a much more effective political mechanism today than it was fifty years ago. This is due primarily to a number of changes in the local, national, and global political economies that have resulted in: (1) decentralization of the production of expert knowledge; (2) a relative decline in trust and unity among public officials and civic elites that has created greater incentives for individual political actors to provoke public controversy; and (3) a loss of elite control over the structuring of contexts in which problems and solutions are linked.

While the agenda of economic development is still largely set by growth elites, the mechanism of controversy forces policy alternatives to be considered outside elite institutions, in public arenas such as City Council, community-based organizations, and the media. Nonelite influence upon policy is thus quite limited, even today. Nonelite access to government is achieved through protest, usually under conditions of imminent government action, allowing opponents of elite plans little time to mobilize for problem redefinition and rational construction of alternatives. Thus now that neighborhoods, preservationists, progressive planners, and government reformers have achieved de facto incorporation into regime politics in Pittsburgh, the question remains whether the short-term, issue-orientation of political action in the region is sufficient to institutionalize more cooperative arrangements for regular dialogue and bargaining among stakeholders concerned with economic redevelopment in Pittsburgh.

NOTES

1. Thus the approach taken in this paper owes a great debt to the micro-interpretive tradition in sociology. See in particular Garfinkel (1967) and Goffman (1974).

2. In southwestern Pennsylvania the term Renaissance is a conventional way of characterizing large-scale, government-led renewal. The term was first used after the Second World War to describe the transformation of the city from a polluted steel town in decay to what Rand McNally would eventually rank as the nation's "most Livable City".

3. John Kingdon (1984) distinguishes three processes in policy-making - defining an agenda, considering alternative solutions, and making choices - each of which may be controlled by a different set of actors. I assume that at the local level those who define agendas also attempt to determine which alternatives will be considered for public discussion.

4. John Delano and Steven Volk "Who's on First?" In Pittsburgh, May 6, 1998, p. 12.

5. See Tom Barnes and Douglas Heuck, "Growth Alliance Unites 10 Counties," Pittsburgh Post Gazette, April 9, 1998; and John K. Manna, "Region Legion," New Castle News, April 1, 1998.

6. This section is based on data from the following sources: Pittsburgh Post Gazette, Pittsburgh Tribune Review, Pittsburgh Business Times, City Paper, and In Pittsburgh, as well as publications and reports released from government and nongovernmental agencies throughout the western Pennsylvania region, and from interviews with public officials and civic leaders.

7. Mayor Murphy's position was presented largely in press releases and press conferences aimed at the major print and broadcast news media in the city.

8. Deborah Stone (1997, p. 189) uses this term to refer to the ways in which political actors strategically craft arguments, using symbols and numbers, to shape alliances around a favored agenda. The point is that in most cases of policy debate, a politician's theory of causation is tested on the basis of its appeal to the desires or needs of a political audience rather than by an empirical test against evidence, as in some philosophies of science.

9. Analytically different political strategies may be interrelated in practice. For example, by using the rhetoric of responsible government to avoid discussing details of his plan with the public (thus controlling the flow of information), the mayor was engaging in a form of subpolitics; he was making many of the important decisions and resolving many of the practical aspects of his plan (e.g. enlisting a developer, securing financing), before it became an "issue" for policymakers.

10. See Ibarra and Kitsuse (1993) for discussion of various strategies of problem redefinition that utilize rhetorical idioms embodying powerful moral appeal.

11. Ibid.

12. Jeff Stacklin, "Market Place Developer Fills in Vision for Shopping District", Pittsburgh Tribune Review, electronic archives, June 6, 1999.

13. Quoted in, ibid., p. 2.

14. See Suzanne Elliot and Tim Schooley, "Textbook Murphy: Mayor's Method of Operation on Development Has Had Mixed Success," Pittsburgh Business Times, V. 19, No. 30, February 11-17, 2000, p. 1.

ACKNOWLEDGMENT

I would like to thank Kevin Fox Gotham for his helpful comments on an earlier draft.

REFERENCES

Ahlbrandt, R. Jr. (1986). Public-Private Partnerships for Neighborhood Renewal. *Annals of the American Academy of Political and Social Sciences, 488*, 120–133.
Ahlbrandt, R. Jr., & Coleman, M. (1987). *The Role of the Corporation in Community Development as Viewed by 21 Corporate Executives.* University Center for Social and Urban Research. University of Pittsburgh.
Bachrach, P., & Baratz, M. S. (1973). *Power and Poverty.* New York: Oxford University Press.
Baumgartner, F. R. (1987). Parliament's Capacity to Expand Political Controversy in France. *Legislative Studies Quarterly, 12*(1), 33–54.
Callon, M. (1985). Some Elements of a Sociology of Translation. In: J. Law (Ed.), *Power, Action, and Belief: A New Sociology of Knowledge?* (pp. 196–233). New York: Routledge.
Clark, G. L. (1989). Pittsburgh in Transition: Consolidation of Prosperity in an Era of Economic Restructuring. In: R. A. Beauregardan (Ed.), *Economic Restructuring and Political Response* (pp. 41–67). Newbury Park: Sage.
Cobb, R. W., & Elder, C. D. (1972). Participation in American Politics: *The Dynamics of Agenda-Building.* Baltimore: The Johns Hopkins University Press.
Coleman, M. (1983). *Interest Intermediation and Local Urban Development.* Doctoral Dissertation, Department of Political Science, University of Pittsburgh.
Cox, K. R. (1999). Ideology and the Growth Machine. In: A. E. G. Jonas & D. Wilson (Eds), *The Urban Growth Machine: Critical Perspectives Two Decades Later* (pp. 21–36). Albany: State University of New York Press
Elkin, S. (1987). State and Market in City Politics: Or, The 'Real' Dallas. In: C. Stone & H. Sanders (Eds), *The Politics of Urban Development* (pp. 25–51). Lawrence: University Press of Kansas.
Feldman, M. M. A. (1996). Spatial Structures of Regulation and Urban Regimes. In: M. Lauria (Ed.), *Reconstructing Urban Regime Theory: Regulating Urban Politics in a Global Economy* (pp. 30–50). London: Sage.
Ferman, B. (1996). *Challenging the Growth Machine.* Lawrence: University of Pittsburgh Press.
Fischer, F., & Forester, J. (1993). *The Argumentative Turn in Policy Analysis and Planning.* Durham: Duke University Press.
Garfinkel, H. (1967) [1984]. *Studies in Ethnomethodology.* Reprint. Cambridge: Polity Press.
Goffman, E. (1959). *The Presentation of Self in EveryDay Life.* New York: Doubleday.
Goffman, E. (1974). *Frame Analysis.* New York: Harper & Row.
Goodwin, M., & Painter, J. (1996). Concrete Research, Urban Regimes, and Regulation Theory. In: M. Lauria (Ed.), *Reconstructing Urban Regime Theory* (pp. 13–29). London: Sage.
Gusfield, J. (1981). *The Culture of Public Problems.* Chicago: University of Chicago Press.
Hajer, M. (1995–1996). Politics on the Move: The Democratic Control of the Design of Sustainable Technologies. Knowledge and Policy: *The International Journal of Knowledge Transfer and Utilization, 8*(4), (Winter), 26–39.
Hall, P. (1972). A Symbolic Interactionist Analysis of Politics. *Sociological Inquiry, 42*, (3-4), 35–75.
Hardin, R. (1982). *Collective Action.* Baltimore: Johns Hopkins University Press.
Jessop, B. (1996). A Neo-Gramscian Approach to the Regulation of Urban Regimes: Accumulation Strategies, Hegemonic Projects, and Governance. In: M. Lauria (Ed.), *Reconstructing Urban Regime Theory* (pp. 51–73). London: Sage.
Jezierski, L. (1988). Political Limits to Development in Two Declining Cities: Cleveland and Pittsburgh. In: M. Wallace & J. Rothschild (Eds), *Research in Politics and Society* (Vol. 3, pp. 173–189). Greenwich CT: JAI Press.

Jonas, A. E. G., & Wilson, D. (1999). The City as a Growth Machine: Critical Reflections Two Decades Later. In: A. E. G. Jonas & D. Wilson (Eds), *The Urban Growth Machine Machine* (pp. 3–18). Albany: State University of New York Press.

Kitsuse, M., & Spector, M. (1973). Toward a Sociology of Social Problems. *Social Problems, 20*, 407–419.

Latour, B. (1991). Technology is Society Made Durable. In: J. Law (Ed.), *Sociology of Monsters* (pp. 103–131). New York: Routledge.

Lauria, M. (1996). Introduction: Reconstructing Urban Regime Theory. In: M. Lauria (Ed.), *Reconstructing Urban Regime Theory* (pp. 1–9). London: Sage.

Lauria, M. (1999). Reconstructing Urban Regime Theory: Regulation Theory and Institutional Arrangements. In: A. E. G. Jonas & D. Wilson (Eds), *The Urban Growth Machine*. Albany: State University of New York Press.

Lubove, R. (1969). *Twentieth-Century Pittsburgh*. New York: John Wiley & Sons.

March, J., & Olsen, J. P. (1984). The New Institutionalism: Organizational Factors in Political Life. *American Political Science Review, 78*(3), 734–749.

Martin, P. (1952). The Allegheny Conference: Planning in Action. American Society of Civil Engineers. *Proceedings, 78* (May).

Muller, E. K. (1989). Metropolis and Region: A Framework for Enquiry into Western Pennsylvania. In: S. Hays (Ed.), *City at the Point* (pp. 181–211). Pittsburgh: University of Pittsburgh Press.

Perry, D. C. (1990). Recasting Urban Leadership In: D. Judd & M. Parkinson (Eds), *Buffalo. Urban Affairs Annual Review, Vol. 37, Leadership and Urban Regeneration*. Newbury Park, CA: Sage.

Peterson, P. (1981). *City Limits*. Chicago: University of Chicago Press.

Quattrone, G. A., & Tversky, A. (1988). Contrasting Rational and Psychological Analyses of Political Choice. *American Political Science Review, 82*(3), 719–736.

Rochefort, D. A., & Cobb, R. W. (1993). Problem Definition, Agenda Access, and Policy Choice. *Policy Studies Journal, 21*(1), 56–71.

Rochefort, D. A., & Cobb, R. W. (1994). *The Politics of Problem Definition*. Lawrence: University Press of Kansas.

Sbragia, A. (1989). The Pittsburgh Model of Economic Development: Partnership, Responsiveness, and Indifference. In: G. Squires (Ed.), *Unequal Partnerships* (pp. 103–120). New Brunswick, NJ: Rutgers Univertsity Press.

Schattsneider, E. E. (1960). *The Semi-Sovereign People*. New York: Holt.

Short, J. R. (1999). Urban Imagineers: Boosterism and the Representation of Cities. In: A. E. G. Jonas & D. Wilson (Eds), *The Urban Growth Machine* (pp. 37–54). Albany: State University of New York Press.

Stave, B. (1970). *The New Deal and The Last Hurrah*. Pittsburgh: University of Pittsburgh Press.

Stewman, S., & Tarr, J. A. (1981). Public-Private Partnerships in Pittsburgh: An Approach to Governance. In: J. Tarr (Ed.), *Pittsburgh-Sheffield: Sister Cities* (pp. 141–182). Pittsburgh: Carnegie Mellon University.

Stone, C. (1989). *Regime Politics*. Lawrence: University Press of Kansas.

Stone, C., Orr, M. E., & Imbroscio, D. (1991). The Reshaping of Urban Leadership in U.S. Cities: A Regime Analysis. In: M. Gottdiener & C. Pickvance (Eds), *Urban Life in Transition* (pp. 222–239). Beverly Hills: Sage.

Stone, D. (1997). *Policy Paradox*. New York: W. W. Norton.

Waste, R. J. (1986). Community Power and Pluralist Theory. In: R. J. Waste (Ed.), *Community Power* (pp. 117–137). Newbury Park, CA: Sage.

Weber, M. P. (1988). *Don't Call Me Boss*. Pittsburgh: University of Pittsburgh Press.
Wilson, D. (1996). Metaphors, Growth Coalition Discourses and Black Poverty Neighborhoods in a U.S. City. *Antipode*, 28(1), 72–96.

ON FRAGMENTATION, URBAN AND SOCIAL[1]

Judit Bodnár[2]

INTRODUCTION

Fragmentation is a powerful, evocative metaphor commonly used in describing contemporary modern societies. Social analysts and even politicians talk about marginalization, the underclass, (bi)polarization, people left behind and society coming apart. In France, the term 'social fracture' (*fracture sociale*) was made famous when it entered French political discourse in current President Jacques Chirac's election campaign in 1995 – a center-right echo of the left social critique of the racial, ethnic and class alienation of the *banlieue*.[3] These concerns gain an additional, spatial, meaning in the context of the city among urban professionals who have been alarmed by the disintegration of the urban form, by the possibility that one of our best collective projects, the city, should become increasingly fragmented. Urban space is assuming a form that has no center and lacks urbanity, where social groups live further and further apart, and racial poverty is ever more confined spatially. Walls are erected around frightened urban middle-class neighborhoods at their citizens' own will. Public space where various social groups could encounter one another is extinguished or privatized or, on a more optimistic note, it is becoming compartmentalized, the argument goes.

Concern with social fragmentation has been, however, a recurrent theme of modernity. The tension between the heterogeneity of citizens and their quest for unity has been at the core of the city from the very beginning. Arguably we have always lived with a sense of fragmentation; fragmentation is a general principle of collective life. Although this piece of historical and theoretical knowledge makes us wiser and more skeptical to those crying the end of city and society, the meaning of fragmentation has varied historically. Our millennial concern with urban and social fragmentation is not quite the same as it was during the advent of modernity. This tension will be a recurring theme of this essay – showing in various ways that fragmentation is not a novelty but also arguing that some aspects of the current physical and social landscape are sufficiently new to warrant a contemporary discussion. I propose a theoretically and historically informed critical examination of today's fragmentation through a set of phenomena that are constitutive of it. In the center of urban fragmentation are the spatial transformation of the city and the metamorphosis of urban society that can be read as a fragmentation of their old forms. Social and spatial fragmentation are intertwined, and one of their configurations, urban fragmentation, is the focus of this essay.

UNDERSTANDING MODERN SOCIETY IN FRAGMENTS

Fragmentation is an implicitly conservative term, it is retrospectively oriented and suggests loss: a unity that had existed is now fragmented. Social fragmentation has been a defining theme of modernity but its peculiarity is precisely that the fragmentation of the old world order and world view makes the new unity possible. Modern social science is conceived exactly at the time when sensitive observers lament the breaking up of communities and start talking about a new totality, modern society. This duality can be found in most major social thinkers of the time – e.g. in the contrast of 'community' and 'society' in Tönnies, 'mechanical' and 'organic solidarity' in Durkheim, etc. – with the emphasis being on the transition from one type of collectivity to another instead of the loss felt over the disappearance of old life forms.

This gained most complete expression in Georg Simmel's work. According to a contemporary of his, as remarks David Frisby, Simmel was "the only genuine philosopher of his time, the true expression of its fragmented spirit" (Frisby, 1986, p. 39). He was the first sociologist to grasp modernity in the sense Baudelaire understood it: "By 'modernity' I mean the ephemeral, the fugitive, the contingent, the half of art whose other half is the eternal and immutable" (Baudelaire, 1964, p. 13). Simmel's analysis of modernity concentrates on the transitory nature of time, the fleeting character of space, and the fortuitousness of causality so that "it is possible to relate the details and superficialities of life to its most profound and essential movements" (Simmel, 1978, p. 56).

How do we grab the fugitive sociologically? Simmel proposes to do so by the analysis of relational concepts that express adequately the flux of life. He suggests, as Frisby points out, that "interaction between individuals and society as a whole constitute a totality that is only apprehendable aesthetically" (Frisby, 1986, p. 49). Modern society can be grasped in flux and in fragments.

This theme is developed more expressly in later modernist thinkers. Walter Benjamin's philosophy of history is a radical reinterpretation of the idea of fragments. As opposed to the usual archeological idea of fragments, holding that once the fragments are dug out, the past that had been once the text of the whole, can be deciphered; Benjamin's fragments need, and ought, not to be dug out. "History is manifest on the surface; everything we know is a knowledge of ruins. No meaning can be rendered to these ruins by human recollection, for they are entirely void of the kind of sense we could possibly understand," points out Agnes Heller (1993, p. 38). "Moreover, recollection, as an attempt to render meaning to the senseless, to bring the speechless to speech, is transgression" (*ibid.*).

Without solving the problem of history either in the modern or postmodern vein, and likewise without offering a fair treatment of the fragmentation of the narratives of history and society, we should stay a bit longer with Simmel who has provided sociologists with an exemplary analysis of the fragmented nature of the modern individual and whose intimate linking of city and modernity has paved the way for any analysis of fragmentation in urban modernity. In his seminal essay, *The Web of Group Affiliations*, Simmel (1955) outlines the emergence of modern personality and freedom in the simplest yet most plausible form.[4]

The thrust of the argument is that the hierarchical and nested pattern of the group affiliations of medieval man was replaced by a network of group affiliations in which one group does not subsume the other; they may not even touch one another.[5] They meet in a single point – the individual. The modern individual is part of many groups but does not owe his soul to any one of them. The proportion between 'competition' and 'socialization' varies with every single group involvement. The individual gives a fragment of himself to many of these collectivities and creates himself precisely out of those fragments. His rewards and his obligations are limited but many. For this reason, "the structure of these connections excluded the possibility that a second individual would occupy exactly the same position within the same context" (Simmel, 1955, p. 152). What is entirely unique to him is this specific *combination*. I differ therefore I am. That combination is him, the individual, and not more than that. But that is not to be underestimated. That combination is his *individuality* and *freedom*. Individual freedom is a truly modern concept; it rests in the *infinite fragmentation of obligations*. Its onset may be accompanied by loss; the loss of ties, depth, intensity, etc., but it could not have been conceived of in any other way, in any period other than modernity.

For Simmel, the emergence of the modern personality is framed by the development of the metropolis. In his sociology of space, space is filled socially; through interaction and sociation (Simmel, 1997). The city is a distinctive social space; an intersection of diverse social groups, a focal point of networks and collectivities. The modern individual reaches into the multiplicity of these groups. The fragmentation of affiliations breeds a new totality that is fragmented by definition. Yet, it produces a meaning unknown before. The city is a sociological entity and an open form. It is a fixed spatial location but one whose networks are in a constant flux. It is the breeding ground of modern personality and freedom; it provides the fragments with a sense of unity. The city is a synecdoche of modern society not only for Simmel but for most social theorists of the time. It is not by accident that the city and society become the central points of inquiry – in fact, the defining themes – of modern social science at the same time. If there is anything called society, then it is the totality that frames fragmented individual experience and social existence. The old world and its view, the exclusive loyalties of the past, may have fallen apart but new connections are created, however minimal(ist) they are. The advent of a mature money economy that is based on a minimalist definition of exchange widens the sphere of interaction and, consequently, our world. It brings together the diversity of people and things; it is the lowest common denominator amidst a great multitude of them. Simmel contends that the metropolitan personality parallels the money economy.

PUBLIC SPACE AS AN EXEMPLARY INSTANCE OF THE CITY

If the city is an exemplary instance of modernity, then public space is at the core of the notion of the modern city. Public space is peculiar to cities. The co-presence of many people in cities renders the emergence of an exhaustive, comprehensive network of personal communication impossible. The predicament of the metropolitan situation, incarnated in urban public space, is the tension between intense physical proximity and moral remoteness. The urban dweller lives increasingly among strangers. This is the source of the typical urban mentality which Simmel describes as a fundamental indifference to distinctions, to every instance of unfamiliarity or difference – the *blasé* attitude. Erving Goffman (1963) accords a more specific social psychological meaning to fundamentally the same attitude, and finds a more virtuous aspect of it by focusing on 'civil inattention' which "makes possible co-presence without co-mingling, awareness without engrossment, courtesy without conversation" writes Lynn Lofland later (1989, p. 462). Building on both Simmel and Goffman, Lofland (1973) strikes a positive tone claiming that the specificity of urban life is precisely that this type of social-psychological situation and its main character, the Stranger, create the very basis of public space where civility toward diversity rules.[6] Public space is where people gather; they represent different levels of wealth, come from different social backgrounds, have different features and bodies – look different – behave and dress differently, strangely. True, interaction in public space is rather limited and fortuitous; it is not about being together. Civility, or indifference toward difference, however, represents a new dimension; public space gathers diverse people and things, and presents this diversity. It gives a new totality to fleeting and fragmented urban life.

THE TRANSFORMATION OF PUBLIC SPACE AND URBAN FRAGMENTATION

Much has been said about the contemporary decline of public space, its privatization and semi-privatization (Sorkin, 1992; Davis, 1992). Optimists reverse the argument and claim that old public spaces become empty – only to be reborn elsewhere. Witold Rybczynski writes in a chapter entitled "The New Downtown" of his latest book (1995) that, for the American citizen who flees the chaos and the challenges of Downtown streets, the shopping mall takes over the function of the street.[7] Regardless of which side one takes in this debate, one thing seems to be clear: public space that once gave a totality to our fleeting and fragmented urban life is going through a profound transformation. This process is partly responsible for an increased sensitivity concerning the fragmentation of the urban public, and the city falling apart.

Urban public space, however, is only one possible location and constituent of the public. Public sphere and public space may overlap in their empirical forms but they need to be distinguished as concepts. Public sphere in the Habermasian sense is "a forum in which the private people come together, readied themselves to compel public authority to legitimate itself before public opinion" (1989, p. 25–6). In a historically less specific and less exclusive tone, Iris Marion Young distinguishes public space from public expression. Expression is public, when besides those expressing themselves at the moment, others can respond and enter into discussion through institutions or the media (Young, 1987). Young's definition of the political is also less exclusive: "Expression and discussion are political when they raise and address issues of the moral value or human desirability of an institution or practice whose decisions affect a large number of people" (Young, 1987, p. 73). This concept of the public is derived from aspects of the modern urban experience, writes Young, acknowledging the importance of urban public space in relation to a heterogenous public life but, correctly, keeping them separate.

The two are linked, even in political theory. Geographers (see Howell, 1993) celebrate Hannah Arendt's spatial imagination in analyses of public space and public sphere: in contrast to Habermas, by 'public space' Arendt means space in the concrete sense as well. It is rarely explained why: that it is the historical period that forms the basis of Arendt's public sphere which allows for her spatial imagination. In the *polis*, the designated place of politics and the most celebrated public space, the *agora*, had both a figurative and an abstract meaning.[8] From Arendt to Habermas there has been a shift from an ocular to an auditory public, from a "space of appearance" – that the *agora* had been – to a desubstantialized public, a virtual community of readers, writers and interpreters that the new publicity of the Enlightenment meant (Benhabib, 1996). The short cohabitation of urban sociability and the political public in the public space of the agora has been replaced by the increasing divergence of urban public space and a public sphere that has neither a body nor a location in space anymore (Benhabib, 1996). With the tremendous expansion of the mass media it has become a truism that "going public today means going on the air" (Carpignano et al., 1993).

A final step in the divergence of public space and public sphere ensues with speculations on the virtualization of public life through the Internet. William Mitchell envisions a new city, the city of bits:

> This will be a city unrooted to any definite spot on the surface of the earth, shaped by connectivity and bandwidth constraints rather than by accessibility and land values. Largely asynchronous in its operation, and inhabited by disembodied and fragmented subjects who exist as collections of aliases and agents. Its places will be constructed virtually by software instead of physically from stones and timbers, and they will be connected by logical linkages rather than by doors, passageways, and streets (Mitchell, 1995, p. 24).

Practically anything becomes accessible through the net; shopping, teaching, healing, chatting, etc. Even *flâneurie* – a pastime made possible by the emergence of the modern metropolis and the urban public – has moved to the virtual city of bits (Featherstone, 1998). If the necessities of daily life can be dealt with from the net and *civitas* will soon be created in cyberspace, as Mitchell maintains, why would one want to leave home and get off the net? And why worry about the decline of urban public space?

Public space offers more than a passive physicality, and its importance goes beyond the pleasure of the urban spectacle. It implies, as said earlier, tolerance for difference and a certain degree of self-limitation, not necessarily of a political nature. The nature of any limitation – should that be exercised by the self, other members of the public, or the authorities – is, however, strongly conditioned by culture and politics. The rules of 'proper conduct' are constructed in constant political debates and cultural interactions. The common denominator for the co-presence of people – tolerance – can be encouraged institutionally. It can be a result of a precarious social and architectural design besides the minimal legal definition of public access. If there exists no credible guarantee of free public appearance, a fear of showing difference emerges. People who have the most to fear – those who diverge from the imagery of the ideal citizen of the public the most – and those who can afford it, start avoiding public spaces. They withdraw into privacy, and the delicate balance in the multi-layered use of public space is offset. When this happens, there is an opportunity to appropriate public space in new, suddenly more exclusive ways. Urban public space is important: it is the place to encounter difference and acquire knowledge about it.

This proposition has a complex relevance for democratic public life. One can, of course, gain this kind of knowledge through various media as well as institutions that are not public. Cyberspace can provide us with extensive knowledge about others, and it increasingly offers interactive possibilities that go beyond the narrow textuality of e-mail communication. Interaction in cyberspace is, however, still a mediated experience. In contrast, urban public space has the unique quality of physical proximity that, even in its most minimalist forms, can be more interactive, more saturating, and richer in surprises than other, de-spatialized (mediated) forms of learning about difference. The public gaze that is a defining constituent of public space implies accountability and, ultimately, responsibility. The interactive experience of difference through physical proximity involves a different degree of *responsibility* than that of any medium, and this immediacy-cum-responsibility remains located in public space. Sending a well-crafted computer virus to one's enemy may well lead to a heart attack, yet, it is different, less direct and less shameful than slapping someone on the face.

Even fierce advocates of a cyberspace-based optimism about the future of the public sphere arrive at the question, as Mitchell does, whether the "shift of social and economic activity to cyberspace mean[s] that existing cities will simply fragment and collapse?" (p. 169). The answer is a reassuring 'no'; cities are flexible, they will adjust as they did earlier, and "we will still care about where we are, and we will still want company" (*ibid.*). As easy as that: we just have to *want* it. Public space becomes a voluntary project, a matter of choice. The elements of *necessity* and constraint are hence removed from public life.

This also explains the contemporary poverty of public space. Public space is not a feasible enterprise anymore when it is not built around the necessity of bringing people together, when it is not the only and natural connection between two meaningful points. This alters the social meaning of public space. Public space by choice is a different undertaking; it is a *theme park*.

FRAGMENTATION OF THE URBAN FORM

The profound transformation of public space may well prompt a rhetoric of fragmentation. The general thinning of publicness in urban life is, however, related to the development of the urban form, the spatial dispersal of city dwellers. This is where the qualifier 'urban' becomes important in the discussion of fragmentation. Whereas the idea of fragmentation has been at the core of modernity, urban fragmentation was not the defining theme of modern social science. In fact, as shown above, the city provided an open and expandable framework that held the fragments of modernity together. True, the city, as any collective project, carries by definition the possibility of fragmentation. Nonetheless, the idea of urban fragmentation is relatively new. When in 1967 historian Robert Fogelson captured a sense of the development of Los Angeles between 1850 and 1930 in the notion of 'fragmented metropolis,' what he had in mind besides the social fragmentation of Los Angeles was also the morphological structure of the city; its lack of a physical center and its diffuse, haphazard growth (Fogelson, 1993 [1967]). This description has since become even truer, and Los Angeles serves clearly as the ideal type of the fragmented city. Motorization and suburbanization figure prominently in the history of the fragmented metropolis.

In fact, there seems to be a consensus that urban fragmentation really starts with the possibility of haphazard growth that is directly linked to the automobile. It is a commonplace by now to blame the fragmentation of the city on cars and suburbs. It is more founded, however, to link urban fragmentation not to cars and suburbs in general but to their scale and class content. It was only the emergence of mass car-ownership that has created the possibility of exit from the city on a massive scale that in turn made serious claims on rearranging urban space.

The Car, the Home and the Search for Happiness

It has become a cliché to point out that American cities are built for cars and for that reason, north American urban life is different from elsewhere. Some think it is only trivial; after all, as opposed to European and Asian cities, most

American cities developed following the age of transportation revolution (among others, see Rybczynski, 1995). Yet, even granting that, the social history of U.S. motorization suggests that the grand victory of the automobile was not inevitable, and transportation technology in itself is not responsible for the fragmentation of the city. Historian Clay McShane (1994) recounts the difficulties and the social resistence the automobile encountered from its early days. When American mechanic Dudgeon attempted to operate his locomotive on ordinary streets at the Crystal Palace Exposition of 1857, a hostile mob attacked him. In 1881, the Milwaukee City Council banned the use of steam engines on railway lines which a real estate developer proposed to use instead of horses that were the common mode of transportation at the time (McShane, 1994). The search for faster transportation prevailed nevertheless. Among all possibilities, the automobile emerged as the most promising means of mobility. For the middle classes it promised the fulfillment of the suburban dream that, according to suburban historian Robert Fishman, was more fundamental than the desire for individual mobility. Even in Los Angeles "[t]he automobile and the highway when they came were no more than new tools to achieve a suburban vision that had its origins in the streetcar era" (Fishman, 1987, p. 15–16). Nevertheless, the suburbs that were built for car transportation were different from the older streetcar suburbs; they did not have to adjust to the pedestrian scale, they were less planned, more amorphous and more fragmented.

Although the suburban detached home may have had its inherent values – such as space, greenery and fresh air – its distance from the social world of the city was equally important. With the rapid industrial growth of American cities, the city started to figure as the repository of social problems: poverty, 'un-American' mores and vice. A move into the suburbs seemed an efficient and nonconflictual individual way of dealing with them. What is quite remarkable is that in pursuing the suburban ideal, the egoistic middle class joined forces with social reformers; the well-conceived self-interests of the former mixed efficiently with the social ideals of the latter.

During the late-nineteenth and early-twentieth-century industrialization of the United States, social problems struck citizens and analysts in the inner city, in the overcrowded and polluted tenements of the industrial metropolis. Relieving congestion seemed an obvious solution. And that became more conceivable than ever as a cheap way of transportation became available. As McShane recounts, Cleveland Mayor John Johnson thought the social benefits of the street railways so great that they justified fare-free municipal ownership: "There was an end of space; an end of the tenement and the slum" (McShane, 1994, p. 29). What turned out to really hold this promise, however, was not mass transportation but the car. In its gradual climb to dominance, the car was pushed by the strong social belief that it would solve the problems of American society. From the introduction of Ford's Model T – the first act that broke the luxury character of the automobile market – the issue was merely making car ownership widely accessible, and that looked feasible. The car was thought to hold remarkably democratic promises; "it democratized social relations by permitting middle-class man to have the same freedom as the more affluent" (McShane, 1994, p. 143). It helped liberate the young and women from family control. Black people also saw the possibility of exit from mass transportation regulated by Jim Crow laws; private or semi-private carpool travel could lift them out of public humiliation (*ibid.*).

The congestion of the inner city did not ease with the advent of massive car transportation. In fact, with the introduction of the skyscraper, inner-city density and traffic have grown considerably. What did change though was the meaning of the street; it became a traffic artery and a storage place for cars while losing its social function as a place of encounters and gatherings. Suburbanization proved to be a rather limited solution of social problems that were seen as rooted in congestion. The ameliorative deficiencies of suburbanization are best shown by the frustrated attempts of the black middle class to pursue the suburban solution that ran into racial and social resistance (Massey & Denton, 1993). In the meantime, the modern American metropolis became a "socially and politically fragmented, gas-guzzling environmental nightmare" (McShane, 1994, p. 228).

Yet, even architects and urban planners were taken by the prospects of a car-driving society. It seemed to offer a solution to congestion that increased precisely due to car traffic and it seemed to realize an old modern utopia, decentralization, that emerged with the modern industrial metropolis. H. G. Wells devoted an essay to "The Probable Diffusion of Great Cities" in 1900. A few decades later Frank Lloyd Wright announced the possibility of the dissolution of the big city and congestion, the epitome of which was the skyscraper. The city in its traditional form was not needed anymore.

> What built the cities that have invariably died? Necessity primarily. That necessity gone and now only a dogged tradition, no sentimental habit can keep any great city alive for a great length of time. Necessity built the city on the basis of 'leg-work or the horse and buggy; wood and coal consumption; food distribution. And when the necessity for communication was to be had only by personal contact.' [. . .] Electric power, electric intercommunication, individual mobilization, and ubiquitous 'publicity' became common denominators and true decentralizing agents of human life (Wright, 1943, p. 318–320).

Congestion he disliked passionately but, for him, that had nothing to do with motor cars, rather, with the overbuilding of cities. Cars were the means for the complete mobilization of the American people and their liberation. "Even now, a day's motor journey is becoming something to be enjoyed in itself" (p. 329). In the new city, "these common highway journeys may soon become the delightful modern circumstance, an ever-varying adventure within reach of everyone" (p. 330). To make sure he drives his point home, Wright concludes with Cervantes's allegory that he takes quite literally: "the Road is always better than the Inn" (*ibid.*). Horizontal development, this "expanded extended sense of place" (*ibid.*) will bring beauty and true freedom into every citizen's life. If the skyscraper stands for capitalist greed and land rent, and a hugely unequal society, horizontal expansion captures the flattening of society, the end of a growth-driven inegalitarian system. Democracy is to be achieved through democratic living arrangements; "there now goes the vexing problem of the tenement" (Wright, 1943, p. 323). To be remembered that however decentralized, the new settlement is still a city: "The new city will be nowhere, yet everywhere. Broadacre City" (Wright, 1943, p. 320). That is, Wright's city will have no history as it will not be built upon contemporary metropolises which, not needed, will simply fragment and disappear. A cornerstone of the new city's democratic potential is its discontinuity with previous cities and its decentralized nature. This, however, does not imply fragmentation but a new way of integration. Broadacres would still have a center. Ominously, in Wright's car-driving society, the roadside gas station may be the embryo of such a functional and symbolic core that would serve as a neighborhood distribution center "naturally developing as meeting place, restaurant, restroom" (p. 328), not very different from what one would call a public space today. Indeed, this intricate linking of individual transportation, parking lot, distribution center and social activities is best seen as a premonition of today's shopping malls, the new locations of public space.

The history of the automobile can be read both as the privatization of transportation and as preparing the ground for spatial fragmentation. More important, however, is the political economy of proportions. A choice with a liberating potential may become a constraint once alternatives are eliminated, or made excessively difficult. The car itself does not fragment space and society; it is one-dimensionality that has that potential. Especially if it means the hegemony of the car – one of the most individualistic means of transportation. The emergence of monocultural car transportation was not an inevitable process even in American cities; it required a major pooling of resources. In fact, it could not have happened without significant federal involvement favoring both car transportation and suburbanization (Jackson, 1985). The car is to liberate from the constraints of space but surely one could appreciate the freedom of fixed tracks when thousands of people seek such individual liberation at the same time. Also, if active support for more collectivistic or vulnerable means of individual transportation – scooters, bicycles, passersby – is neglected, those who do not drive – minors, elders and other misfits – are invariably eliminated from the bliss of individual mobility. The one-dimensionality of social interactions of which transportation is a part, does have the potential of fragmenting urban space and its dwellers. And the possibility of massive exit from the city in the current form of suburbanization that has become available with close to universal car ownership and affordable new construction is the end of the necessity of living amidst diversity that includes unpleasant encounters with dirt and poverty. This has profound social consequences. If one can talk about urban fragmentation today more than before, it is in this sense. The end of the necessity of living with diversity seems possible in some parts of the world.

POLITICAL AND SOCIAL FRAGMENTATION OF THE CITY

Historian Jon Teaford refers to the modern American metropolis as a "fragmented mass" (1979). In fact, in his account the modern American metropolis developed as a "motley patchwork of demands, desires, interests, and schemes" where haphazard development has prevailed:

> the twentieth-century American metropolis. . . has been an uncoordinated mass of clashing social and ethnic fragments. American urban history is a tale of unwilling accommodation rather than harmonious cooperation (Teaford, 1993, p. 5–6).

It has been common to attribute this to governmental fragmentation. A logical extension of that was the blaming of the political division between suburb and city for the 'urban crisis' of the 1960s. Social problems were urban problems – but not in the sense Henri Lefebvre meant it at the time.[9] This *pars pro toto* representation of the city had been rather perverse and social reformers had long targeted urban congestion as the main problem in the U.S. Still, expressing the nation's problems in urban pathology reached a new dimension after World War II and especially during the 1960s. Although the political fragmentation of the city was made possible by a legal structure that developed in the nineteenth century and maximized the opportunity for self-rule, Teaford (1979) shows that the history of urban government has been a continuous struggle between the forces of fragmentation and consolidation. In more general terms, the pendulum kept swinging between the exit and the voice options (Hirschman, 1970) with a legal system that made exit, and the attendant, rampant institutionalization of social identity, relatively easy. Hirschman himself notes the "extraordinarily privileged position" exit has been accorded in American ideology, tradition and practice (Hirschman, 1970, p. 106). The suburban flight was in accordance with this "preference for the neatness of exit over the messiness and the heartbreak of voice" (p. 107) which was arguably a fundamental characteristic of the liberal tradition in America. Yet, until about 1910, claims Teaford, the forces of exit and social segregation met a counterveiling desire for superior infrastructure – better sewage, water, lighting, paving, and fire department – that could only be provided by the pooling of resources that only a larger unit – a city – could afford (Teaford, 1979). The necessity of cooperation in this area kept urban governments together. With the lessening of the dependence of suburbs on cities for quality municipal services, the social, moral and political split between the big city and the suburbs ran deeper. The spread of the automobile only intensified the isolation of white middle-class Americans among their own kind, and the postwar suburbanization of the country – the combined effect of the Federal Highway Act of 1956 and the construction revolution by the Levitts – expanded the category of the middle class enormously. There were several waves of governmental reform and federative plans to change the legal system that framed the urban-suburban division but these attempts failed largely as elite enterprises (Teaford, 1979). This only expressed the deep fissures of society. "Americans preferred the autonomy of the social fragment to some unifying civic ideal preached by starry-eyed reformers" (Teaford, 1993, p. 169). In Teaford's interpretation, the metropolis remained fragmented (1979, p. 173). On second thought developed in a later book, he adds that today's metropolis that remains divided is markedly different from that of 1900 when different groups "shared one umbrella government and a centripetal transit system made the downtown the focus of all commerce" (Teaford, 1993, p. 168). That mixed lot of Americans all did politics and shopping in the city because there was no real alternative, and their differences did lead to conflicts but were also the source of the city's diversity and vitality. The city was multivocal as long as exit was not feasible on a massive scale.

Although it is spatial fragmentation that frames urban fragmentation, the essence of urban fragmentation is the coming apart of urban society, the fission of the urban public. The sociological question is whether spatial fragmentation promotes that and whether there are counterveiling forces that can create a new unity. Even the archetypical story of Los Angeles is about more than its uniquely decentralized urban growth. Its development had been driven by a major tension: how to build a strong democratic consensus in a fragmented society that, to some extent, insisted on remaining so. Strong it had never been, but the democratic consensus that had existed in Los Angeles's development concerned a rather private definition of the 'public interest.' City planning worked only when it coincided with the interests of the subdividers, when its principles did not question the priority of economic growth, low taxes and car traffic, points out Fogelson (1993). What held the city together was the minimalist common denominator of private economic interests – promoting efficiency within the unquestionable premise of decentralization – devoid of any elevated purpose. Collective urban projects and aesthetics certainly suffered from that. With this kind of development Los Angeles has made urban history; it has become the forerunner of a new urban form – a city that defies classic expectations and blurs the distinction between city and countryside, between city and suburb. Its structure does not fit the concentric model of density and social prestige of either the American or the traditional European city, it exhibits very low density in general, and the only thing that reminds one of a city is constant traffic congestion and, upon a closer look, the diversity of cultural consumption.

In the 1980s, urbanists started noticing a change in the landscape. Suburbia was not what it used to be. It had given way to a low – density cityscape that defied the old notion of the suburb as a residential unit. It was a new kind of city; a decentralized one that "nevertheless possesses all the economic and technological dynamism we associate with the city" (Fishman, 1987, p. 184). Fishman calls it *technoburb*, Joel Garreau popularizes the concept as *edge city*, Teaford captures it as *post-suburbia* (Teaford, 1997). The new form emerged historically from suburbia that has tried to maintain a delicate balance between the quiet rusticity of suburban life and the practical necessity to relieve local tax burdens. Its origins also explain the resolutely anti-urban mentality that marks the edge city off from the city. It exhibits some diversity; office complexes, schools, hospitals, single-level detached houses along high-rises, shopping malls and roads give it quite a varied constitution. The most important characteristic of the technoburb, or edge city, upon which its novelty seems to be built, is the renewed linkage it provides between work and residence. The new city, however, is much more private than older cities. It conspicuously lacks mass transit as that would not make sense.

> The true center [. . .] is not in some downtown business district but in each residential unit. From that central starting point, the members of the household create their own city from the multitude of destinations that are within suitable driving distance (Fishman, 1987, p. 185).

For urban dwellers, the true centers of their lives have always been their homes, so in this regard living in a technoburb is not radically different from urban experience at any time. In the technoburb, however, these individual circles intercept less frequently than ever. Increased privatism, of course, is not limited to suburban or postsuburban experience. The non-communicative character of all venues of life is reinforced by architectural and technological design. Air-conditioned family houses are isolated from the street, air-conditioned car travel is a lonely experience anywhere, and urban office space can just as well separate people as its suburban equivalent. Still, there are settings that do more so. Longer highway travel with no sense of orientation is a more solitary experience than city driving, and destinations in edge cities are more isolated from one another than elsewhere. Campuslike office parks – the locations of economic dynamism in the technoburb – as well as shopping malls and elevated plazas, where public life goes on, are institutions that stand on their own, they are not meant to communicate with their environment. Office density does not mean social density, economic growth and dynamism do not translate into vibrant urban public life. These outposts of economic development have generated urban diversity only relative to suburbia.

Even proponents of the new urban form admit to some criticism. Fishman notes the disastrous effect of this development on old cities and on the natural landscape that is eaten up in the standardized sprawl. He also agrees that non-commercial public space is a rare bird, and that these "can be no more than fragments in a fragmented environment" (p. 203). However, he claims that these "deficiencies are in large part the early awkwardness of a new urban type" (p. 204).

Indeed, the edge cities of post-suburbia can be more fragmented than either the cities or their suburbs. Historically, the suburbs still have had an orientation by articulating a sense of center to which they were attached. In the post-suburban sprawl orientation is lost, there are no *urbs* or suburbs any more. One feels constantly on the edge of something – that something else is, however, never reached. It is not only that there is no single center, or there are no centers, to hold the city or an area together physically and symbolically; it is also the fact that city limits have become blurred. Our sense of unity has thus been challenged. Those 'deficiencies' of the new urban form are not temporary, they will not fade away with maturation. These cities will not get old, they are new by definition. They will relocate before they could turn old. Edge cities do not have meaningful boundaries or serious limits to sprawl, their governance usually belongs to several municipalities. Exit is extremely easy for both capital and people. When economic development cuts too deep into the tranquility of residential life, suburbanites move on to areas less touched by economic growth. Businesses also move when they find rents too high. Edge cities, however, cannot defy the logic of capitalism, just as suburbia cannot: blight is not confined to the inner city. Those who celebrate the dynamism of the edge cities or the decency of suburban life often forget that they are embedded in a very unequal development. In fact, decay can be even worse when the downward cycle hits places that emerged in a moment of instant economic need and does not have much to fall back upon.[10] Edge cities even lose their defining characteristic – economic dynamism – and thus their name in a downward cycle. Often popular political proposals that aim to resolve the 'urban problem' by redrawing municipal boundaries – to unify city and suburb, blight and good life – seem outright futile in the face of the new urban form that defies such division and makes the spatial fixes (Harvey, 1982) of capitalist urbanization ever less fixed and solid.

In the new cities there is no necessity of compact development; they are easy to come to and easy to leave. They are rather inconsequential entities. They have no historicity, no permanence. Their landscape is defined by minimalist, 'throw-away' architecture. Houses are high-tech barracks with little initial investment and high maintenance costs that are built for hardly more than a generation. Part of the permanence that makes the city is eliminated by these practices. Cities are built for future generations, they transcend the present. They go beyond immediate necessities and short-term profitability. With David Harvey (1996), cities are spatial permanences that are produced in the process of urbanization. These "distinctive mixes of spatial permanences" (Harvey, 1996, p. 419), while constantly challenged and rearranged, do have some inertia, and that inertia is a crucial part of their worth. The physical landscape connects generations, grounds historical memory, the necessity of adjustment breeds collective and individual creativity, and diversity. It is a constraint and a possibility. Just as the experience of living in a city, among strangers, being part of the urban public, see others in public and being seen in public. The end of this constraint makes for a different city that is based on the denial of constraints and necessities. The city and its exemplary instance, public space, become one of the possible theme parks.

THE SOCIAL CONSTRUCTION OF DISSOLUTION

There are, clearly, forces other than physical design associated with the breaking up of the unity of the city – although, as argued earlier, it is the spatial aspect that is peculiar to urban fragmentation. The lamented decline of urban public space is not only the result of non-communicative urban architecture, the lack of planning, or the virtualization of programmed consumption. The differentiation and the supposed dissolution of the urban public are thoroughly social phenomena in several regards. Anything that increases the dissimilarity of the public as well as how the similarity and the dissimilarity of that public is constructed should be taken into account in analyses of fragmentation. Increasing social inequalities, social polarization as an extreme case of the latter, the insertion of newcomers, strangers, who add to the diversity of the citizenry – all may pull various social groups apart. It is certainly true that, according to research results, social polarization has become more marked recently in big cities but, in all fairness, throughout their history cities had coped with more striking inequalities than today. So, why a growing sense of fragmentation now?

More to the point are arguments that emphasize either the decline of official policies that assert universal principles including the right to a certain standard of living, or the disappearance of some common ideal or purpose that binds citizens together. Fragmentation is striking after an exceptional interlude of affluence, full employment, and planning that produced an image of a 'well-ordered' society that characterized the post World War II period up to the 1970s, especially in Europe. In fact, the relative lack of fragmentation in that period was more exceptional than the heightened sense of fragmentation today (Mommaas, 1996).

The rhetoric of fragmentation often comes as a backlash against increased demands on representation by previously unrecognized social groups. Historian Arthur Schlesinger, Jr. writes in *The Disuniting of America* as a reaction to extreme proposals to adjust school curricula to identity politics:

> What happens when people of different ethnic origins, speaking different languages and professing different religions, settle in the same geographical locality and live under the same political sovereignty? Unless a common purpose binds them together, tribal hostilities will drive them apart. Racial and ethnic conflict, it seems evident, will now replace the conflict of ideologies as the explosive issues of our times (Schlesinger, 1992, p. 10).

The emergence of a 'cult of ethnicity' threatens to segment, resegregate and tribalize American society.

> The multiethnic dogma abandons historic purposes, replacing assimilation by fragmentation, integration by separatism. It belittles *unum* and glorifies *pluribus* (p. 16–7).

What happens to unity and, ultimately, the totality of society? – asks Schlesinger. The validity of his question can be appreciated in all political aisles, though Schlesinger's worries may not be shared by those who have never felt truly part of this unified American society.[11]

Historian Peter Buckley's examination of the emergence of the new narrative of urban degradation, luxury and dividedness following a strong unified republican rhetoric in antebellum New York captures the tension that is inherent in every analysis of the kind. He argues that before the dissolution of republican New York in the 1850s "despite sharp and growing inequalities, perhaps because of them, all parts of the social spectrum engaged in constructing, and in some cases sharing, this culture of difference" (Buckley, 1988, p. 27). On the one hand, he appreciates, and is even fascinated with, the republican understanding of the city that included self-restrictions on visual demonstration of power and wealth. On the other, he notes that "the public sphere appears relatively seamless because the elite, figuratively and geographically, was at the center of its construction" (p. 33). The entrance of middle-class women in masses onto the urban scene and the concomitant expansion of the urban public were also accompanied by a rhetoric of fragmentation, and worries concerning order and the decline in the quality of the public. "Urban unity" is thus redefined as a precarious balance between public order and creeping democratization.

The rhetoric of fragmentation always uses the past as referent. Its implicit belief is always that the city's streets and public spaces had *once* been legible, and they are *not* any more. Students of the public sphere – among them critics of Habermas's homogenous portrayal of the bourgeois public sphere – have argued that the unity of the public has always been an appearance held together by force and the suppression of difference (Robbins, 1993). In this light, any form of the expansion of the public, any form of democratization produces a sense of loss of unity. Some lament it, others hold that what is lost in common understanding may be gained in individual and group freedom and diversity.

This is certainly an optimist reading of both social fragmentation and the profound transformation public space is going through. The loosening of social control that makes public space a difficult enterprise should not be mourned by any rhetoric that holds the publicness of space dear. The idea of public space has always been utopian, and has always dwelt on power and suppression. The present segmentation of the institution by class and function is only honest, realistic, and even democratic. In this reading, the rhetoric of fragmentation is a natural and conservative corollary of democratization.

Yet, there still remains something specific about today's concern with fragmentation. The totality of limited involvements that, in a Simmelian manner, was framed by the modern metropolis and modern society has come under attack. It is challenged by the present disjuncture of nation, state and society that have been instrumental in constructing citizenship even though their conjuncture has always been more of a political project than reality. Their reconfiguration is taking place in an increasingly global context. Fragmentation and globalization are complementary processes; globally connected urban nodes are created along with a multitude of fragmented and increasingly powerless locales.[12] What is new about recent change is the expansion of the social horizon. Connections become routinely produced on the global level at the price of the increasing fragmentation of lower levels. Most prominent are the losses in the integrating role of the city and of the old notion of society. We have moved to a new level of totality; more minimal involvements and more limited meanings than in Simmel's time, but a great multitude of them. Our modern, or postmodern, personality is still created in a web of group affiliations that have become truly global and increasingly virtual. This produces an unforeseen degree of fragmentation and an increased sense of freedom but this restructuring of ties prompts a sense of loss that is ever more unsettling, sometimes even frightening. The dominant experience of the loneliness of today is not what one feels in the urban crowd but the deeper solitude of sitting at home, completely wired, connected to the world wide web.

NOTES

1. An earlier version of this paper was presented at the session on Social Theory at the Annual Meetings of the American Sociological Association, Washington D.C., August 2000.
2. Part of this essay was written while the author was a postdoctoral fellow in the Urban Fragmentation Project of GUST, the Ghent Urban Studies Team, at the University of Ghent, Belgium.
3. Portrayed very forcefully, among others, in the 1995 movie *Hate* by director Mathieu Kassovitz, who, according to a journalist, understood better than the strategists of the presidential campaign what that 'famous *fracture sociale*' meant.
4. Simmel's *Metropolis and Mental Life* also touches upon the topic while a deeper analysis is to be found in *The Philosophy of Money*.
5. For a recent resurrection of Simmel's web of group affiliations that recasts his formula in current network terms, see Pescosolido and Rubin, 2000.
6. See a whole range of work by Richard Sennett (1970, 1974, 1990).
7. See more on this debate in Bodnar, 2001.
8. All the usual restrictions of Greek politics apply: foreigners, slaves, women and minors were excluded from the 'public' space and the politics of the agora. Their exclusion did not include the commercial section of the agora.
9. Lefebvre understood it in the context of his general critique of the production of space in neocapitalism (Lefebvre, 1970).
10. See, among others, Mike Davis's description of Fontana, California (Davis, 1992) and Didion, 1993.
11. See, among others, Massey and Denton, 1993; Portes, 2001.
12. This duality often lingers in accounts of globalization. It has been expressed most clearly perhaps by Manuel Castells (1996).

REFERENCES

Baudelaire, C. (1964 [1863]). The Painter of Modern Life. In: C. Baudelaire (Ed.), *The Painter of Modern Life and Other Essays*. Translated by Jonathan Mayne. London: Phaidon.
Benhabib, S. (1996). *The Reluctant Modernism of Hannah Arendt*. London: Sage & Thousand Oaks.
Bodnár, J. (2001). *Fin-de-Millénaire Budapest: Metamorphoses of Urban Life*. Minneapolis: University of Minnesota Press.
Buckley, P. G. (1988). Culture, Class, and Place in Antebellum New York. In: J. H. Mollenkopf (Ed.), *Power, Culture, and Place. Essays on New York City* (pp. 25–52). New York: Russell Sage.

Carpignano, P., Andersen, R., Aronowitz, S., & DiFazio, W. (1993). Chatter in the Age of Electronic Reproduction: Talk Television and the Public Mind. In: B. Robbins (Ed.), *The Phantom Public Sphere* (pp. 93–120). Minneapolis: University of Minnesota Press.

Castells, M. (1996). *The Information Age: Economy, Society and Culture. Vol. 1. The Rise of the Network Society.* Oxford: Blackwell.

Davis, M. (1992). *City of Quartz. Excavating the Future in Los Angeles.* New York: Vintage Books.

Didion, J. (1993). Letter from California: Trouble in Lakewood. *The New Yorker,* 69(23), July 26, 46–65.

Goffman, E. (1963). *Behavior in Public Spaces.* New York: Free Press.

Featherstone, M. (1998). The Flâneur, the City and Virtual Public Life. *Urban Studies,* 35(5-6), 909–925.

Fishman, R. (1987). *Bourgeois Utopias: The Rise and Fall of Suburbia.* New York: Basic Books.

Fogelson, R. M. (1993 [1967]). *The Fragmented Metropolis. Los Angeles, 1850–1930.* Foreword by Robert Fishman. LA, London: Berkeley.

Frisby, D. (1986). *Fragments of Modernity. Theories of Modernity in the Work of Simmel, Kracauer and Benjamin.* Cambridge, MA: MIT Press.

Garreau, J. (1991). *Edge City: Life on a New Frontier.* New York: Doubleday.

Habermas, J. (1989 [1962]). *The Structural Transformation of the Public Sphere.* Cambridge: MIT Press.

Harvey, D. (1982). *Limits to Capital.* Oxford: Blackwell.

Harvey, D. (1996). *Justice, Nature and the Geography of Difference.* Oxford: Blackwell.

Heller, A. (1993). *A Philosophy of History in Fragments.* Oxford: Blackwell.

Hirschman, A. O. (1970). *Exit, Voice, and Loyalty: Responses to Decline in Firms, Organizations, and States.* Cambridge, MA.: Harvard.

Howell, P. (1993). Public space and public sphere: political theory and the historical geography of modernity. *Environment and Planning D: Society and Space,* 11, 303–322.

Jackson, K. T. (1985). *Crabgrass Frontier: The Suburbanization of the United States.* New York: Oxford University Press.

Lefebvre, H. (1970). *La révolution urbaine.* Paris: Gallimard.

Lofland, L. (1973). *A World of Strangers: Order and Action in Urban Public Space.* New York: Basic Books.

Lofland, L. (1989). Social Life in the Public Realm. *Journal of Contemporary Ethnography,* 17(4), 453–482.

Massey, D., & Denton, N. (1993). *American Apartheid: Segregation and the Making of the Underclass.* Cambridge, MA: Harvard.

McShane, C. (1994). *Down the Asphalt Path. The Automobile and the American City.* New York: Columbia University Press.

Mitchell, W. J. (1995). *City of Bits. Space, Place, and the Infobahn.* Cambridge, MA: MIT Press,

Mommaas, H. Modernity, Postmodernity and the Crisis of Social Modernization: A Case Study in Urban Fragmentation. *International Journal of Urban and Regional Research,* 20(2), 196–216.

Pescosolido, B. A., & Rubin, B. A. (2000). The Web of Group Affiliations Revisited: Social Life, Postmodernism, and Sociology. *American Sociological Review,* 65(February), 52–76.

Portes, A. (2000). Immigration and the Metropolis: Reflections on Urban History. *Journal of International Migration and Integration,* 1(2), 153–175.

Robbins, B. (1993). *The Phantom Public Sphere.* Minneapolis: University of Minnesota Press.

Rybczynski, W. (1995). *City Life. Urban Expectations in a New World.* New York: Scribner.

Schlesinger, A. M., Jr. (1992 [1991]). *The Disuniting of America.* New York: Norton.

Sennett, R. (1970). *Uses of Disorder*. Harmodsworth: Penguin.
Sennett, R. (1974). *The Fall of Public Man*. New York: Norton.
Sennett, R. (1990). *The Conscience of the Eye. The Design and Social Life of Cities*. New York: Norton.
Simmel, G. (1978 [1900]) *The Philosophy of Money*. Translated by T. Bottomore and D. Frisby. London: Routledge & Kegan Paul.
Simmel, G. (1955 [1922]). The Web of Group Affiliations. In: K. Wolff & R. Bendix (Trans.), *Conflict and The Web of Group Affiliations*. New York: The Free Press.
Simmel, G. (1997 [1903]). The Sociology of Space. In: M. Featherstone & D. Frisby (Eds), *Simmel on Culture. Selected Writings*. London: Sage.
Sorkin, M. (1992). *Variations on a Theme Park. The New American City and the End of Public Space*. New York: Farrar, Straus and Giroux.
Teaford, J. C. (1979). *City and Suburb: The Political Fragmentation of Metropolitan America, 1850–1970*. Baltimore: The Johns Hopkins University Press.
Teaford, J. C. (1993 [1986]). *The Twentieth-Century American City*. Baltimore: The Johns Hopkins University Press.
Teaford, J. C. (1997). *Post-Suburbia: Government and Politics in the Edge Cities*. Baltimore: The Johns Hopkins University Press.
Wright, F. L. (1943). *An Autobiography*. New York: Duell, Sloan and Pearce.
Young, I. M. (1987). Impartiality and the Civic Public: Some Implications of Feminist Critiques of Moral and Political Theory. In: S. Benhabib & D. Cornell (Eds), *Feminism as Critique. On the Politics of Gender*. Minneapolis: University of Minnesota Press.

HISTORIC PRESERVATION, GENTRIFICATION, AND TOURISM: THE TRANSFORMATION OF CHARLESTON, SOUTH CAROLINA

Regina M. Bures

INTRODUCTION

In the United States, historic preservation is frequently associated with gentrification: the incursion of middle-class "gentry" on an urban frontier, resulting in the displacement of lower income residents (Smith, 1996). However, historic preservation need not be a consequence or cause of gentrification. The term historic preservation implies the maintenance of both the social environment and the physical environment. Gentrification, on the other hand, implies the improvement of the physical environment at the expense of the existing social environment. Both historic preservation and gentrification are social processes that evolve over time. Historic preservation reflects the social perception that single structures or entire neighborhoods are culturally significant. Older neighborhoods may also experience stages of gentrification through a "chain of gentrifiers" who have a cumulative effect on the social environment (Appleyard, 1979).

The relationship between culture and capital is an important consideration in the redevelopment process. The capital-intensive nature of gentrification often requires "cultural validation" in the form of a historic designation (Zukin, 1987,

p. 143). Historic designation raises property values and displaces less affluent residents. Historic designation may also lead to increased tourism, another form of urban redevelopment. In this context, the history and culture of a city serve as leverage for attracting capital investment and tourists (Hoffman, 2000).

Understanding the consequences of urban redevelopment is a central theme in urban sociology. Historic preservation, gentrification, and tourism are three related, yet distinct, redevelopment processes. All three can lead to improvements in the physical environment of a city but will have different consequences for the social environment. This paper examines the consequences of these redevelopment processes by presenting a case study of the city of Charleston, South Carolina. The redevelopment process in Charleston can be described in three stages. These stages are characterized by different consequences for the social environment, which are illustrated by examining the city's changing racial composition. In describing these stages, I draw from two theoretical perspectives: human ecology and political economy.

Both the human ecology and political economy perspectives contribute to understanding urban redevelopment processes. The ecological perspective describes the effects of redevelopment in terms of housing supply and the residential preferences of specific groups. This perspective is useful for understanding why communities and individuals support renewal through historic preservation. In the case of Charleston, an ecological perspective offers insights into why residents formed the first historic preservation group in the United States: Families organized to preserve the historic, residential character of their neighborhoods. As historic preservation spread in Charleston, a political-economy perspective, framing redevelopment as a process shaped primarily by political and economic forces (Friedenfels, 1992), is useful for understanding the significant expansion of areas designated as Historic Districts. The second and third stages of Charleston's redevelopment, gentrification and tourism, reflect this perspective: Economic incentives in the form of federal funding and tax incentives fueled the expansion of the Historic District, encouraging gentrification. Motivated by the pace and extent of the gentrification, political and economic interests worked to revitalize Charleston's traditional commercial district. This revitalization sought to capitalize on the growing tourism market.

In Charleston, gentrification emerged as a consequence of historic preservation. The preservation movement originated as a strategy to control development in downtown Charleston. Over time, the expansion of Charleston's Historic District led to gentrification, increased property values and taxes, and increased racial and economic segregation. Recently, economic and political support for historic preservation have accelerated the gentrification process and contributed to the growth of tourism as a major industry in the city. By examining these changes for the period 1920–1990, I present a case study of ongoing urban redevelopment that illustrates both a strengthening and a disruption of the social environment.

CHARLESTON AS A CASE STUDY

Case studies enrich our understanding of urban processes by focusing their inquiries on a single city to explore the dynamics of a specific problem. Although the findings from a single case may not necessarily generalize to a wider context, the richness of detail provided by a case study can help to illuminate new dimensions of an issue (Orum & Feagin, 1991). The case study method serves as a tool that both "permits the grounding of observations and concepts about social action and social structures in natural settings studied at close hand" and "furnishes the dimensions of time and history to the study of social life, thereby enabling the investigator to examine continuity and change in life-world patterns" (Feagin, Orum & Sjoberg, 1991, pp. 6–7). In this endeavor, I draw from local publications as well as historical statistics from the U.S. censuses of 1920 through 1990. The combination of these sources allows me to present a broader and more detailed description of the stages of redevelopment as they have occurred in Charleston than would be possible using a single data source.

Charleston represents a compelling case study of multiple stages of the urban redevelopment process for several reasons. These include a long history of preservation efforts, a clearly defined geographic area, and a legacy of racial integration. Charleston had experienced a long-lasting economic depression following the Civil War during which very little new construction occurred. As the preservation movement emerged in the 1920s, it actively sought to limit the development of new structures. At the same time, development was limited by the geography of the city itself. Downtown Charleston is situated on a peninsula, bounded on two sides by rivers that join to form the Charleston Harbor. The peninsula is less than two miles across at its widest point, which ensures stable boundaries over time and limits expansion. Figure 1 presents a map of the city of Charleston, illustrating the boundaries of the "Old and Historic District" defined by the zoning ordinance of 1931. The industrial and commercial areas of the city are darkened; the Historic Districts are outlined (Stoney, 1944, p. 136).

Examining the relationship between the development and expansion of the Historic District, gentrification, and the rise of tourism in Charleston provides insight into the long-term effects of preservation on community change. In Charleston, this change is reflected by the changes in the racial distribution that

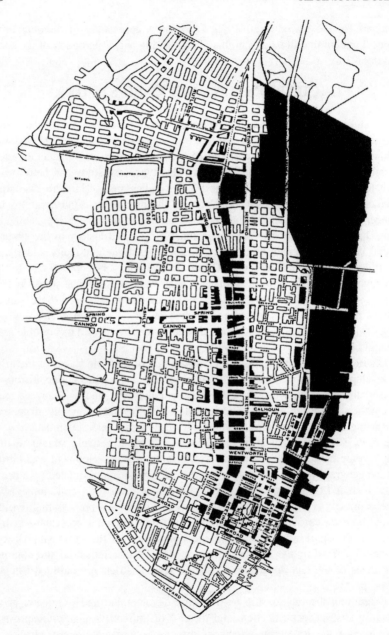

Fig. 1. Map of Charleston, South Carolina (Stoney, 1944, p. 135).

have occurred over time. Early population statistics documented low segregation in peninsular Charleston, but sharp increases in segregation between 1940 and 1960 (Taeuber & Taeuber, 1965). As I describe changing patterns of population distribution in Charleston over the 1920–1990 period, I divide Charleston into two sections, north and south of Calhoun Street, to examine changes in the racial composition of the city. This distinction is useful for two reasons. First, by 1984, the area south of Calhoun was almost entirely within the Old and Historic District. Second, these two areas can be defined consistently over the entire period under study. Before 1960, the area south of Calhoun was represented by 6 census wards. After 1960, the area was redefined as 5 census tracts. A small section of the area below Calhoun was included in a tract that is predominantly above Calhoun. This area is notable in that it was once a Black neighborhood but now houses the Gaillard Auditorium, the city's main public venue, and contains little residential property (though it abuts the Ansonborough neighborhood). Since 1960, the changes in tracts between census years have been relatively minor.

I. Preservation and Community Change

The historic preservation movement in Charleston originated in efforts to preserve the residential character of the downtown peninsula. In 1920, the Society for the Preservation of Old Dwellings was founded to save an historic home (the Manigault House, c. 1803) that was in danger of being razed to make room for a gas station (Stockton, 1985). The Society played a major role in the formulation of the first historic zoning ordinance in the United States and in shaping population movement in peninsular Charleston in the decades to follow. The Planning and Zoning Ordinance, enacted in 1931, designated the city's first Old and Historic District and established the Board of Architectural Review. A 23-block area containing 400 buildings (south of Broad Street from Logan Street to East Bay Street) was designated as Old and Historic (See Fig. 1). This zoning resulted in the rehabilitation of an area known as Rainbow Row, which consisted of pre-Revolutionary homes that had fallen into slum conditions following the Civil War. The primary role of the Board of Architectural review was to regulate demolition, renovation, and new construction within this district. Figure 2 summarizes the Historic Preservation movement and major events in city development.

In the 1920s, the white population increased slightly and the Black population decreased slightly. This period was characterized by much movement within the Black population: Many Blacks migrated to northern cities in search of better jobs while others moved into the city of Charleston from rural areas

I. Preservation and Community Change

1920	Society for the Preservation of Old Dwellings established.
1929	The Cooper River Bridge opens, connecting Charleston, with the eastern suburbs and the beaches.
1931	First historic zoning ordinance in the U.S.
1947	Historic Charleston Foundation founded. First annual Festival of Houses and Gardens.
1958	Ansonborough Rehabilitation Project begins.
1963	The Silas Pearman Bridge opens, a double-span bridge offering greater access to suburbs east of the city.

II. "Classic" Gentrification

1966	Expansion of the Historic District.
1966–1968	Construction of Interstate 26, the cross-town expressway and the Gaillard Auditorium displaces Black homeowners.
1970s	A decade of urban renewal; the city defines its business district from Calhoun Street south to Market Street.
1975	Expansion of the Historic District.
1976	East Side Redevelopment and Planning Program, an early indicator of attempts to expand the Historic District into Black neighborhoods north of Calhoun.
1977	First annual Spoleto Festival USA held.
1984	Expansion of the Historic District.

III. The Rise of Tourism

1986	Charleston Place opens at the corner of King and Beaufain Streets.
1989	July: Historic Zoning proposal defeated by East Side homeowners. September: Hurricane Hugo causes millions of dollars of damage in Charleston.
1991	Visitor Reception and Transportation Center opens north of Calhoun.

Fig. 2. Summary of the Historic Preservation Movement and Major Events in City Development: Charleston, South Carolina, 1920–1991.

(Rosen, 1997). These newcomers would settle north of Calhoun Street in chiefly Black neighborhoods.

Between 1930 and 1940, major preservation efforts began and public housing was introduced. The enactment of the historic zoning ordinance in 1931 served as a catalyst for renovation throughout the newly created Old and Historic District. Segregated public housing followed in 1937 (Carolina Art Association, 1949). While white housing projects were located south of Calhoun Street and near the commercial districts, Black projects were concentrated on the East Side of the city and north of Calhoun. It is likely that the introduction of public housing contributed to population stability in Charleston between 1940 and 1950.

But change was underway: The city expanded north northward, up the peninsula, where white neighborhoods flourished. Meanwhile in the southern tip of the peninsula, the Historic Charleston Foundation, a second preservation organization, formed in 1947. The primary goal of this foundation was to purchase, restore, and resell historic buildings (Stockton, 1985). In 1958, the Foundation began the Ansonborough Rehabilitation Project. Ansonborough is bound by Calhoun Street to the north, King Street to the west, Wentworth on the south, and the Cooper River to the east. It had been a lower-middle class white neighborhood during the 1940s but had fallen into slum conditions by the late 1950s. Some of Charleston's oldest homes can be found in this area, and its restoration triggered interest in restoring the remaining areas below Calhoun Street. In 1966 the Historic District tripled in size to include a large portion of the residential area between Broad and Calhoun Streets (Stockton, 1985). As a result of this rezoning, the entire southern tip of the peninsula was designated as historic.

Preservation efforts intensified as suburbanization of the areas surrounding the city occurred in the 1950s and 1960s. As preservation efforts began to spread down the streets and reach back into alleys, Black residents moved up the peninsula, north of the affluent neighborhoods that occupy the southern tip. Many middle-class white residents moved off the peninsula, and the Black population became increasingly concentrated in an area north of Calhoun Street on the east side of the city.

In 1959, Charleston's public housing was still racially segregated (Taeuber & Taeuber, 1965). Many of the white families that lived in public housing worked either on the docks or in the shipyard. In the 1960s, these workers moved out of the city, and Black residents took their place. By the late 1960s, nearly all of the public housing on the Charleston peninsula would be occupied by Blacks. As the Black population increased in the previously white housing projects, it also increased in the surrounding neighborhoods, and these areas fell into slum conditions.

The first forty years of preservation in Charleston occurred in a constantly changing social environment. Between 1920 and 1930, the Black population dropped slightly in Charleston, as many Black families moved North. Other Black families relocated from the smaller streets and alleys south of Broad Street, and ultimately south of Calhoun Street. Figure 3 illustrates the trend in the racial composition of peninsular Charleston over time. Despite substantial population movement within the city, the racial composition of Charleston remained stable during the period 1930–1950. Between 1950 and 1960, the proportion of whites residing in the city began to drop significantly, a decline that would continue through the late 1960s. This decline is attributable to the

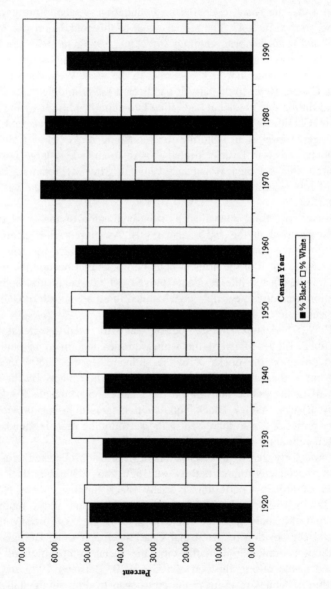

Fig. 3. The Racial Composition of Charleston, SC: 1920–1990.

significant outmigration of middle and lower class whites from downtown Charleston to the suburbs. The movement of white residents into suburban neighborhoods represents an ecological change. Neighborhood succession occurred as Black families moved into the homes vacated by the movement of white families to the suburbs.

II. "Classic" Gentrification

In the 1960s and 1970s preservation continued despite continuing migration to the suburbs. But the tone changed: Charleston began to experience gentrification. Federally subsidized economic incentives (funding and tax breaks) fueled the gentrification process. White families began to move into Black neighborhoods close to Calhoun Street in anticipation of Historic District rezoning. The Old and Historic District was expanded to include Ansonborough and the surrounding area in 1966.

The city had experienced a 25% decrease in the total population and a 42% decrease in the white population between 1960 and 1970. The declines in the white population were extreme north of Calhoun, where the population decreased by 53%. This dramatic decline reflected the continued movement of white middle-class residents to suburbs outside the city, which had become easier to reach as the number of bridges connecting the city to the surrounding areas increased from two to four. These bridges were connected by a crosstown expressway, which also joined to the newly completed Interstate 26. The development of these roadways, as well as the construction of the Gaillard Auditorium, which occupied the northeastern quarter of Ansonborough, displaced numerous Black families. These families moved into the northern portions of the peninsula, into both traditionally Black areas and traditionally white areas that were experiencing significant outmigration by white residents.

The Historic District was rezoned again in 1975, incorporating two neighborhoods just north of Calhoun: Radcliffeborough and Wraggsborough. This expansion was influenced by the expansion of both the Medical University of South Carolina and the College of Charleston. Many students were soon living in what had been considered Black neighborhoods. In 1984, several small areas adjacent to existing historic zones were officially incorporated into the district. This extended the Old and Historic District from Radcliffeborough and Wraggsborough through the entire area south of Calhoun Street.

Public housing in Charleston was now primarily occupied by Blacks. South of Calhoun Street, gentrification and preservation reclaimed the areas surrounding Robert Mills Manor, isolating the Black population within a predominantly white area. North of Calhoun, the largest declines in the Black population occurred in the gentrifying tracts closest to Calhoun Street, while the population of tracts at the northern-most parts of the city remained stable. Calhoun Street had become the dividing line between areas designated as Old and Historic and areas that had yet to be zoned as such.

In the late 1980s, city planners sought to expand the Historic District into a predominantly Black neighborhood north of Calhoun. A substantial number of the residents in this area were elderly and lived on fixed incomes. They feared that changes in the historic designation of the area would result in higher rent, higher taxes, and ultimately in their being forced to move out by financial pressures. Many of these families had lived in the city for years. There was a real sense of neighborhood in these areas, especially on the East Side. This served as a unifying force and, in 1989, homeowner petitions defeated a proposal to expand the Historic District into the East Side. Black residents, understandably, saw the expansion of the Historic District as a threat to their community.

During this period, the suburbs outside the city of Charleston were experiencing tremendous growth. In the city, speculators began to buy and restore homes in transitional neighborhoods. Because federal funding was often tied to historic designations, the Historic District grew rapidly. While the shifting of the population within the greater Charleston area reflected ecological settlement patterns, the changes that occurred within the city are better understood in terms of the economic and political forces that increasingly supported the maintenance and expansion of the Historic District.

III. The Rise of Tourism

Despite the success of East Side residents in preventing the expansion of the Old and Historic District, the city opted to build a Visitor's Center to the west of their neighborhood. The goal of this development was to relieve tourist congestion and traffic flow from areas south of Calhoun Street (Knack, 1994). Congestion became a growing problem as the city entered its third stage of redevelopment, which began in the late 1980s with efforts to revitalize the downtown commercial district. This stage differed from the two previous periods: It sought to capitalize on the growing tourism market. In 1986, Charleston Place, a luxury hotel and shopping mall, opened between King and Meeting Streets at Beaufain Street. The location of Charleston Place at the center of the city and adjacent to the traditional King Street commercial district (see Fig. 1) was intended to provide a boost for additional retail development. It is likely that the Visitor's Center, also located off King Street, was intended to not only relieve tourist congestion but to serve as a northern anchor for the growing commercial district. This stage of redevelopment had strong political support, and much attention was given to preserving the historic character of the revitalized commercial areas. The combination of well-funded restoration and solid planning led to Charleston's emergence in the 1990s as one of the top domestic travel destinations (Conde Nast *Traveler*, 2000).

Charleston has become a successful tourist destination for a number of reasons. It has hosted an annual arts festival, Spoleto Festival USA, since 1977. The preservation of both the architecture and the "culture" of the Old South mean that Charleston offers a unique blend of history and regional culture. Distinctions such as "most polite" and "best mannered" city contribute to this image and reflect the strong service economy that is a necessary element of tourism. In addition, Charleston is situated on the South Carolina coast, a short drive from a number of barrier islands. For those who tire of the historic atmosphere, beaches, golf and tennis resorts are a 30-minute drive away. The completion of I-26 in the late 1960s facilitated travel to the Charleston area, and investors soon began to develop resorts on two of the larger islands to the south, Kiawah and Seabrook. This was followed by the development of the Isle of Palms to the north in the 1980s.

In the 1970s and 1980s, low interest loans and generous tax credits for rehabilitation accelerated the preservation process. Preservation became a business and ". . ., old Charleston became something she never had been – clean, new, painted, and preserved" (Rosen, 1997, p. 163). Charleston's tourism business boomed. The number of tourists visiting Charleston increased from an estimated 2.1 million in 1980 to 4.7 million in 1990 (Rosen, 1997, p. 164). This increase is particularly dramatic in the context of the extensive damage caused in 1989 by Hurricane Hugo. Yet, this natural disaster led to an infusion of millions of dollars into the city. Repairing the damage wrought by Hugo fueled a new wave of restoration in the 1990s, increasing tourism even more.

HISTORIC PRESERVATION AND RACE IN CHARLESTON

A much less romantic side of Charleston's history is its connection with slavery. Charleston's history is deeply embedded within the Southern slavery system. Although early population statistics suggest minimal segregation in peninsular Charleston, the Black population was, in fact, isolated from the white population. Blacks lived behind the large townhomes, on the smaller streets and alleys, just as their enslaved ancestors had done, in close proximity to the white households. Higher population density in these areas, as compared to the adjacent areas inhabited by whites, helped to contribute to the facade of integration.

While Fig. 3 illustrates the citywide trends in the racial composition of Charleston, the relative increase in the Black population is deceiving. When examining Charleston in the context of its historic preservation movement, it is more useful to study the city in two parts: north and south of Calhoun Street. Figure 4 summarizes the population by race and region of the city for the years 1920–1990. During this period, the black population has become increasingly concentrated north of Calhoun, while the white population has maintained a strong presence south of Calhoun. The neighborhoods north of Calhoun experienced a 70% decrease in their white population, which fell from 30% to 17% of the city's population, while the Black population experienced a moderate increase over time (approximately 7%). The area north of Calhoun represents the typical urban transition scenario: white migration to the suburbs, Black succession, and the expansion of neighborhoods surrounding public housing. On the other hand, the neighborhoods south of Calhoun were all included in the Historic District by the mid-1980s. The proportion of the city's population that was white and living south of Calhoun declined temporarily between 1960 and 1970, but had rebounded by the 1980s. Meanwhile, the proportion of the city's Black population residing south of Calhoun has steadily declined. In terms of net population change, the neighborhoods south of Calhoun experienced a 33% decrease in the white population but an 88% decrease in the Black population during the 1920–1990 period. This pattern reflects the extensive gentrification that has occurred in these neighborhoods.

During the period 1920–1990, the population of Charleston underwent a net decrease of 44%. Much of this decrease was neighborhood-specific. After increasing slightly between 1920 and 1940, the white population of neighborhoods north of Calhoun Street decreased steadily between 1950 and 1980.

	1920	1930	1940	1950	1960	1970	1980	1990
Black, North of Calhoun	34.8	34.5	38.4	39.2	50.7	60.9	60.1	53.9
White, North of Calhoun	29.8	33.3	33.8	32.5	27.7	17.3	14.3	16.7
Black, South of Calhoun	14.3	10.5	6.2	5.9	3.0	3.5	2.9	2.5
White, South of Calhoun	21.1	21.7	21.6	22.4	18.6	18.3	22.7	26.9
Total %	100	100	100	100	100	100	100	100
N	65,768	62,239	71,275	70,174	62,165	46,575	40,795	36,540

Source: U.S. Census of Population and Housing 1920–1990.

Fig. 4. Summary of Population by Race and Region of the City: Charleston, South Carolina, 1920–1990.

Blacks moved into the areas left vacant by the declining white population, expanding the already existing Black neighborhoods north of Calhoun Street. Meanwhile, the Black population south of Calhoun Street declined from 14% of the city's population to less than 3%.

These changes reflect the significant impact of gentrification on the social environment in Charleston. As the preservation movement grew and gentrification became profitable, lower-income homeowners, often Black, were displaced, and the historic neighborhoods became increasingly segregated. The rejection of an historic designation by East Side residents in 1989 was the consequence of this ongoing shift in the racial distribution of Charleston, which had resulted from the historic preservation movement and its legacy of gentrification.

CONCLUSION

Historic preservation and gentrification can change communities. While the early preservation efforts in Charleston sought to maintain community, the long-term consequence of preservation has been the maintenance of the physical façade of the city and the displacement of the social one. An indirect consequence of the continued success of the historic preservation movement in Charleston is the growth of tourism and the displacement of native families to other sections of the city. This represents one of the contradictions often inherent in preservation: present-day communities may be destroyed in order to restore the past. Fitch (1990, p. 404) suggests that this need not be the case: Preservation can appeal to a broader proportion of the population and offer them an opportunity to regain a sense of community identity that is often lost in the process of urbanization.

To fully understand the consequences of urban redevelopment, urban scholars must consider the interdependence between the physical and social environments of a city. Clearly, as preservation and gentrification occur, changes in the physical environment will affect the social environment as well. In Charleston, the establishment of the Old and Historic District has played an important role in shaping both the physical and the social environment within the city. The long history of preservation in Charleston makes it possible to identify three stages in the redevelopment process. These stages are not meant to serve as a template for other cities, but to illustrate the complex relationship between historic preservation, gentrification, and tourism.

It is important to remember that urban redevelopment is a social process that evolves over time. Two of the key issues in the urban renewal debate, neighborhood succession and involuntary dislocation, follow from the ecological and political economy perspectives. The changing ecology of communities leads to neighborhood succession, while changes in the political economy of an area may result in the dislocation of residents. The case of Charleston, South Carolina, illustrates multiple dimensions of the impact of restoration efforts on the social environment, particularly the social and the political contexts of historic preservation. While change is an important part of the urban environment, we need to consider more innovative approaches to maintaining community and social environment while preserving the physical environment. As we learn more about the relationship between maintaining communities and restoring communities, urban sociologists and planners should seek to balance preservation and redevelopment with the interests of the community.

REFERENCES

Appleyard, D. (1979). *The Conservation of European Cities*. Cambridge, MA: The MIT Press.
Carolina Art Association (1949). *Charleston Grows: An Economic, Social and Cultural Portrait of an Old Community in the New South*. Charleston, SC: Walker, Evans, and Cogswell Company.
Expanded Historic Zone Blocked By Petitioners (1989, July 6). *The Post and Courier*, pp. 1A+.
Feagin, J. R., Quinn, A. M., & Sjoberg, G. (Eds) (1991). *A Case for the Case Study*. Chapel Hill: The University of North Carolina Press.
Fitch, J. M. (1990). *Historic Preservation*. Charlottesville: University of Virginia Press.
Friedenfels, R. (1992). Gentrification in Large American Cites from 1970 to 1980. *Research in Urban Sociology, 2*, 63–93.
Hoffman, L. M. (2000). Tourism and the Revitalization of Harlem. *Research in Urban Sociology, 5*, 207–223.
Knack, R. E. (1994). Charleston at the Crossroads. *Planning, 60*(9), 21–27.
Orum, A. M., & Feagin, J. R. (1991). A Tale of Two Cases. In: J. R. Feagin, A. M. Quinn & G. Sjoberg (Eds), *A Case for the Case Study* (pp. 121–147). Chapel Hill: The University of North Carolina Press.
Rosen, R. N. (1997). *A Short History of Charleston*. Columbia: University of South Carolina Press.
Smith, N. (1996). *The New Urban Frontier*. New York: Routledge.
Stockton, R. P. (Ed.) (1985). *Information for Guides of Historic Charleston*. Charleston, SC: Tourism Commission.
Stoney, S. G. (1944). *This Is Charleston: A Survey of the Architectural Heritage of A Unique American City Undertaken by the Charleston Civic Services Committee*. Charleston, SC: Carolina Art Association.
Taeuber, K. E. & Taeuber, A. F. (1965). *Negroes in Cities: Residential Segregation and Neighborhood Change*. Chicago: Aldine Publishing Company.
U.S. Bureau of the Census (1922). *Fourteenth Census of the United States: 1920*, Vol. III, Table 13. Washington D.C.: U.S. Government Printing Office.
U.S. Bureau of the Census (1932). *Fifteenth Census of the United States: 1930*, Vol. III, Part 2, Table 23. Washington D.C.: U.S. Government Printing Office.
U.S. Bureau of the Census (1943). *Sixteenth Census of the United States: 1940*, Population, Vol. II, Table 34. Washington D.C.: U.S. Government Printing Office.

U.S. Bureau of the Census (1952). *United States Census of Population: 1950*, Vol. I, Table 8. Washington D.C.: U.S. Government Printing Office.
U.S. Bureau of the Census (1952). *United States Census of Population: 1950*, Vol. II, Part 40, Table 33. Washington D.C.: U.S. Government Printing Office.
U.S. Bureau of the Census (1952). *United States Census of Housing: 1950*, Vol. V, Part 21, Table 3. Washington D.C.: U.S. Government Printing Office.
U.S. Bureau of the Census (1961). *United States Census of Population and Housing: 1960*. Census Tracts. Final Report PHC(1)-23, Table P-1. Washington D.C.: U.S. Government Printing Office.
U.S. Bureau of the Census (1972). *United States Census of Population and Housing: 1970*. Census Tracts. Final Report PHC(1)-39. Washington D.C.: U.S. Government Printing Office.
U.S. Bureau of the Census (1983). *1980 Census of Population and Housing*. Census Tracts, PHC80-2-114, Table 2. Washington D.C.: U.S. Government Printing Office.
U.S. Bureau of the Census (1993). *1990 Census of Population and Housing*. CPH-3-107, Table 8. Washington D.C.: U.S. Government Printing Office.
Zukin, S. (1987). Gentrification: Culture and Capital in the Urban Core. *Annual Review of Sociology*, *13*, 129–147.

GENTRIFICATION, HOUSING POLICY, AND THE NEW CONTEXT OF URBAN REDEVELOPMENT

Elvin K. Wyly and Daniel J. Hammel

1. INTRODUCTION

Recent years have brought a dramatic transformation of the North American inner city. The decade of the 1990s certainly began on a dystopian note: the recession punctuated two decades of deindustrialization and working-class job losses, while widespread drug use and spiraling crime rates inspired vivid metaphors of "adrenaline cities," "shooting gallery cities," and "kill zones" in urban areas written off as "beyond the point of no return" (Waste, 1998; Rusk, 1994). Heightened racial tension focused attention on the lack of progress in reducing segregation since the late 1960s, lending an eerie, contemporary flavor to the warnings of the Kerner Commission – the authoritative image of urban inequality that derived "its most devastating validity from the fact that it was drawn by representatives of the moderate and 'responsible' establishment." (Wicker, 1968, p. v; cf. Boger & Wegner, 1996). But by the end of the 1990s, metaphors of catastrophe and crisis had been pushed aside. The nation's longest economic expansion finally brought a wave of reinvestment that rivaled the brief interludes of urban "renaissance" that were occasionally glimpsed during the long downward slide of American cities in the 1970s and 1980s. And despite deep cuts in social programs and devolution of federal responsibilities to states and local governments, several national policy initiatives endured. Efforts to expand homeownership enjoyed broad, bipartisan support, while structural changes in the financial services industry and the institutional framework of the housing finance system helped to funnel capital to urban neighborhoods. Even the perpetual neglect of the nation's public housing stock had finally inspired a bold, innovative response under the banner of a suite of programs dubbed "Homeownership and Opportunity for People Everywhere" (HOPE). The sustained expansion of the 1990s held out the promise that twenty-five years of worsening inequality in American cities might finally be reversed.

Such hopes proved unrealistic. The new urban revival was born of widening inequalities of income, wealth, and opportunity, and the most recent round of urban reinvestment magnified the class polarization that once inspired daring pronouncements of a metropolis turned inside out:

> If present trends accelerate, the social geography of the nineteenth century city may appear to urban scholars as a temporary interlude to a more historically persistent pattern of higher status segregation adjacent to the downtown core (Ley, 1981, p. 145).

Such predictions added fuel to a vociferous debate over the magnitude, extent, and significance of gentrification for the future of the inner city (Ley, 1980, 1981; Berry, 1980, 1985; Smith & Williams, 1986). More fundamentally, the debates of a generation ago foreshadowed the imprint of public policy and private profit on the landscapes of poverty and wealth inscribed at the heart of the city. Today, sifting through the voluminous policy literatures on urban homeownership, community development, and the reinvention of public housing leaves the reader with a sobering reminder of the historical continuity of class polarization in the North American city:

> ... in its interaction with private interests, particularly in the land market, the reform movement was perhaps too naive, not recognizing that its humane philosophy might be coopted by the calculus of the marketplace and lead to an inequitable outcome where the vulnerabilities of the poor would be exposed. (Ley, 1980, p. 258).

Two decades on, urban redevelopment is a story of ambitious reform, entrepreneurial reinvention, and an ideology of disciplined privatization meant to encourage the "personal responsibility" of the urban poor alongside heated competition among the newly-wealthy professional classes searching for the very best urban residential experience. Private interests in the land market have never been so sophisticated in exposing the vulnerabilities of the poor under the banner of work, opportunity, and empowerment.

In this chapter, we offer three hypotheses on the changing nature of inner-city redevelopment in the U.S. First, we suggest that the economic expansion of the last decade has revived inner-city housing markets – and that the force, depth, and focus of resurgent capital investment have been sufficient to invalidate the predictions of "degentrification" voiced in the early 1990s. Second, we suggest that the neighborhood mosaic of wealth and poverty created by three decades of gentrification has coalesced into a distinctive urban social context for the reinvention of low-income housing policy; this shift is apparent in many local policies and public-private redevelopment efforts, but is most starkly illustrated by local endeavors under the federal HOPE VI program. Third, we hypothesize that restructuring of the national system of housing finance has altered key facets of the gentrification process itself.

Our analysis proceeds in six sections. We begin with a brief review of the recent history of gentrification research to establish the context for our argument. We then outline our methods of mapping and measuring inner-city neighborhood change. In the subsequent sections we draw on a hybrid of quantitative and qualitative evidence to evaluate each of our hypotheses in turn: that gentrification has rebounded, that it has collided with assisted housing policy, and that it has itself been altered by developments in housing finance. In the final section we venture concluding remarks on the implications of our findings for continuity and change at the metropolitan core.

2. THE DISAPPEARANCE AND RETURN OF GENTRIFICATION RESEARCH

Research on gentrification expanded dramatically during the 1970s and 1980s, yielding a diverse, interdisciplinary literature with its own debates over methodology, theoretical significance, and epistemology. Even the definition of the process itself eluded consensus, and inspired considerable disagreement over the magnitude and relevance of a turnaround in the inner city. Our research is premised on a conceptualization of the process that is cognizant of these long-standing debates, and that is at once social and spatial. Gentrification is the class transformation of urban neighborhoods that were devalorized by previous rounds of disinvestment and outmigration amidst metropolitan growth and suburbanization – a problem that grew especially severe in the twenty-five years after the Second World War. Class transformation is rooted in long-run changes in the social and technical division of labor, as well as the distribution of income, wealth, and educational opportunities; but it is the geographical intersection of these demand-side changes with the supply-side forces of urban land markets that imbues gentrification with theoretical significance. The process is nothing less and nothing more than the reconstruction of urban space to serve those of a "higher" social class than those currently using a particular part of the built environment (Gottdiener, 1985; Hutchison, 1992; Smith, 1979a; Zukin, 1982). This conceptualization emphasizes the necessary relations among place, class, and uneven development – and not the contingencies of specific types of land uses or population flows. An influx of middle- and upper-class professionals into an urban neighborhood, for example, usually involves the displacement of working-class renters who face rising costs and harassment by landlords, speculators, and developers eager to capitalize on increased market potential. Yet displacement is often indirect or gradual, and some of the most spectacular inner-city luxury redevelopment schemes involve upscale retail construction and "festival marketplaces," or the adaptive reuse of industrial, warehouse, or vacant waterfront parcels. In these circumstances, the reconstruction of urban space to serve middle-class interests is manifest in public subsidies with opportunity costs for affordable rental housing, aggressive "quality of life" policing and the criminalization of homelessness, and strategic privatization of public spaces to maintain the exclusivity required to sustain and protect housing investments.

Causes and Controversy

If definitional concerns have been subject to dispute, underlying causes have inspired even greater disagreement. In the first decade after Ruth Glass (1964) coined the term to describe changes in parts of London, it was imported to the United States and pressed into service as a descriptive, uncritical celebration of a curious trend that seemed to be limited to a few cities and neighborhoods. And yet the class connotations could hardly be more explici – making it all the more remarkable that the word entered the popular and policy lexicon "without the least hint of squeamishness" (Smith & Williams, 1986, p. 1). These class connotations, along with the rapid growth of the process itself, gave rise to a burgeoning literature in urban sociology, geography, and allied fields in the 1970s and 1980s (e.g. Castells & Murphy, 1982; Holcomb & Beauregard, 1981; Laska & Spain, 1980; Ley, 1980, 1981; Marcuse, 1986; Palen & London, 1984; Zukin, 1982; for a comprehensive bibliography, see Smith & Herod, 1991). This literature ranged widely through alternative theoretical traditions and methodological frameworks, and thus defies any simple categorization. For our purposes, however, one central thread running through most of the research in the 1980s helped to shape the point of departure for a new generation of investigators. As early as 1980, explanations of gentrification could be divided along a demand/supply axis, with accounts usually focusing *either* on the changing consumer preference of middle-class households *or* the strategies of speculators, developers, and state institutions. Demand-side theories ranged from economic attempts to rework bid-rent models in light of changes in household structure or fuel costs, to demographic analyses of household formation in the baby boom cohort, to sophisticated narratives of new occupational structures and class formations with distinctive consumption aesthetics and political sensibilities favoring diverse urban locales (Berry, 1985; Ley, 1980, 1981; Long, 1980; Zukin, 1982). Supply-side accounts, by contrast emphasized the explicit political-economic strategies used in the days of urban renewal to subsidize luxury redevelopment, and the fundamental processes of the circulation of capital in the built environment in creating opportunities for profitable redevelopment (Harvey, 1973, 1985; Smith, 1979a, b; Smith & Williams, 1986; Logan & Molotch, 1987).

Much of the attention devoted to gentrification, therefore, was spent on the conceptual battleground provided by an unexpected empirical phenomenon viewed through the lens of supply- and demand-side assumptions. The prospects for a general urban revival in the 1970s, for instance, immediately rekindled urban ecology debates inherited from the Chicago School. Gentrification could be seen as a repudiation of conventional theory, or it could be explained as the necessarily limited backwash from the overwhelming preference for decentralization – which necessitated the expansion of a fully-articulated nucleus to coordinate an increasingly complex, dispersed metropolis (Berry & Kasarda, 1977; Hawley, 1981; Hoover & Vernon, 1959; cf. Gottdiener, 1985). The functionalist and organicist tendencies of the ecological literature, of course, were challenged in Smith's (1982, 1996) work on uneven development and by Logan and Molotch's (1987) rendering of the multifaceted coalitions of public and private institutions in urban growth machines, which exposed downtown redevelopment as anything but natural (see also Jonas & Wilson, 1999). From other quarters, social production of space arguments drew on the work of Henri Lefebvre (1991) and, to a lesser extent, structuration theory (Giddens, 1984) to explicate the relations among social practices, institutional ideology, spatial form, and representations of place (Gottdiener, 1985; Soja, 1980, 1996). Gentrification can be understood as one particular facet of a new spatiality of material practices and representations that create "designer neighborhoods" as residential elements of the society of the spectacle, in which commodified, alienated consumption is born out of the postfordist erosion of alienated industrial production (Dear & Flusty, 1998; Debord, 1994; Knox, 1991, 1993; Pinder, 2000; Soja, 1980, 1996).

Several prominent theorists sought to reconcile conflicts among these alternative explanations (Hamnett, 1984; Smith & Williams, 1986; Clark, 1992). Yet gentrification was used as a lens through which to view some of the richest (and most hotly disputed) dichotomies of urban research – production/consumption, economy/culture, structure/agency – and thus eluded any broad consensus (e.g. Bondi, 1991; Clark, 1992; Hamnett, 1984; Rose, 1984). By the end of the 1980s these debates seemed to exhaust themselves with no prospect of a final "resolution," and on empirical grounds gentrification appeared ready to collapse under the weight of every demand- and supply-side factor cited to explain its emergence. Perhaps the hard facts of "degentrification" (Bourne, 1993a, b) on the urban landscape would be mirrored in a corresponding theoretical retreat.

We contend that debate over the causes of gentrification does not exhaust the array of important and relevant questions. Regardless of its underlying causes, gentrification has become a durable feature of the urban ecology, uneven development, and social production of cities – with all of the heavy theoretical baggage that arrives with each of these labels – and it has itself assumed an important role in how urban fortunes are understood and planned. Whether a product of capital or culture, production or consumption, urban redevelopment has been woven into wider public, scholarly, and policy portrayals of urban futures in ways that are simultaneously new and historically precedented. It is now tied integrally into policy discussions over fair lending, racial discrimination in urban housing markets, the privatization of community development, the potential alliance between city residents and suburban advocates of neotraditional design and smart growth, the drive to disperse the concentrated urban poverty created by metropolitan restructuring and public policy, and the devolution of certain federal housing programs to states and cities. How did we arrive in the thick of the gentrification debates again?

The Renaissance of Theory in the 1990s

By the middle of the 1990s it was possible to discern a revival of theoretical and empirical research on gentrification, and it is from this literature that we distill our central hypotheses. Three avenues of inquiry proved most fruitful. First, prominent theorists engaged the "degentrification" debate in order to strengthen the connections between visible symptoms of neighborhood change and underlying processes of suburbanization and uneven development (Lees & Bondi, 1995; Smith, 1996; Smith & DeFilippis, 1999). The degentrification hypothesis was ultimately premised on a demand-side view of the process, and ignored the critical function of recession and disinvestment in creating the conditions for new rounds of investment and class turnover. We strongly suspect that gentrification has recovered from the recession of the early 1990s, and we marshal recent anecdotal evidence as well as a more rigorous, field-verified database to verify this claim. Second, analysts turned once again to the consequences of what is now recognized as a durable, dynamic, and theoretically vital phenomenon. The process is here to stay, and it provides a lens through which to scrutinize broader trends affecting many kinds of urban and suburban environments. Van Weesep's (1994) astute review proposed a renewed emphasis on "the effects of gentrification rather than its causes," as a way to "put the gentrification debate into a policy perspective." We contend that a critical analysis of public policy is essential to understand the contemporary relevance of gentrification: three decades of reinvestment into favored quarters of the city have altered the social composition and spatial geometry of many neighborhoods – in some cases driving capital infusion and wealth to the boundaries of public housing landscapes created at the zenith of liberal urban policy during the 1960s. When viewed at the scale of the urban core, Berry's (1985) famous metaphor of "islands of renewal in seas of decay" has been replaced by islands of decay in seas of renewal.

Three "Waves" of Gentrification

A third stream of research represents an attempt to reconstruct the relations between gentrification and broader changes in the urban political economy (Hackworth, 2000a, b; Hackworth & Smith, forthcoming; Smith, 1996, 1997; Smith & DeFilippis, 1999). Our work follows in this line of inquiry, and thus it merits careful consideration. Several decades of turbulent change in disinvested urban neighborhoods now invite, for the first time, a full-fledged historiography of the evolving links among class turnover, real estate cycles, public policy, and community activism. Jason Hackworth (2000a, b) and Hackworth and Smith (forthcoming) delineates three waves of gentrification in cities of the U.S., Western Europe, and Australia – each wave tied to "a particular constellation of political and economic conditions nested at larger geographic scales" (Hackworth & Smith, forthcoming, p. 5). First-wave gentrification emerged in the 1950s and lasted until the economic crisis of 1973; this phase was marked by widespread but sporadic reinvestment, and relied heavily on public subsidy and urban renewal schemes explicitly designed to counter the private market decline wrought by rapid suburbanization. The subsequent recession of the 1970s eviscerated many urban programs and city budgets, but its effect on gentrification was ambiguous, especially in light of Harvey's (1973, 1985) theory linking crises of overaccumulation to a switching of capital out of industrial production into the speculative realm of real estate investment and downtown construction. A second wave began in the late 1970s, and gathered momentum through the 1980s. Intensified reinvestment in these years involved aggressive entrepreneurial spirit and speculation by developers and owner-occupiers, while local-state efforts shifted away from direct orchestration of redevelopment to public-private partnerships and *laissez-faire* subsidies. But the signal feature of second-wave redevelopment was its "integration into a wider range of economic and cultural processes at the global and national scales." (Hackworth & Smith, forthcoming, p. 7). This integration was evidenced in a nascent internationalized property development industry, the rise of global cities and an expanded transnational professional elite, and the diffusion of a sophisticated planning ideology constructed around the virtues of public-private cooperation committed to giving each city its own version of Boston's Faneuil Hall, Baltimore's Inner Harbor, Philadelphia's Society Hill, or New York City's Battery Park City.

Ultimately, the second wave crashed on the rocks of a short but sharp recession in the early 1990s, which deflated urban property markets and gave rise to pronouncements of a "post-gentrification" era after the historical aberration of the baby boom and sudden bouts of restructuring (Bourne, 1993a, 1993b). But gentrification recovered quickly and flowered in scores of cities after the early 1990s, and several analysts argue that this third wave departs considerably from earlier rounds (Hackworth, 2000a, 2000b; Hackworth & Smith, forthcoming; Smith & DeFilippis, 1999; Wyly & Hammel, 1999, forthcoming). Reinvestment expanded beyond the confines of neighborhoods transformed in the first and second rounds; large developers gained a broader role in the process after the shakeout of the early 1990s took out their competition; and neighborhood resistance subsided as the gutting and privatization of city services forced once-militant community organizations to assume the role of housing and social service providers. Moreover, the state undertook a newly-interventionist stance towards gentrification, drawing selective precedent from the history of urban renewal while advancing new goals, methods, and alliances. The endeavor came just in time. By the mid-1990s, private market reinvestment in many cities had exhausted the supply of easy targets, bringing developers and speculators to the edge of mixed-use land parcels, remote locations, and public housing projects. State assistance was essential to remove these barriers, and public initiatives were launched amidst a broader emphasis on the devolution of social reproduction burdens (welfare provision, housing assistance) to localities, while economic autonomy was elevated to ever-higher scales. The results have been dramatic, as the federal agency once seen as a barrier to gentrification (the Department of Housing and Urban Development) has been "reinvented" in a way that emphasizes the privatization of public functions, the spatial integration of assisted and market-rate housing units, and the virtues of market processes adapted to socially desirable public goals (Lane, 1995; Gotham, 1998; Varady et al., 1998). It is crucial to recognize at once the Janus-faced character of this new local-federal axis in housing policy. It represents a comprehensive and sophisticated attempt to atone for the failures of previous generations of public intervention, which left a legacy of concentrated poverty and desolate public housing reservations (Epp, 1998; Goetz, 2000; Gotham & Wright, 2000; Nenno, 1998; Varady, 1994; Wilson, 1987, 1996). Yet the reinvention of assisted housing is based on a narrow foundation of social science research that reveals more of the problems than of alternative solutions, and the initiative proceeds in the context of reduced housing assistance in an overheated rental market that HUD itself believes is in crisis (U.S. Department of Housing and Urban Development, 2000). As such, the new local-federal state entrepreneurialism exposes a broad swath of the inner city to gentrification pressures in new and troubling ways.

Our work falls squarely within this line of inquiry, but attempts to extend and refine it in two ways. First, as noted earlier, we contend that gentrification has been installed as a prerequisite for certain facets of public policy, in that local private market conditions are used to justify and condition the reinvention of assisted housing. Second, we hypothesize that the political economy-gentrification nexus elaborated by Hackworth and Smith (forthcoming) is also manifest in the obscure and Byzantine yet powerful institutions of housing finance – which have undergone their own reinvention in the past decade. We hypothesize that several distinct economic and political trends coalesced in the 1990s – community activism around the issues of redlining and disinvestment, Federal antidiscrimination policy, and financial services restructuring – and altered the relationship between housing finance and neighborhood change. Private mortgage capital, once notorious for its role in disinvestment and the creation of rent gaps in the inner city, has been drafted to lead a secular reinvestment into low- and moderate-income communities now cast as "untapped" or "underserved" markets (Listokin & Wyly, 2000; Wyly & Hammel, 1999, forthcoming). This national shift has brought unprecedented opportunities to racial and ethnic minorities and low- and moderate-income families once excluded from the mainstream mortgage market. But it has also unleashed powerful gentrification forces, and has reshaped the links between local, "private" housing market activity and national, publicly-sanctioned capital investment networks.

3. MAPPING THE NEW URBAN FRONTIER

How much gentrification is there? How much of the urban fabric does it affect? Where is it? Sometimes the simplest questions prove the most contentious. After thirty years of sustained research, a multiplicity of conceptual definitions, methodological traditions, and data sources permit no clear consensus on the magnitude, extent, and precise location of gentrification. Much of the early literature (and a significant fraction of current work) took the form of case studies of one or two specific neighborhoods in a particular city (Fusch, 1980; Laska & Spain, 1980). While yielding a richly-detailed portrait of the localized effects of the process, such an approach provided little information on the broader extent of redevelopment. The few attempts at extensive, comparative measurement encountered difficulties in operationalizing theoretical concepts or distinguishing the phenomenon from other types of neighborhood change (Bourne, 1993a, b; Clay, 1979; Ley, 1986; Lipton, 1977).

Our research began as an attempt to reconcile tensions between the *generalities* of uneven development in the reconstruction of urban space and the specific expression of social and economic change on the ground in particular neighborhoods (Hammel & Wyly, 1996; Wyly & Hammel, 1998). Our goal, quite literally, has been to "ground-truth" the standard statistical series of the census, and to link this information with other data sources. The long-term project involves a database for the thirty most populous U.S. cities, but to date we have assembled complete datasets for eight: Boston, Chicago, Detroit, Milwaukee, Minneapolis-St. Paul, Philadelphia, Seattle, and Washington D.C. While not necessarily representative of American cities as a whole, this sample provides insight on a broad spectrum of contextual variations in the process – from Detroit's tiny trickle of redevelopment in the face of deindustrialization, to old industrial metropoli now thriving with service-sector growth (Chicago, Boston), to advanced postindustrial economies with little history of manufacturing employment (Washington D.C.).

We employ a three-step process to identify gentrified neighborhoods in each city. First, we review scholarly, policy, and popular press sources to identify neighborhoods recognized as existing or potential loci of class turnover (Fig. 1). Second, we conduct a block-by-block field survey. Our methodology relies on visible evidence of reinvestment in the housing stock, an assumption that is certainly not free of theoretical complications, but nevertheless captures certain essential relations between class turnover and the revalorization of the built environment. At this stage we distinguish between heavily-redeveloped "core" neighborhoods (where substantially upgraded or new upscale construction comprise at least one-third of all housing units) and more marginal "fringe" districts (with some redevelopment on each block, and at least one block where renovated or new structures account for more than one-third of all units) (see Hammel & Wyly, 1996; Wyly & Hammel, 1998). Finally, we use stepwise discriminant analysis to identify measures from the decennial census[1] that best distinguish core and fringe neighborhoods from the rest of the central city. Socioeconomic change in established middle-class areas is theoretically distinct from gentrification, so we imposed a criterion that a tract must have

Gentrification, Housing Policy and Urban Redevelopment

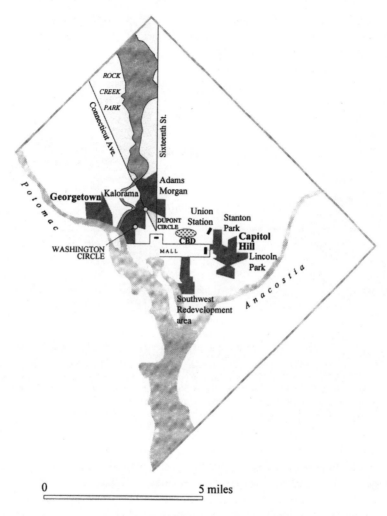

Fig. 1. Gentrified Neighborhoods in Washington D.C.

endured some prior disinvestment in order to qualify for core or fringe designation; this requirement is operationalized with a dichotomy between "inner city" and "rest of city," based on 1960 median household income for each city (see Fig. 2). Over time, as our database of field-verified census tract indicators has expanded, we have been able to use the discriminant models to identify potential sites of gentrification before conducting field surveys;

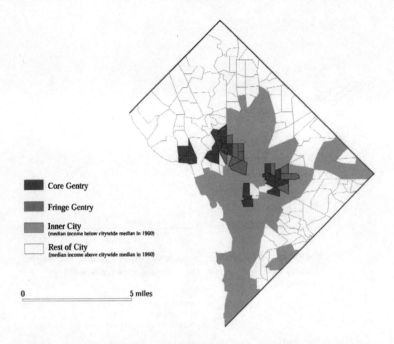

Fig. 2. Neighborhood Classification for Washington D.C.

A combination of qualitative and quantitative methods and field surveys were used to classify census tracts in each case study city. Core gentrified neighborhoods show substantial and widespread evidence of housing investment correlated with class turnover, while dispersed or emerging areas of gentrification are classified as "fringe."
Source: Authors' field surveys and analyses of census data (Wyly & Hammel, 1998).

this technique provides a valuable means of triangulating among qualitative, quantitative, and field-observation techniques (Table 1).[2]

Our field reconnaissance and quantitative analysis yields important findings on the extent of inner-city redevelopment. First, it is clear that suburban growth dwarfs gentrification, and that significant class turnover affects only a small portion of the urban built environment. In the most extreme scenario, Washington D.C.'s intensely-redeveloped enclaves (Fig. 3) account for just over 17% of the total housing stock of the central city – which itself encompasses a steadily shrinking fraction of the metropolitan housing market.[3] Paltry figures like this lead many scholars to readily acknowledge the quantitative irrelevance of the process, and to concentrate instead on the unique theoretical insights gained from its study. Such reasoning is indeed persuasive, and we generally

Table 1. Gentrification Field Surveys, 1994–1998.

City	Initial field survey*	Number of census tracts added to database	
		Core	Fringe
Boston	March, 1998	14	11
Chicago	August, 1995	37	32
Detroit	July, 1998	3	4
Milwaukee	August, 1995	6	7
Minneapolis-St. Paul	January, 1994	7	19
Philadelphia	July, 1998	13	7
Seattle	July, 1998	9	12
Washington, DC	July, 1995	17	11
Total		106	103

Notes:
* Subsequent visits and informal field inspections were conducted in Chicago, Philadelphia, and Washington, D.C. to document changes in neighborhood conditions in 1998 and 1999.
Source: Authors' field surveys.

concur with the broad outlines and implications of the argument. Nevertheless, one cannot dismiss the social-geographic realities of gentrification or the well-established theoretical traditions developed to understand it – unless, that is, we are prepared to sweep aside other, similarly partial accounts of the urban condition. Taken as a group, the core and fringe gentrified neighborhoods of our eight case study cities encompass a total of 570,408 residents, equivalent to 21.6% of the total "inner city" population as defined in our classification. Consider that Ricketts and Sawhill's (1988) operational definition of so-called "underclass" neighborhoods for these same cities is 526,566, while Kasarda's (1993) more restrictive definition of underclass *population* comes in at 419,780. Just as the voluminous body of research on the ills of concentrated urban poverty is justified given the high stakes of structural versus culture-of-poverty explanations (Goetz, 2000; Wilson, 1987, 1996), gentrification demands careful attention for what it tells us about social polarization at the urban core. And it also requires close scrutiny of the relations between "private" market processes and public policies that reflect and reinforce the restless remaking of favored quarters of the inner city. In the rest of this chapter, we present an empirical analysis of these processes – the return of gentrification, its effects on certain types of policies, and its vulnerability to other kinds of policy interventions – in an attempt to discern what is new and what is not.

Fig. 3. Florida Avenue and R Street, Northwest Washington D.C.

Gentrification activity in the District has a long history, beginning with the influx of a large cohort of professionals during the expansion of federal government employment in the New Deal. Postwar reinvestment focused on four distinct clusters: Georgetown, Capitol Hill, the Southwest Redevelopment Area, and (shown here) a swath of neighborhoods in the city's Northwest quadrant between Dupont Circle, Adams Morgan, and the wealthy enclave of Kalorama. Taken as a group, these neighborhoods fell well below citywide income levels in 1960, and lost population until the 1980s. By 1990 these neighborhoods posted income levels substantially above the citywide median, and two-thirds of all residents were college graduates. In the expansion of the 1990s, private mortgage capital investment in gentrified areas grew at three times the pace of Washington's suburbs. *Photograph by Elvin Wyly.*

4. THE RESURGENCE OF GENTRIFICATION

The early 1990s were a bleak time for urban America, and the returning specter of the forgotten urban crisis of the 1970s seemed to have suddenly erased popular fascination with urban pioneers, trendy warehouse districts, and an ultimately illusory "back to the city" movement. Smith (1996, p. 210) put it most memorably: "After the stretch-limo optimism of the 1980s was rear-ended" by recession and collapse, "real estate agents and urban commentators quickly began deploying the language of '*de*gentrification' to represent the apparent reversal of urban change in the 1990s." Barely a decade later, jaded critics and effusive boosters alike are searching for ever more evocative superlatives to describe the long economic boom and its effects on local property markets. The evidence is anecdotal but striking. In Boston, the influx of high-tech and high-finance professionals into the city's old-money enclaves is driving what Alan Wolfe calls "turbo-gentrification" (quoted in Goldberg, 1999). In the San Francisco Bay area, dot-commodified housing inflation has finally brought growth pressures to Oakland, where community activists lament Mayor Brown's overtures to developers as a sign of "Jerryfication" (Neives, 2000, p. A16). In Chicago, market-rate housing is under construction steps away from the notorious Cabrini-Green public housing project, now almost completely surrounded by gentrification and described by a local broker as "some of the best land in the entire city" (Sharoff, 2000, p. NJ RE 7). One of the first buyers told the *New York Times*, "We're excited to be part of the experiment" of a mixed-income community created by an innovative public-private partnership; yet he also clearly understands the real estate equivalent of a lucrative stock option: "It's an area with a lot of potential. We didn't jump into this for altruistic reasons. It was a financial decision. We wanted the best house we could find for the money" (Sharoff, 2000, p. NJ RE 7).

Perhaps it comes as no surprise that New York generates the strongest torrent of real estate lore – by turns horrific or boastful, depending on who is buying, selling, renting, or subleasing – and at what price. Signs of class turnover are remarkable throughout a city that has always been considered paradigmatic of restructuring and inequality (Mollenkopf & Castells, 1991; Sassen, 1991, 1994; Smith, 1997). Manhattan brokers compete in a "helium-filled" market (Hevesi, 1998) infused with a speculative frenzy, where recent revisions to the rent control laws fuel co-op to condominium conversions while thousands of rentals cross the magical "luxury decontrol" threshold of $2,000 per month. Just as Soho and the East Village were touted as the cutting edge of urban chic for a previous generation priced out of Greenwich Village, "fashion's new frontier" has shifted to Loho – a stretch of the Lower East Side between East Houston and Canal Streets, where a loft recently sold for $1.1 million and newly-renovated 450-square foot studios are renting for $3,000 per month (Hamilton, 2000, p. F1). Press accounts are suffused with a frontier mythology that echoes the 1970s (Smith, 1996; Cole, 1987). The "trendy frontier" of artist pioneers pushes "Eastward Ho!" from Soho to the distant quarters of Bushwick, Brooklyn (Barnes, 2000).

As house prices surge well beyond $1 million in the established outposts of gentrification in Manhattan and Brooklyn, the city's last frontiers for affordable historic brownstones have reached into Bed Stuy and Harlem. Bedford-Stuyvesant, once an icon of crime and decay, is now "a gold mine, just waiting to be discovered by newcomers who have no idea of its charms, of its thousands of historic homes just 15 minutes on the A train from downtown Manhattan" (Hall, 2000, p. NJ RE 1). Harlem is changing too. In the 1980s, Schaffer and Smith wrote an article titled with a question ("The Gentrification of Harlem?"), which they answered with a compelling and provocative analysis of the early stages of reinvestment as white middle-class homebuyers and private capital began "to perceive Harlem as a viable and even lucrative investment especially when backed by public funds" (Schaffer & Smith, 1986, p. 362). Today the rhetoric and reality of gentrification is clear to everyone in the neighborhood. A Second Harlem Renaissance has been proclaimed, with a push for a Sugar Hill Historic District, surging brownstone prices, and construction of Harlem USA, a $65 million entertainment and retail complex with Disney and other high-profile tenants lured by city, state, and Federal Empowerment Zone subsidies (Pataki, 1998). And "if Starbucks, the Gap, and the Disney store open up in a neighborhood, can residential real-estate offices be far behind?" (Brozan, 2000, p. B9). The *Times* now issues regular dispatches from "The Battle of the Brokers" in Harlem (Rozhon, 2000, p. F1).[4]

And finally, there are scores of horror stories for small businesses and industrial concerns. When the *Times* featured a ten-year retrospective on the 1988 gentrification protests and police riot in Tompkins Square Park, they sought out commentator and talk-show host Ed Koch, mayor at the time of the uprising. Koch joyfully declared that "gentrification has definitely returned to the East Village. And like Martha Stewart says, 'It's a good thing'" (Jacobs, 1998). Koch's playful soundbite was soon reflected in the realities of Manhattan's competitive market. In early 2000, Martha Stewart Living Omnimedia joined several other Silicon Alley tenants signing up to pay $20 to $30 per square foot for bare-bones space in a giant 1930s building in the remote corners of Chelsea near the Lincoln Tunnel; a dozen of the building's industrial and arts-related tenants with current leases for $6 per square foot have filed suit, charging the new owners with harassment in their attempts to get new tenants willing to pay for what brokers now call "shabby chic" (Bagli, 2000; Dunlap, 2000).

No matter how vivid, all of these images can be dismissed as nothing more than anecdotal evidence of a localized, highly visible, and ultimately quixotic process of neighborhood change (Bourne, 1993a, b). A more sober assessment requires that we use an indicator that is rigorous, conservative, and quantifiable. For this purpose we turn to a dataset assembled as part of the requirements of the Home Mortgage Disclosure Act (HMDA) (Federal Financial Institutions Examination Council, 2000). HMDA records come out of a rich history of community activism and Federal legislation designed to eliminate lending industry discrimination and redlining. Through the late 1970s and 1980s, HMDA records were not widely available, and only provided aggregate information on the number, value, and geographic distribution of mortgage loans made by banks and savings institutions. But in the wake of a high-profile exposé in the *Atlanta Journal-Constitution* (Dedman's [1988] "Color of Money" series), a provision for expanded disclosure was inserted into the 1989 savings and loan bailout legislation. Since 1990, institutions have been required to report individual records for all loan applications received, including information on the applicant (sex, race, income), the loan (amount requested, conventional or FHA-insured, single or multifamily dwelling,[5] whether the house will be occupied by the borrower), and the decision rendered on the application (approved and granted, declined by the borrower, denied, withdrawn, or closed as incomplete). Due to changes in the criteria under which institutions are required to report, we can safely assume comparability for the annual releases from 1993 onward.

These data are by no means ideal: they only capture a specific segment of housing demand and residential investment, and they include terribly little information on the household circumstances or class character of prospective homebuyers.[6] Nevertheless, this segment of the housing market provides an extremely rigorous test of the reinvestment hypothesis. In light of the deeply entrenched social and cultural traditions of homeownership and the intricate web of public subsidies designed to support it, households face considerable pressures to maximize the value of a house as an investment vehicle – and thus to seek out neighborhoods with opportunities for healthy price appreciation. Banks and investors, acting on the basis of economic theories that portray delinquency or default as rational when the value of collateral falls, are likewise concerned about future trajectories of neighborhood change. If gentrification is indeed a fundamental urban process and not simply a trivial imprint of the postwar baby boom, we should expect a renewed willingness of middle-class buyers and lenders to accept the long-term investment risks of the inner city.

To test this hypothesis, we focus here on a tabulation of mortgage commitments by the most conservative buyers and investors in the market: those negotiating over conventional, market-rate loans reserved for applicants with prime ('A'-rated) credit histories.

The results of this analysis are striking (see Table 2). Although suburban areas account for the vast majority of investment flows, the boom of the 1990s brought a pronounced recentralization of capital towards the urban core. Taken as a group, the suburbs of our eight case study metropolitan areas enjoyed a growth rate of 52% in conventional home purchase mortgage commitments between 1993 and 1999. By decade's end, prime, conventional mortgage loans in the suburbs topped $53 billion annually. Investment in the central cities, however, surged by 82%, and in turn this growth displayed a clear bias towards the center. The disinvested ring of neighborhoods that stood below the city-wide median in 1960 (labeled here as the "inner city") attracted capital growth more than double the rate of suburban expansion, and posted the highest rate of increase in median loan amounts (a growth of 38%, to $124,000 in 1999). But it is in the gentrified districts of the inner city where capital markets, regulatory changes, and household demand have funneled the most dramatic expansion of investment. These neighborhoods account for a tiny fraction of the metropolitan housing stock, but they now compete with the upper echelons

Table 2. Aggregate Mortgage Reinvestment in Selected Metropolitan Areas, 1993–1999.

	Total value of conventional home purchase loans originated		Median loan amount	
	1993	1999	1993	1999
Core gentrified	456,193	1,161,508	144	160
Fringe gentrified	229,554	697,701	121	158
Inner city*	592,756	1,281,425	90	124
Rest of central city	3,294,187	5,184,849	93	115
Suburbs	34,920,703	53,045,820	134	142

Notes:
1. All values are expressed in thousands of constant, 1999 dollars, adjusted for inflation with the MSA-level CPI series for housing expenditures.
2. Tabulations are restricted to conventional, home-purchase loans made by lenders not classified by Scheeselle (2000) as subprime or mobile home lenders. Tabulations exclude records failing validity or quality edit checks, with missing geographic information, or with extreme values for applicant income (less than $10,000 or more than $500,000).
* Tracts with median household income below respective citywide median income in 1960.
Data Source: FFIEC (1994, 2000).

of the suburban market. In the boom of the 1990s, residential investment in core gentrified areas advanced by 155%, three times the suburban growth rate. As gentrification has remade significant, although localized parts of the urban core, class turnover has pushed out into adjacent neighborhoods. Mortgage commitments in fringe gentrified areas ballooned by 204% between 1993 and 1999. By the close of the 1990s, both core and fringe gentrified areas posted median loan amounts well above the suburban figure.

Disaggregating the analysis adds detail to the picture, but does not fundamentally alter the results (see Table 3). Chicago's vast swath of redeveloped neighborhoods drives the results when the cities are grouped together; the pace of reinvestment in Chicago's core gentry districts is more than twice as fast as the suburbs, and fringe investment growth is more than four times the suburban rate. For the other cities, however, three distinct groups emerge. First, fringe reinvestment outpaces the core in the long-established outpost of urban chic in Boston, Philadelphia, and Washington, D.C.; these markets grew briskly in the 1990s, but priced out many middle-class buyers who then sought homes in the advancing reinvestment frontier. Chicago also conforms to this pattern. Although these fringe neighborhoods began from a small base in the recession of the early 1990s, subsequent growth has been staggering – in Philadelphia, more than seven times faster than the stagnant suburban rate (Fig. 4). In a second group of cities the opposite situation prevails: growth rates remain tightly

Table 3. Growth in Mortgage Capital Reinvestment in Selected Metropolitan Areas, 1993–1999.

City	Total value of conventional home purchase loans originated, 1999		Growth 1993–1999, as ratio of suburban growth rate	
	Core	Fringe	Core	Fringe
Boston	216,404	72,948	1.82	3.62
Chicago	508,177	377,100	2.21	4.18
Detroit	978	5,709	. . .*	8.01
Milwaukee	13,938	21,512	7.42	1.62
Minneapolis-St. Paul	22,309	43,635	0.92	2.18
Philadelphia	117,860	10,532	6.47	7.24
Seattle	164,335	96,220	3.83	1.51
Washington D.C.	117,507	70,045	3.04	3.75

Notes:
1. See footnotes 1 and 2 on Table 2.
* Decline of 5.3%, compared with suburban expansion rate of 68%.
Data Source: FFIEC (1994, 2000).

circumscribed in core neighborhoods in Milwaukee and Seattle. In a third group of cities reinvestment has stalled, providing mixed evidence on the fate of local gentrification processes. Redevelopment in Detroit remained tenuous for most of the decade, and mortgage data for core neighborhoods reflected the construction of several mixed-use condo developments that had the misfortune of landing on the market between 1991 and 1993. Still, a small trickle of investment has accelerated in Detroit's fringe neighborhoods – a mixture of Victorian and interwar single-family units around New Center (near the General Motors headquarters complex) and Corktown, and new developments along parts of the riverfront (Fig. 5). Lending in these areas grew eight times as fast as in Detroit's suburbs. Finally, the Twin Cities stand out as anomalous, with stagnant trends in core neighborhoods. We strongly suspect that this result stems from the aggressive and entrepreneurial role of the local state, which established large concentrations of market-rate, highrise condominiums on the edge of downtown Minneapolis and St. Paul (Hammel, 1999). These developments remain popular among professional singles and childless couples as well as empty nesters (Adams & VanDrasek, 1993), but if turnover is low this demand will not be detected by the home purchase mortgage data. In the blocks of attractive Victorian homes in fringe gentry neighborhoods of Minneapolis and St. Paul, however, lending activity has been brisk (see Table 3).

These results provide clear evidence of a surge of investment into gentrified neighborhoods in eight large cities. When measured with the conservative and partial benchmark of prime, conventional mortgage loans for home purchase, the turnaround of parts of the inner city is remarkable: core gentrified areas rebounded from the recession of the early 1990s twice as fast as the suburbs, while fringe areas exploded at four times the suburban rate. There can be little doubt that predictions of a "post-gentrification" era were premature (Bourne, 1993a, b). But the last decade has involved far more than a simple restoration of the yuppie-driven housing markets of the 1980s. In many large cities, the latest wave of urban reinvestment has washed up to the edges of landscapes inscribed by the stalled public policies of the 1960s. As a consequence, the rhetoric, profitability, and politics of gentrification have collided with the reinvention of assisted housing policy and the demolition of the "projects."

5. THE REINVENTION OF PUBLIC HOUSING

The rebound of urban housing markets did not take place in a vacuum. The 1990s opened with a recrudescence of the euphemistically-termed "fiscal discipline" of the previous recession, with deep cuts in social services and

Fig. 4. Reinventing Gentrification in Philadelphia.

Deindustrialization and the recession of the early 1990s eviscerated housing market activity in older centers of the Northeast, but subsequent years brought vibrant reinvestment. In Philadelphia, prime capital commitments for home purchase loans in gentrified neighborhoods ballooned from $55.8 million in 1993 to $128.4 million in 1999 (in constant, 1999 dollars), a growth rate more than six times that of the suburbs. Home mortgage data underestimate the resurgence of gentrification by exluding retail development and investment in rental properties. The Locust properties along Philadelphia's Schuylkill River, shown here in the Summer of 1998, illustrate the contrasts and similarities between second- and third-wave gentrification as private and public markets are reinvented (see Walsh, 1997; Historic Landmarks, Inc.). At center-left is Locust Point, a publishing facility built in 1912 and converted to a 110-unit luxury apartment building with a $9 million refurbishment in 1986. The building was redeveloped and is still owned by Historic Landmarks for Living, a local firm specializing in historic rehabilitation and the "urban pioneer" rental market. At right is Locust on the Park, a current project targeted to "knowledge workers" employed in the Market Street office corridor and the hospitals and classrooms of University City. Spearheaded by the now-independent former president of Historic Landmarks, the project was slated as the first rehab after the Philadelphia City Council approved a ten-year tax abatement program for conversions of historic buildings. Three-quarters of the $20 million is financed by a 40-year loan backed by HUD multifamily mortgage insurance, along with private commitments raised from Wall Street and the sale of historic preservation tax credits. *Photograph by Elvin Wyly.*

Fig. 5. Motown Rebound.

Three decades of industrial hemorrhage installed Detroit as a national icon of the fall of urban fortunes. But by the late 1990s, the long-running economic expansion had finally reached the vast swaths of abandoned blocks in the city, and the Detroit Free Press published "Motown Rebound," a series coinciding with a tour by Vice President Al Gore as part of a White House Community Empowerment Conference (Dixon & Solomon, 1997). The Free Press tabulated $5.74 billion in new private and public development planned or under construction, including General Motors' takeover and facelift of the Renaissance Center to accommodate 9,000 workers, Chrysler's expansion plans for six plants across the city, and new stadia for the Lions and the Tigers; riverfront casino development promised an additional $1.2 billion, led by a $220 million MGM Grand that opened on a temporary site in July, 1999 (Meredith, 1999). Detroit's nascent landscapes of gentrification, born in spite of chronically weak demand for city housing, were battered by the recession of the early 1990s but have subsequently begun a tentative expansion. Harbortown, a gated community of retail shops, apartments, and condominia anchored by a marina east of downtown, expanded with a new, 172-unit luxury apartment building set in one of the city's warehouse districts (shown here). A few blocks away new single-family homes went on the market in 1998 at asking prices of $450,000. *Photograph by Elvin Wyly.*

intergovernmental transfers (Staeheli et al., 1997). Far from a replay of the cyclical budget swings of most of the postwar years, however, this wave of cuts signified the culmination of a second round of the "new federalism" initiated during the Nixon years, and was eventually carried forward during the Clinton Administration (Goetz, 2000; Staeheli et al., 1997; Waste, 1998). The mid-term elections of 1994 sealed the fate of this shift, and added weight to the secular movement to privatize government functions, to reinvent the remnants of public institutions in line with private-sector models, and to devolve social welfare provision and other social reproduction functions to units of local government. Assisted housing policy proved particularly adaptive to these policy experiments, and its chronically weak and powerless constituency ensured no effective resistance to creative interventions. Few scholars or policymakers anywhere on the political spectrum were prepared to defend public housing in the form it took in the late 1980s (Marcuse, 1998; Varady et al., 1998).

Chicago provides a vivid and nationally-recognized icon of the failures of American public housing. The Chicago Housing Authority (CHA) was formed in 1937 to help provide suitable housing for a rapidly-growing working class population, which included a large cohort of recent immigrants and African-Americans newly arrived from the rural South. Initially lured to the city by the opportunities in its burgeoning industrial economy, these groups suffered the most during the Great Depression. Deplorable housing conditions grew even worse, and inequalities were magnified by virulent racism, extreme segregation, incipient suburbanization, and entrenched machine politics. Hirsch (1998) provides a comprehensive history of the CHA's efforts between 1940 and 1960, but for our purposes the key point of departure is the landscape produced by this tortured institutional and political history. Between the early 1950s and the early 1960s, the CHA built thirty-three projects comprising some 21,000 housing units; ninety-eight of all units were located in virtually all-black neighborhoods (Hirsch, 1998, p. 243). Many American cities compete for the dubious distinction of most segregated in the nation, but Chicago's vast, desolate landscapes are certainly among the most populous and nationally-recognized.

Public housing problems were not entirely of Chicago's making. Chicago and all major U.S. cities were handicapped by a general neglect of public housing nationwide (Varady et al., 1998; Lane, 1995). The recent history of public housing in America is one of perpetual crisis. Federal regulation attempted to balance the conflicting goals of providing for the poorest residents while minimizing capital expenditures and operating subsidies (Quercia and Galster, 1997). A series of Congressional amendments beginning in 1969, for example, placed ceilings on the share of public housing residents' income devoted to rent, and subsequent legislation instituted admissions priorities for progressively poorer groups. When combined with perpetual shortfalls in capital allocations, these decisions sent most Public Housing Authorities (PHAs) into an immediate cycle of deficit, deferred maintenance, and cuts in supportive services for residents. The virtual halt in construction in the early 1970s tightened the supply of low-rent housing, exacerbating the results of new admissions criteria and concentrating the very poorest and most vulnerable populations into public housing developments. Given local ambivalence or outright disdain for public housing coupled with a series of increasingly contradictory federal mandates, it is not surprising that by the late 1980s public housing was universally denounced as hopeless, isolated, and dangerous. One *Washington Post* reporter dubbed them "high-rises to hell" (quoted in Committee on Government Reform and Oversight, 1995, p. 4). The strongest rhetoric came from the federal government itself during the infamous 104th Congress: "Most Americans believe that public housing is a failure and a waste of their hard-earned taxpayer dollars" (U.S. House of Representatives, 1996, p. 66).

In the context of an expanding economy with a conservative Congress committed to privatization and government cuts, the Clinton Administration crafted a radical attempt to save the Department of Housing and Urban Development with "reinvention" along the lines of private-sector models (Osborne & Gaebler, 1992; Varady et al., 1998). Public housing figured prominently in this effort, with an ambitious new program crafted from a series of local "best practices" experiments across the country. It began as an attempt to solve the problems in the worst public housing communities (National Commission on Severely Distressed Public Housing, 1992), but was then recast as a high-profile, pre-emptive measure to save public housing. HOPE VI is built on the twin themes of income mixing and lower densities with a healthy dose of privatization. These themes appeal to both liberals and conservatives, and are based loosely on policies developed from a sometimes-simple reading of several complex arguments – especially Newman's defensible space (1972, 1980) and Wilson's underclass hypothesis (1987, 1996; cf. Goetz, 2000; Marcuse, 1998). The effectiveness of mixed income public housing is a subject of much academic debate (Bennett & Reed, 1999; Miller, 1998; Nyden, 1998; Rosenbaum, Stroh & Flynn, 1998; Schill, 1997; Vale, 1998). It still may be too early to gauge its long term effects, but income mixing has become a fundamental principle of public housing reform.

HOPE VI comprises only a small part of HUD's budget, but provides the potential for a substantial influx of capital for local PHAs who are successful in the competitive grant process. In Chicago, for example, the CHA and HUD signed an agreement in February of 2000 for $1.5 billion over ten years to assist in the demolition and reconstruction of all of the CHA's major projects

(Chicago Housing Authority, 2000a; Claiborne, 2000; U.S. Department of Housing and Urban Development, 2000a). This level of funding for a public housing authority, even one as large as the CHA, is substantial. Perhaps more importantly, the guarantee of federal money also enables local authorities to leverage more funds through bond issues.

The origin and goals of HOPE VI are complex, but we suggest that the form of the development plans have been influenced by the new geography of the inner city. Public housing projects in many of the nation's cities are no longer located in "seas of decay" (Berry, 1985). As we have argued elsewhere (Wyly & Hammel, 1999), some are now surrounded by rapidly gentrifying neighborhoods and are now islands of decay in seas of renewal. It seems obvious to suggest that a public housing project surrounded by gentrified neighborhoods might tempt local officials to submit a HOPE VI redevelopment plan calling for a mix of incomes with some market rate housing. That is, after all, one of tenets of the program. But this must be understood in an academic context where the extent and significance of gentrification is still a topic of debate, and within a geographical context where some of the nation's most infamous neighborhoods, the dystopian icons of the ills of urban America, have become hot real estate.

At this point we suggest only that gentrification has changed the context of the inner city to such an extent that it has influenced the formation of public housing policy. We do not yet suggest that HOPE VI itself will encourage or accelerate gentrification. However, if the mixed income projects realize their goals, some will spur nearby market activity, and in doing so make the neighborhood more palatable for investors and developers – especially with low density, architecturally contextualized projects. While the Department is concerned about the potential for displacement, resident mistrust, and private abuse of public funds, the agency also finds it necessary to use the program as a banner for its new image of efficiency and entrepreneurialism. The rhetoric sometimes descends into bizarre twists of logic to promote the effort. HUD's Fiscal Year 2001 Budget Summary highlights the Centennial Place Project in Atlanta (located near downtown and the Georgia Tech campus) as a "true mixed-income community where public housing residents earning less than $3,000 per year live next door to professionals earning more than $125,000 per year" (U.S. Department of Housing and Urban Development, 2000b, p. 30). Nonetheless, at this point our primary inquiry regards the effect of gentrification upon public housing policy, and not the reverse.

To test our hypothesis we examined all of the successful HOPE VI grants from fiscal years 1992–1998 in each of our eight cities.[7] Local PHAs must submit a site profile for each project seeking HOPE VI funds, and while the profiles vary in comprehensiveness, they all specify the redevelopment plans and how the supposed ills of high-rise projects will be remedied. Typically the plans suggest rehabilitation, or construction of new units in combination with vouchers and certificates to be used for private market housing. The profiles do not provide a complete picture of the new development, but they do provide a general view of the plan, which we can then incorporate with our gentrification database.

Gentrification and Public Housing Redevelopment

There were twenty-four grants in all totaling $652.8 million delivered or promised by the federal government (Table 4). These grants represent slightly less than a quarter of all HOPE VI funds awarded so far, and provide a striking variety of plans. PHAs are faced with a broad range of projects needing assistance, and they must prioritize them based on local context. In addition they must seek as many opportunities as possible to "partner" with other local institutions to develop a range of initiatives. Within this variety born out of complex local contexts there is qualified support for our hypothesis.

Ten grants involve no provision for income mixing, and six of them have no nearby gentrification. Two more are in cities with limited amounts of gentrification (Detroit and Milwaukee), but two others occur near gentrified areas in cities with substantial gentrification pressures (Boston and Philadelphia). For policy makers operating in a difficult political landscape overlaid on a rapidly changing social landscape, complex local contingencies are often of primary importance in attempting to balance the challenging tasks of improving the worst public housing conditions, and securing federal funding. Given the complexities, the presence of gentrification is clearly not sufficient in itself to induce PHAs to include a mixed-income component in the plans.

Ten other plans call for mixed income developments. Four of these plans include mixes of poor and moderate-income residents and three of these four are surrounded by core or fringe gentrified tracts. Six plans call for market-rate housing in the mix. All but two are *in* or adjacent to gentrified tracts, and in all of these cases the PHAs explicitly identify the importance of the surrounding areas in the site profile. The two exceptions provide indirect support for our hypothesis. Both are in locations outside the inner city and almost suburban in character, and the less impoverished local context seems to have been influential in the plans. In Washington D.C., the Valley Green/Skytower development is at the boundary between the District and suburban Prince George's County. In Seattle, Holly Park is a low-density project started as war-worker housing lying between solidly working class neighborhoods and middle class areas near Lake Washington.

Table 4. Profile of HOPE VI Revitalization Plans in Case Study Cities, 1993–1998.

City and Development	Grant	FY	Type	Demolition	Rehab	New	Mix	Description	Tract	Location
Chicago										
Cabrini Homes Extension	50,000	1994	I	660	65	493	Market	Mixed with market rate units. Most replacement units will be off-site in surrounding community.	819	Surrounded by core and fringe tracts.
ABLA Homes/Henry Horner/Rockwell Gardens	400	1995	P							
Robert Taylor Homes	25,000	1996	D-R	790	125	125		No replacement units on-site; site to be developed as light industrial park.	3817	Surrounded by poor tracts; more than 1 mile from nearest gentrification activity (fringe tract).
Brooks Extension	24,483	1996	D	300		200	Moderate	Mixed with moderate: half of units will be open to families earning 50%–80% of AMI; the remaining half will be reserved for those earning less than 30% of AMI. Adjoining parcels to be acquired; 54 units to be built on-site, 146 on adjacent parcels.	2820	In fringe tract.
Henry Horner Homes	18,435	1996	I	743		150	Moderate	Mixed with moderate: half of units will be open to families earning 50%–80% of AMI; the remaining half will be reserved for those earning less than 30% of AMI.	2804	Adjacent to fringe tract.
ABLA Homes	35,000	1998	I	2,776		2,598	Market	Site replacement includes 1,052 public units; 580 affordable rental/ownership units, and 966 market rate rental/ownership units.	2820	In fringe tract.
Boston										
Mission Main	49,992	1993	I	807		850	Market	Plan to build 850 mixed-income condominium, ownership, and market-rate rental units, and constructing 4–6 story public housing units.	808	Adjacent to fringe tract.
Orchard Park	30,400	1995	P,I	441		280	Market	On-site development of 280 public housing units, 40 tax credit rental units, 10 ownership units; 224 public units to be constructed in surrounding area.	803/804	Approx. 0.4 miles from core tract.

Table 4. Continued.

City and Development	Grant	FY	Type	Demolition	Rehab	New	Mix	Description	Tract	Location
Detroit										
Parkside Homes	48,120	1994 1995	P,I3	92	350	180		Plan to demolish about half of the 60 buildings; reconfigure remaining units; construct 180 townhouses in adjacent neighborhoods; and acquire and rehabilitate 345 single-family homes throughout EZ to be sold to public-eligible households.	5122	No gentrification within 2 miles.
Jeffries Homes	49,807	1994 1996	D,I	612				Construction of low-rise replacement units both on and off-site.	5207	Approx. 0.5 miles from downtown core tract.
Parkside Addition/Herman Gardens/Gardenview	400	1995	P							
Herman Gardens	24,224	1996	D-R	1,223	274	672		Plan to increase income of current residents; 176 ownership units to be constructed for residents completing self-sufficiency program.	5454	No gentrification within 3 miles.
Milwaukee										
Hillside Terrace	45,689	1993 1995 1996	I*		119			Goal is reduction of density and revitalization of existing development.	141	Approx. 0.5 miles from downtown core and fringe tracts.
Parklawn	34,230	1998	I	138	380	40		Demolition of 138 public units, to be replaced by 40 new single-family lease-to-purchase homes, half on-site and half off-site. Remaining public units to be rehabilitated.	40	No gentrification within 5 miles.
Philadelphia										
Richard Allen Homes	50,000	1993	I	129	314	80		Plan for reconfiguration of three existing quadrants to 314 townhouse units, and construction of new five-story building with 80 elderly units.	131, 132	Approx. 0.25 miles from fringe and core tracts.
Martin Luther King	25,630	1995 1998	P,I	537		330	Moderate	Plan to build 85 new public units, 93 affordable rental units, and 152 home-ownership units for "a range of incomes"	15	Surrounded by core tracts on north and east.

Table 4. Continued.

Site	Amount	Year(s)	Type	Units1	Units2	Units3	Goal	Description	#	Gentrification
Schuylkill Falls	26,401	1997	I	266		300	Moderate	Plan to build 330 new rental and home-ownership units for a cross-section of incomes. Families with incomes up to 120% of AMI will have opportunity to purchase home through lease-to-own program.	207	No gentrification within 4 miles.
Seattle										
Holly Park Apartments	48,617	1993 1995 1996	P,I*	893		1,200	Market	A complex development with multiple financing sources; consists of three phases. Plan for 800 total rental units, 40 market-rate; 400 total ownership units, 300 at market rates.	110	No gentrification within 3 miles; between poor and middle-class tracts.
Rainier Vista/High Point	400	1995	P							
Roxbury House/ Roxbury Village	17,810	1996 1998	D,I	60	151	60		Plans to rehabilitate Roxbury House, a 151-unit elderly highrise; Roxbury Village to be demolished and replaced with 60 family townhomes.	114	No gentrification within 4 miles.
Washington, D.C.										
Ellen Wilson Dwellings	25,076	1993 1995	I*	134		153	Market	Construction of townhouse units to integrate development into surrounding historic district; 19 market-rate units, remaining 134 organized in cooperative. Half of coop units open to families earning 50%–80% of AMI; provision for up to 20 units at no more than 115% of AMI.	70	In core gentry tract.
Sheridan Terrace	400	1995	P							
Fort Dupont	1,995	1996	D	133				Demolition only.	99.02	No gentriciation within 3 miles.
Valley Green/Skytower	20,300	1997	I	312		314	Market	Plan for 48 new public units; 100 elderly public housing; 30 public lease-to-purchase; 32 market-rate rentals; and 104 for-sale homes.	97	No gentrification within 3 miles.

Notes: Grant awards are expressed in thousands of dollars, not adjusted for inflation; awards are conditional Federal commitments spread over several years, and do not include leveraged funds from local government or private sources.
Grant type codes: P = Planning; D = Demolition; R = Revitalization; I = Implementation.
Mix: Explicit goals of redevelopment with regard to mixed-income housing on original site. *Includes amendment funds.
Minneapolis and St. Paul were ineligible to apply under the initial HOPE VI criteria, but received $1.8 million in FY 1998 to demolish the Glenwood/Lyndale Towers.

Source: Authors' analysis of site profiles published by Housing Research Foundation (1998).

Two projects are exemplary in their combination of gentrification, mixed income housing, and their attempt to blend the project into the existing social, economic and even architectural context of their neighborhoods. The Ellen Wilson Dwellings were opened in 1943 on a 5.3-acre site less than a mile south of the U.S. Capitol. The area is somewhat isolated by the I-395 corridor, but within a short walk of a Metro station near several blocks of redeveloped row houses. They also sit on the fringes of the Capitol Hill district, one of the most intensely gentrified neighborhoods in the country. By the late 1980s all 143 units had deteriorated to the extent that the local PHA declared them uninhabitable. An abandoned project in such a prime location made an excellent candidate for HOPE VI sponsored mixed income development. Demolition commenced in April of 1996.

The plan calls for a complete redevelopment of the site with 153 townhouse units designed to resemble mews typical of the historic district of which the complex is part. Nineteen of the units scattered throughout the site will be sold at market rates. The remainder will be divided into three income brackets with one quarter of the housing open to residents earning no more than 25% of the area median income (AMI); one quarter open to those earning between 25 to 50% of the AMI; and the remainder going to moderate income families earning between 50 and 80% of the AMI. However, 20 units in the top income group will be open to residents earning up to 115% of the AMI "to enhance marketability and expand the range of household incomes in the new community" (Housing Research Foundation, 1998).

Chicago's Cabrini-Green

The Cabrini-Green development in Chicago provides perhaps the single best example of the influence of gentrification upon public housing policy. Cabrini-Green consists of three adjacent developments. The first, the Cabrini Homes, were completed in 1942 and include 55 two and three story buildings in "regimental rows ... giving the appearance of army barracks" (Chicago Housing Authority, 2000b). Despite this description the Cabrini Homes are similar in design to the low density housing style favored in HOPE VI redevelopment plans, and are thus slated for rehabilitation while the newer towers will meet with the wrecking ball. Completed in 1959, the Cabrini Extension consisted of fifteen buildings of seven, ten and nineteen stories. The William Green Homes built soon afterward consist of eight monolithic concrete towers of 15 and 16 stories. Although initially an ethnically diverse development, Cabrini-Green was largely an African American ghetto by the end of the 1960s.

Despite the location, just northwest of the Loop and within a short walk of the Gold Coast, the area remained unsuitable for residential use for decades. Proximity to the North Branch of the Chicago River dictated an industrial character in surrounding areas. Intense fumes from a nearby gas works lent the area the nickname "Little Hell", and it served as the original slum in Zorbaugh's classic ethnographic study, *The Gold Coast and the Slum* (1929). As commercial activity expanded out of the Loop and deindustrialization killed many of the firms along the Chicago River, gentrification spread northward and surrounded the public housing projects. Intense local market pressures allow the CHA to contemplate the cross-subsidies between market-rate and assisted housing units that are the linchpin of current management strategies for public housing authorities (Lane, 1995; Quercia & Galster, 1997) (Fig. 6). Mixed-income strategies have little chance of success where local demand will not support market-rate redevelopment (Fig. 7). Residents of Cabrini-Green, and other Chicago public housing communities, understand these contextual factors better than anyone. Both the CHA and HUD are well-aware that there is a long history of mistrust in Cabrini-Green, and that any redevelopment plans have to be viewed in light of previous broken promises (U.S. General Accounting Office, 1998). Thus, when the CHA drafted its plans for redevelopment it encountered widespread suspicion that the land would be sold to developers. The income-mixing component of HOPE VI provided a more subtle means to accomplish the goal of eliminating a major barrier to further reinvestment on the Near North Side.

HUD rejected the CHA's original HOPE VI application in 1993, but changes in the program's requirements the following year forced HUD to approve all 1993 applicants, and the CHA was awarded a $50 million grant in 1994. The situation was complicated in 1995 when HUD took over direct control of the CHA. Eventually a revised plan was approved in September of 1997 (U.S. General Accounting Office, 1998). The original plan called for the demolition of 660 units of the Cabrini Extension and the construction of 493 replacement units, most of them in the surrounding neighborhood. The revised plan calls for the demolition of 1,324 units. Ultimately, the new development will contain a mix of 30% public, 20% moderate income, and 50% market-rate units (Salama 1999; U.S. General Accounting Office, 1998). The mix in the development is only part of the story. The CHA has plans to relocate many of the current residents through use of Section 8 housing units. When the location of the approved housing was made public, controversy erupted. The neighborhoods were largely African American, and poor. It appeared to many residents that the real goal of the HOPE VI plan was not to ease barriers of entry in more desirable neighborhoods, but to relocate residents from a development site with lucrative profit potential (Smith, J., 1999).

Fig. 6. Million Dollar Views of the Gold Coast and the Slum.

North Cleveland Avenue, Chicago, July 1999. The urban fabric nestled between Lake Michigan and the North Branch of the Chicago River has always been a sharp gradient between wealth and poverty, providing the setting for Zorbaugh's (1929) classic. After several generations of social change and public policy, however, reinvestment has swept across the city's Near North side and has turned "islands of renewal in seas of decay" into "islands of decay in seas of renewal." The towers in the background at left are slated for demolition under the HOPE VI revitalization plan for Cabrini-Green, "a pocket of isolated and concentrated poverty surrounded by wealth" (Chicago Housing Authority, 1997, 1.2). The redevelopment plan sharply reduces the supply of public housing units in favor of a mixed-income community to integrate affordable and market-rate units. A few steps from the building in the foreground, construction is nearing completion on a 4,000 square foot single-family home with a rooftop deck and a posted price just shy of $1,100,000. *Photograph by Elvin Wyly.*

The form of HOPE VI redevelopment plans is dependent upon a myriad set of influences, but neighborhood context is an explicit principle of the program. Mixed income developments are feasible in gentrified areas, and the policy record provides clear evidence of the linkage between private reinvestment and

Fig. 7. Reinventing the Second Ghetto.

Demolition of the Robert Taylor Homes, July 1999. Hirsch (1998) renders a detailed history of the making of the "second ghetto," the deeply-segregated public housing reservations on Chicago's South Side. Local entrepreneurialism and Federal support have joined forces in an effort to erase the geographical imprint of thirty years of public housing policy. The Robert Taylor Homes, a line of 28 buildings with more than 4,300 units built between 1959 and 1963, are being demolished with no on-site replacement units; residents are receiving portable Section 8 certificates, while the parcel will be redeveloped as a light industrial park. Most of the south side has endured massive outmigration and disinvestment in the last twenty years, and HOPE VI principles emphasize the careful integration of assisted housing sites with nearby market conditions. Assisted housing policy is now strongly conditioned and justified by localized geographies of disinvestment and reinvestment. *Photograph by Elvin Wyly.*

public redevelopment strategies. The last three decades of uneven urban development have altered parts of the inner city, conditioning attempts to redefine public assistance and private profitability in ways that imbue "public housing reform" with new and disturbing connotations.

6. THE TRANSFORMATION OF GENTRIFICATION

The third strand of our argument focuses on the role of housing finance in processes of neighborhood change. Even as favored quarters of the inner city have become popular destinations for middle-class professionals to live, work, and play, a host of less visible trends have reshaped the links between local neighborhoods and national (and global) capital markets. Structural changes in the mortgage lending industry and the expansion of a sophisticated secondary mortgage market have intersected with stepped-up regulatory oversight and hard-fought successes by the community reinvestment movement. These forces have altered the magnitude and terms of capital investment in urban neighborhoods, even as entrenched features of the urbanization process reinforce longstanding patterns of discrimination and disinvestment. The result is a more complicated terrain of disinvestment and reinvestment, as shifts in the risks and rewards of mortgage lending have channeled capital to parts of the inner city undergoing demographic and class turnover. Part design and part accident, the acceleration of selective reinvestment stands in sharp contrast to the prevailing postwar relationship between housing finance and neighborhood change.

The foundations of the American housing finance system were laid out in the 1930s, when Depression-wracked housing markets required broad Federal intervention to reduce foreclosures and resuscitate construction activity. Although many features of the system were altered in subsequent years, the underlying principles of Federal involvement have endured – encouraging homeownership as a national policy goal, reducing barriers to capital investment in the housing sector, and developing standards to unify fragmented business practices in construction, appraisal, and mortgage lending. This evolving system proved enormously successful in promoting the twin goals of homeownership and a healthy construction industry, but there is near-universal agreement that the norms established by government precedent and private practice helped to accelerate suburbanization and its counterparts – redlining, white flight, and neighborhood abandonment (for the most comprehensive synthesis, see Jackson, 1985). Oddly enough, when early signs of gentrification in the 1960s seemed to promise an urban renaissance, it soon became clear that banks wanted no part of the risks. In the urban renewal efforts of the 1950s and 1960s, the risks of capital investment at the urban core were almost invariably borne by the public. Risks and rewards were negotiated among financial institutions, federal, state, and local governments, and quasi-public development and investment authorities; the shared goals were to "take advantage of the useful effects of agglomeration" and to "exploit the social, interactive creation of value in space based on the externalities of growth." (Gottdiener, 1985,

p. 64). But private financial institutions were reluctant without extensive public guarantees and deep subsidies, and thus preferred large redevelopment projects involving carefully-planned mixtures of luxury housing, retail and entertainment facilities, and downtown office construction (e.g. Leinsdorf et al., 1973; Smith, 1979b). In established urban neighborhoods, banks were usually content to remain on the sidelines until maverick developers, city governments, or residents proved that an area could defy the received wisdom of the neighborhood life cycle (Hoover & Vernon, 1959; Metzger, 2000; cf. critiques by Downs, 2000, Temkin, 2000, and Galster, 2000). Even wealthy, white gentrifiers could not obtain conventional financing in many of these neighborhoods, and the emphasis on "urban pioneers" and "sweat equity" in the early literature is revealing. Well-educated, sophisticated professionals enjoying the fruits of white privilege were nevertheless forced into creative schemes to circumvent the barriers established by decades of redlining and disinvestment.

This situation grew more complicated in the 1980s, setting the stage for a fundamentally new process of neighborhood change after housing markets recovered from the early 1990s recession. The story involves community activists, Federal regulators, scholars and investigative journalists, and lending institutions searching for new profit opportunities in a turbulent and competitive industry (Listokin & Wyly, 2000; Schwartz, 1998a, 1998b; Squires, 1992; Stegman et al., 1991; Vartanian et al., 1995). Not all of the efforts of these groups yielded expected results, and several unanticipated developments proved decisive at key turning points. But the net effect of public and private initiative over the past fifteen years has been a new relationship between housing finance and neighborhood change in the inner city.

After the stagflation of the 1970s, the next decade brought a housing market recovery in the context of falling inflation and declining interest rates. Federal restructuring of the secondary mortgage market, intended to reduce geographical and institutional bottlenecks in the availability of mortgage funds, reinforced industrywide trends towards standardization and automation. Loans meeting clearly-defined financial parameters could readily be sold in pools of mortgage-backed securities, freeing capital for further rounds of lending and creating new markets and new pricing mechanisms for many of the contractual obligations embedded in the ordinary home loan. The secondary market shifted certain risks from lenders to investors, even as increasingly sophisticated behavioral models further enhanced the safety and profitability of mortgage pools that were already comparatively low in risk. Downpayment requirements began to edge downward, permissible debt burdens were increased, and shifts in the composition of lenders themselves encouraged competition for residential mortgages (Listokin and Wyly, 2000). Bank consolidation continued, while savings and loans were battered by disintermediation in the 1970s and mounting commercial loan losses in the 1980s. All of these changes signalled a more complex relationship between local housing markets and national and global capital markets. In the 1970s, total single-family mortgage debt outstanding grew briskly (53.8% in real terms), but the next decade brought an acceleration (74.2%) and a shift away from savings institutions, whose share of total holdings fell from half to less than a quarter. Mortgage pools and trusts accounted for only 12.9% of all single-family loans outstanding in 1980, but this share tripled in the next decade. By 1996, mortgage pools accounted for just under half of the $3.9 trillion of single-family mortgage debt outstanding (Simmons, 1997, p. 225).[8]

Federal policy translated these economic circumstances into specific changes in urban investment patterns. Most of the relevant legislation had been in force for many years – including the Fair Housing Act (1968), the Equal Credit Opportunity Act (1974), the Home Mortgage Disclosure Act (1975), and the Community Reinvestment Act (1977) – but Vartanian et al. (1995, p. 1–4) conclude that "it is probably fair to say that until very recently the implementation and the enforcement of these acts was not a high priority of the bank regulatory agencies or depository institutions." Several developments in the late 1980s changed this situation. In May, 1988, the *Atlanta Journal-Constitution* published "The Color of Money," a four-part series that documented severe redlining in Atlanta's African-American neighborhoods (Dedman, 1988). The series won Bill Dedman the Pulitzer Prize for investigative reporting, inspired similar studies in other cities, and triggered a landmark Department of Justice investigation into the lending practices of a local savings and loan. It also led to the inclusion of seemingly innocuous amendments to HMDA and CRA as part of the savings and loan bailout, otherwise known as the Financial Institutions Reform, Recovery, and Enforcement Act (FIRREA) of 1989. Amendments to HMDA required institutions to collect and report data on the race, gender, and income of all applicants and the action taken by the lender, and these loan-level data led to a new wave of research on individual and geographical disparities in mortgage lending (Canner & Gabriel, 1992; Fishbein, 1992). New examination procedures and public disclosure of CRA ratings gave ammunition to the community reinvestment movement in its drive to challenge redlining and urban disinvestment. Banking industry consolidation also helped, because the key leverage point of CRA has always been the provision requiring regulators to solicit public comments and consider institutions' record of lending to low- and moderate-income neighborhoods in applications for mergers and acquisitions.

Two key events in the early 1990s completed the infrastructure that now governs the relationship between housing finance and urban neighborhoods.

First, Congress passed the Federal Housing Enterprises Financial Safety and Soundness Act (FHEFSSA) of 1992. The act established new regulations to reduce the risk of failure of Fannie Mae and Freddie Mac, the Government Sponsored Enterprises (GSEs) accounting for the lion's share of secondary market purchases of residential mortgages. The act also reaffirmed the GSEs' obligation "to facilitate the financing of affordable housing for low and moderate income families in a manner consistent with their overall public purposes, while maintaining a strong financial condition and a reasonable economic return" (Sect. 1302.7). The legislation established "affordable housing goals" for acquisitions of mortgages to low and moderate income families as well as borrowers in central cities, rural areas, and "other underserved areas." This last provision is pivotal: "underserved" areas are defined at the census tract level on the basis of income and racial composition as measured in the 1990 census, and the thresholds used to define underservice yield a broad swath of urban neighborhoods.[9] Moreover, the designation was incorporated into scores of urban policy initiatives and mortgage assistance programs at the local, state, and Federal levels. Fannie Mae's flagship affordable lending product, for example, relaxes downpayment requirements and household debt ceilings to help low and moderate income buyers; but in underserved census tracts, these incentives are available to buyers of all income levels. We contend that all of the methods used to define underserved neighborhoods leave many parts of the inner city vulnerable to accelerated gentrification: in areas where gentrification was just beginning after the recession of the early 1990s, the "underserved" designation allowed housing policy to lubricate capital flows to middle-income homebuyers.[10]

The second factor grew out of Bill Clinton's election in 1992. The new administration's urban policy agenda included strong commitments to promote fair housing, combat racial discrimination in lending, and expand access to homeownership (Stegman, 1995, 1996). Equally significant, however, was the broad, durable, and bipartisan consensus that the administration managed to forge with successive Congressional sessions: most Senators and Representatives agreed on the virtues of homeownership as a policy goal worthy of public subsidy even in an age of privatization and lean government. The 1990s were marked by a proliferation of public, private, and quasi-public initiatives to encourage homeownership among underserved *populations* and underserved *neighborhoods*. The uncertain distinction between these two categories, we suggest, has allowed gentrification to mediate the local effects of the national system of housing finance and homeownership policy. This shift marks an important transformation of the relationship between capital investment flows and processes of neighborhood change.

Modeling the Spatiality of Mortgage Capital

Testing the transformation hypothesis is extremely difficult. To suggest that selective reinvestment has replaced redlining is to venture a broad institutional and socio-historical assertion; sustaining the argument demands detailed and comparable records of mortgage market transactions in specific neighborhoods through several business cycles. With few exceptions, the data meeting these requirements are lost forever – they simply were not collected on a systematic or consistent basis until the late 1970s, and provided no individual-level information until 1990. It is possible, however, to examine one economic recovery in sharp detail, and to test the hypothesis that middle-class homebuyers and private lending institutions no longer systematically avoid urban neighborhoods. It is even possible to test a more fundamental proposition: that the institutions and agents of capital investment are more important in the transformation of the inner city than are the locational preferences of middle-class homebuyers. To evaluate these hypotheses, we return to the HMDA data for a more intensive multivariate analysis.

Consider an underwriter's decision to approve or deny a loan request, based on the applicant's financial profile (A_i), the requirements of the mortgage product, (M_i), and the distinctive market specialization of the lending institution (I_i):

$$\ln\left[\frac{P_{approve}}{1-P_{approve}}\right] = b_0 + b_A A_i + b_M M_i + b_I I_i + e_i \tag{1}$$

This framework mimics the economic calculus of profit and risk in the mortgage lending industry, and serves as the basic foundation for a voluminous housing finance literature. Since the first release of loan-level records in the early 1990s, much of this research has sought to determine whether a borrower's race (R_i) has any effect on the loan decision:

$$\ln\left[\frac{P_{approve}}{1-P_{approve}}\right] = b_0 + b_A A_i + b_M M_i + b_I I_i + B_R R_i + e_i \tag{2}$$

In this framework, a statistically significant coefficient estimate for the b_R term is normally interpreted as discrimination – i.e. as lender decisions that cannot be justified on the basis of economic considerations captured in the rest of the model. But public-release HMDA files do not include all of the borrower and loan characteristics considered by underwriters, prompting critics to dismiss claims of discrimination on the basis of "omitted variable bias." If the model excludes legitimate underwriting concerns that differ across racial groups, then the b_R coefficient will be biased by these omitted variables – rendering it impossible to tell how much of the race effect is discrimination, and how much results from minorities' generally lower assets and higher debt burdens.

Omitted-variable bias creates serious problems in discrimination research, but for our purposes it presents an analytical opportunity. Consider an expanded model that controls for varied conditions across the eight case study metropolitan areas (MSA_i), and for distinctive lending decisions in gentrified neighborhoods:

$$\ln\left[\frac{P_{approve}}{1-P_{approve}}\right] = b_0 + b_A A_i + b_M M_i + b_I I_i + b_R R_i + b_1 MSA_i + b_2 CORE_i + b_3 FRINGE_i + e_i \quad (3)$$

This approach exploits omitted-variable bias to test the hypothesis that geographical shifts in class turnover are linked to mortgage capital investment decisions. Statistically significant coefficient estimates for the b_2 and b_3 terms would suggest that there is something distinctive about mortgage market transactions in the gentrified neighborhoods identified in our field surveys. Affluent professionals in these neighborhoods may be more qualified than homebuyers elsewhere in the city, even after accounting for income and other variables included elsewhere in the model. Conversely, the transformation of housing finance may have altered lenders' perceptions of risk and profit in the new urban frontiers inscribed in the gentrified inner city.

These models allow a rigorous and conservative analysis of residential preference, capital investment, and public policy in the transformation of the inner city. By synthesizing our fieldwork with loan-level records for the 1990s, we are able to measure specific aspects of reinvestment that were difficult or impossible to examine in the 1970s and 1980s. We estimate models for Eqs 2 and 3 with 1993 and 1999 loan records for home purchase loan applications; observations with missing or questionable data are excluded on the basis of generally-accepted cleaning standards.[11] A standard set of applicant, loan, and institutional variables is defined to control for prevailing variations in applicant risk and loan underwriting procedures (see Table 5).

Results

Mortgage lending is a complex synthesis of economic and social processes, embodied in institutionalized practices and norms of professional responsibility among executives, loan officers, and underwriters. Nevertheless, broad regularities govern these complex transactions, allowing one to model the process with a fair degree of accuracy. Consider first a baseline analysis of loan decisions, pooling applications filed in 1993 and 1999 (see Table 5). Results generally mirror those that recur throughout the vast literature on mortgage lending and race discrimination (cf. Holloway, 1998; Munnell et al., 1996). Denial is less likely for higher-income applicants requesting larger loans, so long as the amount does not exceed the conforming limit for mortgages salable to Fannie Mae and Freddie Mac: jumbo loans are 1.7 times as likely to be rejected as otherwise identical conforming notes. Striking institutional variations prevail, an indication of the deep market segmentation of different types of lenders targeting specific categories of borrowers: FHA products incorporate several underwriting flexibilities that make them ideal for loan officers trying to maximize a borrower's chances for approval, while subprime lenders operate in ways that depart radically from mainstream, A-paper institutions.[12] There are also significant contrasts between independent mortgage companies, credit unions, banks, and savings institutions. Finally, note that racial and ethnic disparities remain severe even after controlling for income, loan amount, and institutional characteristics. Latinos are 1.4 times as likely to be rejected compared with otherwise identical non-Hispanic whites, and African Americans are twice as likely to be denied. Borrowers choosing not to report racial information are also twice as likely to be rejected, raising troubling questions for research and policy on racial disparities in credit markets.[13]

The second model adds contextual variables to test for metropolitan variations and gentrification (see Table 5, Model 2). Compared with Chicago, Boston and Detroit are distinguished by higher rejection probabilities, while more favorable market outcomes appear in Milwaukee, Minneapolis-St. Paul, Philadelphia, and Washington, D.C. Loan denial models of the sort used here, of course, are extremely sensitive to pre-application screening. These regional contrasts may simply be the product of market processes that influence the home search and households' ultimate decision to file a formal loan application.[14] Nevertheless, accounting for metropolitan context does not fully capture the distinctive market outcomes observed in gentrified neighborhoods. Even after controlling for all applicant, loan, and institutional characteristics, mortgage investment shows a slight but statistically significant preference for applicants in core and fringe gentrified neighborhoods. Prospective borrowers in these areas are 0.9 times as likely to be rejected compared with similarly qualified borrowers elsewhere in the city. Although this effect remains small, it provides a critical test of the hypothesis that mortgage capital no longer sees all of the inner city as an unacceptable risk.

Calibrating separate models for 1993 and 1999 provides further insight (see Table 6). The context for gentrification shifted rapidly during the 1990s with a vigorous economic recovery and restructuring in the financial services sector, and contrasts in model results for the two periods bear this out. Between 1993 and 1999, changes in conditional denial likelihood suggest a magnified

Table 5. Logistic Regression of Application Denial, 1993 and 1999.

	Model 1		Model 2	
	Parameter estimate	Odds ratio	Parameter estimate	Odds ratio
Intercept	−1.3716****	.	−1.1814****	.
Applicant income (× $1,000)	−0.00921****	0.991	−0.00975****	0.990
Applicant income squared	0.000018****	1.000	0.000019****	1.000
Loan amount (× $1,000)	−0.00075****	0.999	−0.00126****	0.999
Loan exceeds GSE Limit	0.5205****	1.683	0.5866****	1.798
Loan for owner occupancy	−0.2575****	0.773	−0.2699****	0.763
FHA-Insured Loan	−0.3553****	0.701	−0.373****	0.689
Female (1 = yes)	−0.1178****	0.889	−0.1171****	0.889
Traditional white family	−0.1333****	0.875	−0.1179****	0.889
Black	0.7502****	2.117	0.6918****	1.997
Hispanic	0.3548****	1.426	0.2964****	1.345
Other race	0.1767****	1.193	0.1585****	1.172
Race unreported	0.7159****	2.046	0.7069****	2.028
Regulator OCC	0.2099****	1.234	0.2431****	1.275
Regulator FRB	−0.0652***	0.937	−0.0385	0.962
Regulator FDIC	−0.1532****	0.858	−0.1306****	0.878
Regulator OTS	−0.2048****	0.815	−0.213****	0.808
Regulator NCUA	0.1268*	1.135	0.1963***	1.217
Subprime Lender	1.5916****	4.911	1.5737****	4.824
Boston			0.083**	1.087
Detroit			0.1319****	1.141
Milwaukee			−0.3935****	0.675
Minneapolis-St. Paul			−0.3566****	0.700
Philadelphia			−0.2104****	0.810
Seattle			−0.0307	0.970
Washington, D.C.			−0.1536****	0.858
Core gentry tract			−0.0802**	0.923
Fringe gentry tract			−0.0816*	0.922
Number of observations	154,558		154,558	
−2 Log Likelihood	118,082		117,665	
Chi-square vs. null model	12,914****		13,332****	
Chi-square vs. Model 1			417****	
Percent correctly classified	70.5		70.9*	

* Significant at $p \leq 0.10$; ** $p \leq 0.05$; *** $p \leq 0.01$; **** $p < 0.001$.
Traditional white family is defined as white male applicant, white female co-applicant.
Data Source: FFIEC (1994, 2000).

divergence between the underwriting standards of conventional and FHA-insured loans, a greater willingness of lenders to accept the risks of jumbo loans, and a realignment in the varied lending practices of banks, thrifts, credit unions, and independent mortgage companies (compare the Model 1 odds ratios for 1993 and 1999, Table 6). Market segmentation among mainstream institutions pales in comparison to subprime lenders, however, whose rapid growth has been accompanied by dramatic changes in credit allocation: all else constant, a borrower is five times as likely to be rejected for a home purchase loan at a subprime lender in 1999, up from a factor of 2.1 in 1993. There is also evidence of a slight reduction in racial disparities: by the end of the decade, an African American applicant was 1.77 times as likely to be rejected compared with an otherwise identical non-Hispanic white; six years earlier, this ratio stood at 2.43. It is certainly tempting to seize upon this finding as the long-awaited fruits of community activism, government regulation, and industry attempts to find new markets and eliminate discrimination; yet this apparent evidence of improvement is vulnerable to the same long litany of theoretical and methodological criticisms leveled against studies that show persistent or worsening discrimination. What we can conclude is that the last years of the nation's longest economic expansion were still marked by severe racial and ethnic inequalities in mortgage market *outcomes* in these eight cities. Unobserved variations in borrower credit profiles may explain most of these inequalities, but disparities also reflect the durable housing and labor market segregation entrenched by half a century of urbanization processes, institutionalized bigotry, and Federal policy (Jackson, 1985; Teaford, 2000; Yinger, 1995).

Gentrification is central to these polarizing forces in the inner city, and the evidence suggests that its role has changed in recent years. As central city land markets began to recover from the short, sharp recession of the early 1990s, gentrified neighborhoods were seen as no riskier than other parts of the city – but they also enjoyed no particular advantage. The parameter estimates for core and fringe tracts are not statistically significant, and the income interaction terms imply that higher-income applicants in gentrified areas faced more careful scrutiny than their affluent peers elsewhere in the city (see Model 2, 1993, in Table 6). But after six years of sustained growth and restructuring of housing finance, a more complex picture emerges. Loans in core and fringe tracts are significantly less likely to be rejected than similarly qualified applicants elsewhere in the city, and higher-income buyers are evaluated no differently if they choose a *pied-a-terre* in the outposts of wealth inscribed by thirty years of gentrification. Again, the magnitude of this effect remains small, but nevertheless persists as highly significant after controlling for an extensive set of borrower and lender characteristics. Moreover, the boundaries between wealth and poverty appear to be shifting rapidly in the inner city, with important consequences. Note that a borrower in a fringe neighborhood is less than three-quarters as likely to be denied as an otherwise identical applicant elsewhere, a bonus effect that exceeds the margin in core gentry areas. Yet lenders are more likely to reject nonconforming loans in fringe tracts than otherwise identical jumbos in other city neighborhoods: the clear implication is that lenders perceive considerable opportunity on the expanding frontier of gentrification, so long as someone else is willing to bear the risk (in this case, investors in mortgage-backed securities). But in spite of the expansion of affordable lending and secondary market acquisition goals, gentrified areas remain landscapes of exclusion. The insignificant parameter estimates for the income-gentry interaction terms could be the result of income mixing, but racial polarization is another matter. African American applicants in gentrified areas are one and one-half times more likely to be rejected than similarly-qualified blacks in other parts of the city – who already face barriers to credit that cannot be explained by income, loan amount, or institutional characteristics. To the degree that gentrification is an inherently polarizing reproduction of urban space, these findings strongly suggest that recent changes in public policy and housing finance have reinforced inequalities in the inner city.

Polarization or Equitable Reinvestment?

The evidence marshalled thus far points to important changes underway at the core of the American city. There is little doubt that gentrification is back on the agenda – for community activists, low-income residents, and middle-class home buyers as well as speculators and powerful institutions of finance and investment. It is also clear that mortgage capital no longer systematically avoids all parts of the inner city. Nevertheless, our interpretation remains vulnerable to critique on the basis of three limitations. First, the recentralization of middle-class preference and capital investment may not have any significance at the metropolitan scale; perhaps we are seeing nothing more than an eddy in central city housing markets that fall farther behind the wave of suburban growth. Second, reinvestment may be a simple, unproblematic response to demand-side forces not captured by applicant variables in the denial models. The apparent preference of lenders for gentrified neighborhoods might reflect a stream of particularly well-qualified professionals desperate for high-quality residential environments close to downtown office districts; maybe housing finance and public policy play no significant role. Finally, it is possible that innovative public policy and sustained economic growth have achieved goals once dismissed as utopian: expanding affordable ownership opportunities through neighborhood redevelopment, while transform-

Table 6. Denial Models with Contingency Tests, 1993 and 1999.

| | 1993 | | | | 1999 | | | |
| | Model 1 | | Model 2 | | Model 1 | | Model 2 | |
	Parameter estimate	Odds ratio	Parameter estimate	Odds ratio	Parameter estimate	Odds ratio	Parameter estimate	Odds ratio
Intercept	−1.3975****	.	−1.3751****	.	−1.0544****	.	−1.0469****	.
Applicant income (× $1,000)	−0.0102****	0.990	−0.0103****	0.990	−0.00955****	0.990	−0.0095****	0.991
Applicant income squared	0.000022****	1.000	0.000002****	1.000	0.000018****	1.000	0.000018****	1.000
Loan amount (× $1,000)	−0.00274****	0.997	−0.00275****	0.997	−0.00086****	0.999	−0.00086****	0.999
Loan exceeds GSE Limit	0.8766****	2.403	0.894****	2.445	0.4936****	1.638	0.4341****	1.544
Loan for owner occupancy	−0.2832****	0.753	−0.2878****	0.750	−0.2592****	0.772	−0.2579****	0.773
FHA-Insured Loan	−0.1368****	0.872	−0.1381****	0.871	−0.5039****	0.604	−0.5018****	0.605
Female (1 = yes)	−0.1688****	0.845	−0.1694****	0.844	−0.0881****	0.916	−0.0885****	0.915
Traditional white family	0.0133****	1.013	0.0106	1.011	−0.1845****	0.832	−0.1866****	0.830
Black	0.889****	2.433	0.8791****	2.409	0.5715****	1.771	0.5492****	1.732
Hispanic	0.3241****	1.383	0.3173****	1.373	0.3003****	1.350	0.2908****	1.337
Other race	0.3126****	1.367	0.3081****	1.361	0.0623	1.064	0.0576	1.059
Race unreported	0.9123****	2.490	0.9127****	2.491	0.6249****	1.868	0.6222****	1.863
Regulator OCC	0.1732****	1.189	0.1716****	1.187	0.2584****	1.295	0.2579****	1.294
Regulator FRB	0.0681	1.070	0.0674	1.070	−0.0838***	0.920	−0.0847****	0.919
Regulator FDIC	0.1645****	1.179	0.1633****	1.177	−0.3819****	0.683	−0.3825****	0.682
Regulator OTS	−0.0549	0.947	−0.0559	0.946	−0.2679****	0.765	−0.2689****	0.764
Regulator NCUA	0.4789****	1.614	0.4764****	1.610	−0.0121	0.988	−0.0163	0.984
Subprime Lender	0.749****	2.115	0.7485****	2.114	1.5952****	4.929	1.5932***	*4.920
Boston	0.2077****	1.231	0.2068****	1.230	0.00532	1.005	0.0101	1.010
Detroit	0.1596****	1.173	0.159***	1.172	0.0817**	1.085	0.0851**	1.089
Milwaukee	−0.3394****	0.712	−0.3433****	0.709	−0.4446****	0.641	−0.4429****	0.642
Minneapolis-St. Paul	−0.2877****	0.750	−0.288****	0.750	−0.4205****	0.657	−0.4218****	0.656

Table 6. Continued.

Philadelphia	0.1747****	1.191	0.1722****	1.188	−0.4526****	0.636	−0.4512****	0.637
Seattle	0.3122****	1.366	0.3174****	1.374	−0.2172****	0.805	−0.2127****	0.808
Washington D.C.	0.0756	1.079	0.0752	1.078	−0.2503****	0.779	−0.2534****	0.776
Core gentry tract	0.0831	1.087	−0.0872	0.917	−0.1483****	0.862	−0.1449*	0.865
Fringe gentry tract	0.0845	1.088	−0.1319	0.876	−0.1757****	0.839	−0.3101****	0.733
Limit * Core			−0.0986	0.906			0.1882	1.207
Limit * Fringe			−0.1662	0.847			0.3699**	1.448
Income * Core			0.00234**	1.002			−0.00092	0.999
Income * Fringe			0.00332**	1.003			0.000098	1.000
Black * Core			0.1027	1.108			0.3782***	1.460
Black * Fringe			0.2021	1.224			0.3921***	1.480
Number of observations	60,229		60,229		94,329		94,329	
−2 Log Likelihood	43,472		43,462		73,499		73,475	
Chi-square vs. null model	2,074****		2,084****		11,463****		11,487****	
Chi-square vs. Model 1			10*				24****	
Percent correctly classified	65.1		65.1		74.3		74.3*	

* Significant at $p \leq 0.10$; ** $p \leq 0.05$; *** $p \leq 0.01$; **** $p < 0.001$.
Traditional white family is defined as white male applicant, white female co-applicant.
Data Source: FFIEC (1994, 2000).

ing gentrification from an elite, polarizing force into a more progressive balance between moderate- and low-income households.

Addressing these criticisms demands several analytical revisions. We can calculate estimates from the models reported in Table 6 to evaluate the tensions between equitable reinvestment and polarization; but the other critiques require more extensive changes, for which we turn to a separate database developed as part of previously published research (Wyly & Hammel, 1999). This dataset is restricted to 1996 and 1997, capturing the middle of the long 1990s boom but excluding the most recent developments in the turbulent urban land markets of the last two years. The database also encompasses the entirety of each metropolitan area surrounding our case study cities. The most crucial enhancement, however, is an attempt to account for the credit histories of prospective homebuyers. We borrow an approach first suggested by Abariotes et al. (1993) and refined by Holloway (1998) to construct an instrumental variable measuring the likelihood that an applicant will be rejected on the basis of bad credit history (see Table 7, Table 8).[15] This instrument relies on a part of the disclosure file that is optional for lenders, and so the approach is by no means perfect.[16,17] Yet it provides the best possible way of modeling the lending decision with publicly available information. Adding the instrument to the loan denial models takes the analysis into new terrain: it is now possible to disentangle the preferences of well-qualified, high-income borrowers from the institutional calculus of housing finance in the renewed investment into urban neighborhoods.

Consider first the possibility that the resurgence of gentrification is nothing more than an ephemeral, demand-side process. The metropolitan models allow an especially rigorous and conservative test of the hypothesis that reinvestment cannot be explained solely in terms of the characteristics of would-be homebuyers. Adding the credit history instrument yields a significant improvement in model fit, and reduces the severity of racial disparities in loan denial (compare Models 1 and 2, Table 8); but considering credit does not come close to eliminating these deeply entrenched racial divisions, which are understated by the fact that the optional denial codes can be used to mask discriminatory behavior. Equally significant, however, is the persistence of statistically significant coefficient estimates for core and fringe gentrified neighborhoods (Model 3, Table 8). Distinctive lending outcomes persist even after accounting for every demand-side characteristic for which we have information: the type of lending institution, the racial and ethnic identity of homebuyers, and the income, loan amount, and credit profile of borrowers. Even after considering all of these factors, conventional mortgage investment displays a clear preference for the new enclaves of wealth inscribed by historical and contemporary rounds of gentrification. From the perspective of gentrification theory,

Table 7. Logistic Regression of Application Denial for Poor Credit History, 1996–1997.

	Parameter Estimate	Odds Ratio
INTERCPT	−3.1391***	
Applicant income (× 1,000)	−0.0075***	0.993
Loan amount (× 1,000)	−0.0056***	0.994
Female?	−0.1529***	0.858
Traditional white family	−0.2514***	0.778
Black (1 = yes)	1.0463***	2.847
Hispanic (1 = yes)	0.2894***	1.336
Other race (1 = yes)	−0.1374**	0.872
Race unreported (1 = yes)	0.891***	2.438
OCC	1.2157***	3.373
FRB	0.7674***	2.154
FDIC	0.7329***	2.081
OTS	1.1196***	3.064
NCUA	0.3231***	1.381
Number of observations	389,274	
−2 LL	101,808	
Chi-square vs. null model	8,191.3 ***	
Percent correctly classified	71.5*	

* Significant at $p \leq 0.05$; ** $p \leq 0.01$; *** $p \leq 0.001$.
Traditional white family defined as white male applicant, white female co-applicant.
Data Source: FFIEC (1997, 1998).

it matters little *who* exercises this preference – underwriters who evaluate the many risk considerations of a loan; appraisers who must assess the value and risk of the collateral home and its surroundings; or insurance companies who indemnify lenders and investors against the losses of delinquency, default, and foreclosures. What matters is that the systematic preference for gentrified neighborhoods cannot be explained solely in terms of middle-class residential preference. The resurgence of gentrification is fundamentally rooted in the economic calculus of capital investment and the institutional framework of housing policy and mortgage finance.

These results also address the argument that gentrification has little significance at the metropolitan scale. On one level, the critique is indeed persuasive: instances of localized class turnover affect only a small segment of the metropolitan housing market, and are dwarfed by the sheer magnitude of decentralized suburban residential growth. Yet this line of reasoning is a fundamental misunderstanding of gentrification theory: the significance of the process lies

Table 8. Metropolitan Models, 1996–1997.

	Model 1		Model 2		Model 3	
	Parameter estimate	Odds ratio	Parameter estimate	Odds ratio	Parameter estimate	Odds ratio
Intercept	−0.0967****	.	−0.2396****	.	−0.5439****	.
Applicant income (× $1,000)	−0.00552****	0.994	−0.00482****	0.995	−0.00509****	0.995
Loan amount (× $1,000)	−0.0116****	0.988	−0.011****	0.989	−0.0104****	0.990
Loan/income ratio exceeds 3.0	0.6901****	1.994	0.6936****	2.001	0.6646****	1.944
Loan exceeds GSE Limit	1.9314****	6.899	1.8844****	6.582	1.8004****	6.052
Female (1 = yes)	−0.1609****	0.851	−0.1468****	0.863	−0.1245****	0.883
Traditional white family	−0.3283****	0.720	−0.308****	0.735	−0.2669****	0.766
Black	0.5526****	1.738	0.3427****	1.409	0.3623****	1.437
Hispanic	0.0905****	1.095	0.041**	1.042	0.1595****	1.173
Other race	−0.0689****	0.933	−0.0568****	0.945	−0.0345**	0.966
Race unreported	0.3629****	1.437	0.2473****	1.281	0.3042****	1.356
Regulator OCC	−0.4417****	0.643	−0.6047****	0.546	−0.5165****	0.597
Regulator FRB	−0.5242****	0.592	−0.6042****	0.547	−0.5793****	0.560
Regulator FDIC	−0.8612****	0.423	−0.9374****	0.392	−0.8615****	0.423
Regulator OTS	−0.8645****	0.421	−0.9964****	0.369	−0.9967****	0.369
Regulator NCUA	−1.2601****	0.284	−1.288****	0.276	−1.2956****	0.274
Loan for owner occupancy	0.101****	1.106	0.0956****	1.100	0.1219****	1.130
Credit history instrument			3.3844****	29.501	4.2862****	72.690
Boston					0.0689****	1.071
Detroit					0.5235****	1.688
Milwaukee					−0.4664****	0.627
Minneapolis-St. Paul					0.00417	1.004
Philadelphia					−0.1825****	0.833
Seattle					0.3748****	1.455
Washington D.C.					0.2248****	1.252

Table 8. Continued.

Core gentry tract		−0.3134****	0.731
Fringe gentry tract		−0.4069****	0.666
Income * Core		0.00639****	1.006
Income * Fringe		0.00685****	1.007
Number of observations	734,437	734,437	
−2 Log Likelihood	504,051	498,719	
Chi-square vs. null model	63,795****	63,980****	69,128****
Chi-square vs. Model 1		185****	5,332****
Chi-square vs. Model 2			5,147****
Percent correctly classified	74.1	74.1	74.4

* Significant at $p \leq 0.10$; ** $p \leq 0.05$; ***P ≤ 0.01; **** $p < 0.001$.
Traditional white family is defined as white male applicant, white female co-applicant.
Data Source: FFIEC (1994, 2000).

not in its numerical magnitude or spatial extent, but in the concrete, place-based expression of broader social and cultural trends. Viewed in this way, gentrified neighborhoods represent the leading edge of socio-spatial change. Our analysis confirms that housing finance plays an important role in pushing this edge deeper into the disinvested inner city, and that the effect stands out even when compared with the suburbs (see Table 8, Model 3). When we account for every demand-side factor influencing the lending decision, we still find that homebuyers in core gentrified neighborhoods are only three-quarters as likely to be denied compared with their peers elsewhere in the metropolitan area – including the suburbs. The reduced denial likelihood is even more pronounced in fringe gentrified neighborhoods, where applicants are only two-thirds as likely to be rejected. Private market reinvestment and public policy have funneled a wave of capital into the inner city, such that the spatially limited but theoretically significant enclaves of gentrification are now viewed as some of the best investment opportunities in the metropolis.

Finally, consider the argument that the "dirty word" of gentrification (Smith, 1996, Chapter 2) has been eclipsed by a more equitable language and reality of homeownership opportunities in "new markets." The metropolitan model results offer a hint that the benefits of reinvestment have not been limited to upper-income households: note that the income-gentry interaction terms are positive and statistically significant, confirming that higher-income borrowers in gentrified areas are scrutinized more carefully than high-income borrowers elsewhere (see Model 3, Table 8). To explore this issue more fully, we used the parameters of the 1999 denial model (Model 2, Table 6) to examine how different groups of would-be homebuyers are evaluated by lending institutions. It is a simple matter to calculate the likelihood that a "reference applicant" will be rejected, and then to measure how this probability varies with applicant income.[18] The results are striking. Note that at all income levels, applicants in core gentrified neighborhoods are significantly less likely to be turned down than their peers elsewhere in the city. Fringe neighborhoods, by contrast, present greater risks when upper middle-class buyers bump up against the conforming loan limit, beyond which risks cannot easily be offloaded into the secondary market. At incomes below $85,000, then, lenders view homebuyers in core and fringe gentrified neighborhoods as better risks (or profit potential) than applicants in other city neighborhoods. The differential is greatest in the fringe, providing some support for the notion that affordable lending initiatives have broadened opportunities for low and moderate income households.

These findings provide encouraging evidence of equitable reinvestment in parts of the inner city, and the benefits should not be underestimated (Listokin & Wyly, 2000; Quercia & Galster, 1997). Nevertheless, three factors suggest that the evidence is insufficient to proclaim the demise of class polarization. First, the favorable treatment of applicants in gentrified neighborhoods is not limited to low and moderate income buyers. Denial probabilities are lowest for upper middle-class buyers earning around $85,000 who seek homes in the reinvested fringe, corroborating our suspicion that the expansion of geographically-targeted lending has unleashed gentrification pressures. Second, the apparent benefit to lower-income households implied by Fig. 8 refers only to a very specific group: non-Hispanic whites who are ready to purchase homes, and who undertake the application process to obtain conventional credit from prime lending institutions. We know that African American homebuyers in gentrified areas face more severe discrimination than in other parts of the city (recall the race-gentry interaction terms in Table 6); and we also know that the expansion of homeownership invariably undermines efforts to preserve the supply of affordable rental units. Finally, even among low and moderate income homebuyers, our models are almost certainly comparing distinct groups that cannot be empirically distinguished in the loan files. The model results are consistent with a stream of middle-aged, working class minorities making considerable sacrifices to qualify for homes in the remaining affordable blocks in fringe gentrified neighborhoods, while adjacent blocks are invaded by young, white professionals in the early stages of their career and earnings trajectory. Some of these young professionals rely on their parents to come up with downpayments, a social reality that repeatedly frustrates economists' attempts to model home-buying behavior; other young professionals will be upwardly mobile single or divorced women (Markusen, 1980; Rose, 1984). Such "proto-yuppies" or "marginal gentrifiers" figure prominently in theoretical debates over the epistemological status of gentrification (Rose, 1984; Smith, 1996, Chapter 5), but there is no reason to suspect that these groups represent a genuine, progressive expansion of opportunities to residents of working-class urban neighborhoods. If mainstream urban theory is correct, these young professionals will move on to the suburbs in a few years when they have children – perhaps to be replaced by a new cohort of college graduates. If other theoretical perspectives prevail, then these "proto-yuppies" who arrive in search of nontraditional diversity will, over time, come to support the wide array of public and private strategies used to replicate the comforts of a controlled suburban life in the untamed lands of the new urban frontier: "quality of life" crime control, the privatization of public space, the criminalization of homelessness, and the ubiquitous expansion of upscale residential and retail development that enhances that most sacrosanct of American values – property values.

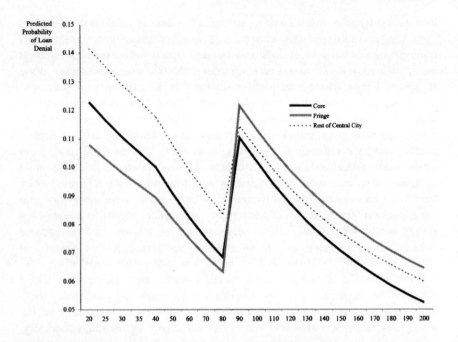

Fig. 8. Conditional Probability Plot of Loan Denial by Applicant Income, 1999.

To illustrate changes in denial probability as applicant income increases, we used the parameters of the denial equation in Table 6 to calculate the likelihood of denial for a "reference applicant" defined to represent the market segment traditionally courted by mainstream financial institutions. We calculated the denial equation for white, non-Hispanic males with white, non-Hispanic female co-applicants seeking conventional loans from prime, national lenders in Chicago. Substituting the core and fringe dummies along with progressive values for the effects of rising income and loan amounts yields the probability curves shown.
Source: Model results shown in Table 6 (FFIEC, 2000).

7. CONCLUSIONS

With time, we may come to see the 1990s as a decade's tidy rhetorical summary of the perennial cycle of decline, crisis, and renaissance that has shaped urban studies in the last half of the twentieth century. The narrative of inner-city devastation that prevailed in the early 1990s was replaced a few years later by an increasingly effusive portrayal of urban fortunes. The image of dysfunctional urban housing markets gave way to a brighter picture of heightened demand for city living, new opportunities for investment and redevelopment, and the long-overdue "discovery" of lucrative new markets in low-income and minority communities. In the realm of urban policy, the pains of recession and fiscal belt-tightening that were justified on the basis of a supposedly permanent federal deficit soon gave way to a frantic race to estimate and exploit the billions and trillions of a mounting surplus. The previous wave of urban cutbacks was largely forgotten in this newfound fiscal abundance, pushing cities farther along the path of privatization, reinvention, and devolution that has guided recent rounds of the "new federalism" in U.S. urban policy.

The empirical economic and rhetorical urban policy shifts of the 1990s have altered the context, process, and outcomes of urban redevelopment. Our research points to three critical facets of the new context of neighborhood change in the inner city. First, we find a clear resurgence of capital investment that lays to rest any doubt that gentrification is an inherent feature of North American urbanization. Measured on the basis of a conservative indicator of investment, the revival of inner-city markets is unmistakable. Conventional, prime investment for home purchases in heavily-redeveloped "core" gentrified neighborhoods expanded at three times the suburban growth rate between 1993 and 1999. Growth has pushed beyond the boundaries of intensely-redeveloped enclaves into "fringe" neighborhoods marked by a juxtaposition of new wealth and persistent poverty; investment in these areas ballooned at nearly four times the suburban growth rate. The resurgence of the 1990s has favored traditional bastions of urban chic that enjoy fully-diversified downtown employment centers, as well as dynamic dot.com economies where suburban housing inflation sustains a recentralization of demand and investment; but resurgence has also reached into seemingly unlikely frontiers, such as Milwaukee and Detroit. Press accounts and fieldwork conducted throughout the 1990s complement the conservative empirical benchmark of mortgage lending data, providing a rich, ground-level view of the localized consequences of inner-city reinvestment.

North American cities are now in the third wave of gentrification (Hackworth, 2000a, b; Hackworth & Smith, forthcoming). Three decades of reinvestment have reshaped the inner city in ways that confirm the durability of a theoretically significant process; and the rebound also carries implications for the choices available to urban policymakers. In the context of selective federal devolution and an ongoing redefinition of public and private institutions, gentrification has come to mediate the design and implementation of local efforts to remake assisted housing policy.

We also find strong evidence of a transformation in the process of gentrification itself. Access to conventional mortgage capital has been eased by the automation and standardization associated with the expansion of the secondary market. Public policy and industry competition have opened new markets among low income and minority borrowers and neighborhoods, blurring the boundaries between sustainable community reinvestment and polarized class turnover in many parts of the inner city. As public policy and private institutions have negotiated a reallocation of risks and rewards, mortgage capital no longer systematically avoids urban neighborhoods. Indeed, there is compelling evidence that structural changes in the housing finance system and public policy have combined to reconcile the once antagonistic goals of "fair lending" and the imperatives of private profitability. Conventional mortgage capital, once responsible for the creation or exacerbation of rent gaps in the inner city, now plays an instrumental role in closing these gaps with a surge of reinvestment. The effect is a simultaneous expansion of affordable homeownership and accelerated class turnover as housing finance unleashes powerful forces of gentrification. Our logistic regression analyses confirm that conventional mortgage capital favors gentrified areas by a significant margin, even after accounting for the varied characteristics of lenders and prospective homebuyers. Housing policy has forged a tight linkage between local trajectories of neighborhood change and national (and global) capital markets, lubricating the mechanisms of investment, disinvestment, and reinvestment that lie at the heart of the production of the urban built environment (Harvey, 1973, 1985; Smith, 1979, 1996).

And so what is new? Our analysis of third-wave gentrification is necessarily limited and provisional, due to the paucity of comparable historical data (cf. Hackworth & Smith, forthcoming; Smith & DeFilippis, 1999; Shlay, 1986). Our use of the overworked language of a "transformation" in urban redevelopment, therefore, might seem exaggerated or premature. And yet some things *do* seem to contrast with earlier incarnations of urban policy, as nearly all realms of public intervention are recast on the foundations of private sector principles. Conceived as a flawed compromise to rectify the failures of private housing activity in the 1930s, assisted housing policy now adapts to and aggressively promotes the imposition of private market principles on poor residents of city neighborhoods. As a manifestation of the circulation of capital in the built environment, gentrification is an inherently unequal and polarizing force – even if public intervention can ameliorate its most severe effects. In a policy environment where every alternative to the market model has been eliminated, however, the reconstruction of urban space to serve the new middle classes promises worsening inequality (Harvey, 2000; Staeheli et al., 1997; Smith, 1984, 1997).

If the historical assessment of third-wave gentrification remains ambiguous, our analysis of its contemporary dynamics is explicitly crafted to be as conservative as possible – in the extent of gentrification, its rate of expansion, and its linkage to broader forces of policy formulation and investment. Our use of a multivariate accept/reject framework to analyze mortgage lending decisions, for example, has the effect of defining a surge of high-income homebuyers in gentrified neighborhoods as a background, demand-side factor – and not as evidence of a resurgence or transformation of the process itself. Yet even after accounting for every borrower and loan characteristic for which we have information, the risks and rewards of mortgage investment in gentrified neighborhoods stand out from the rest of the urban land market. It is our contention that these distinctive geographies cannot be explained solely in terms of the whims of middle-class homebuyers. The reinvention of public and private markets and institutions has transformed the inner city, recasting the disinvested core as a profit opportunity and as a testing ground for a newly-devolved and privatized social policy. There is now a diverse, vibrant, and interdisciplinary body of research that documents separate facets of this complex socioeconomic and geographical change (Bennett, 1998; Bennett & Reed, 1999; Hackworth & Smith, 2000; Lane, 1995; Marcuse, 1998; Nenno, 1998; Quercia & Galster, 1997; Smith, J., 1999; Smith, N., 1997; Smith and DeFilippis, 1999; Vale, 1998). Our contribution, such as it is, simply draws the connections between these separate literatures, and offers a systematic framework to guide analysis in the new urban frontier. And all evidence suggests that further analysis is essential, as private capital and public policy endeavor to reinvent the social production of urban spaces of poverty and wealth.

NOTES

1. Although our primary concern is with conditions measured in the 1990 census, we also incorporate changes from the census enumerations of 1970 and 1980 into the discriminant models. See Wyly and Hammel (1998, 1999).

2. Our methods, of course, are not without limitations. Gentrified areas do not always conform nicely to tract boundaries, and areas of intense redevelopment often straddle two or more tracts. Field research invariably reveals these localized exceptions and allows an assessment of modeling quirks. The time lag between the (now vintage) 1990 census and our fieldwork (conducted between 1994 and 1998; see Table 1) also presents some problems, but it is usually not difficult to identify zones of post-1990 expansion. Finally, the reliance on visible housing reinvestment introduces complications in tightly regulated housing markets where wealthy owners take pains to hide their expenditures (especially New York City). In our case study cities, however, physical rehabilitation and new luxury construction provides a fairly reliable surrogate for the consumption and investment decisions associated with the complex economic and demographic forces of gentrification (Knox, 1991, 1993).

3. All of the figures cited in this paragraph refer to population and housing estimates from the 1990 census. Updated measures of reinvestment in the 1990s are provided by a more specialized database, which is described in a section below.

4. And yet the new urban frontier remains a risky wilderness, with reinvestment and wealth never far removed from poverty, real estate speculation, and predatory business practices. The most recent reminder of the risks came when the *New York Times* published a front-page, above-the-fold article on Harlem's nascent realty boom in November, 2000 (Pristin, 2000), at a time when sustained uncertainty in the aftermath of the Presidential election made it exceedingly difficult to find *any* front-page story that was not concerned with Bush, Gore, Florida recounts, dueling lawsuits, or pregnant chads. The article tempered earlier predictions of a boom in brownstone sales and real estate broker competition. After taking a closer look at Harlem, the city's major brokers found a shortage of choice homes in good condition in the right locations – but even more serious was an unfolding web of federal and state inquiries into abuses of HUD's FHA 203(k) rehabilitation mortgage program, which was "causing havoc just when things seemed to be going so well." (Pristin, 2000, p. 51). Designed to foster homeownership and renovation in low-income neighborhoods, the program remains vulnerable to speculators who acquire properties and flip them at inflated prices to inexperienced or fraudulent nonprofit rehab groups with the cooperation of high-risk mortgage lenders or brokers. Harlem now has 184 203(k) loans in default, out of a total of 450 in New York City.

5. HMDA defines "multifamily" structures as those with five or more dwelling units. Homes with four or fewer units are aggregated into a single category, which has the unfortunate effect of precluding analysis of the complex circumstances of newly-mortgaged landlords. Buying a two- or three-unit building requires a borrower to rely on a stream of rental receipts in order to meet mortgage obligations. In urban areas with the right mix of housing stock and increases in homeownership, there is considerable pressure to force out low- and moderate-income tenants in favor of more affluent renters. To the degree that underwriters or lax rent regulations allow such displacement to occur, the result will be capitalized into still higher house prices.

6. HMDA data have been subjected to a violent barrage of criticism over the past decade, primarily owing to their heavy use in attempts to prove (or disprove) racial discrimination. For our purposes, however, three limitations are most significant. First, the loan application records provide little social information on prospective homebuyers, precluding any nuanced analysis of changes in neighborhood class composition. Second, the data are only comparable after 1993, providing no historical perspective; it is not possible to construct a baseline study of neighborhoods that were gentrified in the 1960s or 1970s. Nevertheless, the data do allow us to examine most of the long boom of the 1990s – a period of buoyant housing market expansion, renewed urban redevelopment, and sweeping changes in the nation's housing finance system (Listokin & Wyly, 2000). Third, HMDA capture only part of the changes involved in urban redevelopment. Residential mortgage flows ignore retail investment, commercial development, and rental market trends – all of which reflect and reinforce capital commitments in the owner-occupied sector. In sum, it is safe to conclude that our reliance on home purchase mortgage data provide a conservative estimate of gentrification activity.

7. Obtained from Housing Research Foundation. World wide web page at: http://www.housingresearch.org, accessed September, 1998 and July, 2000.

8. All of the figures cited in this paragraph are adjusted for inflation and refer to residential nonfarm mortgage debt outstanding on 1- to 4-family structures. See Simmons (1997), pp. 224–225.

9. The 1992 act directed the Secretary of Housing and Urban Development to set preliminary goals for 1993 and 1994, which were to be treated as transitional years. After soliciting and considering comments, HUD established the final rule in 1995, under which a metropolitan area census tract is considered underserved if it meets either of two conditions: (a) its median household income is less than 90% of the area median, or (b) its median income is less than 120% of the area median and it has a minority share greater than 30%. Nationwide, a total of 21,598 tracts qualify as underserved, equivalent to 46% of all tracts in metropolitan statistical areas.

10. In no way do we intend to promote a simplistic conspiracy theory in which the agents and institutions of housing finance collude to foster polarization and displacement in viable urban neighborhoods. Intentions are explicit and admirable. Federal and state policymakers, the GSEs, and neoliberal policy analysts all subscribe to a comprehensive theoretical framework that predicts substantial, shared benefits from mixed-income neighborhoods (e.g., Varady, 1994; Varady, Preiser & Russell, 1998; cf. Schwartz & Tajbakhsh, 1997). We contend, however, that the admirable goals of public policy are too easily distorted and redirected in the gritty realities of urban redevelopment and the strategic mobilization of public burdens to enhance private profit (Jonas & Wilson, 1999; Squires, 1989). Moreover, the public policy consensus on inner-city reinvestment masks considerable disagreement over how underserved areas should be defined. In March, 2000, HUD solicited comments on proposed revisions to the rules specifying the share of Fannie Mae and Freddie Mac purchases that must qualify as "affordable," depending on the income of the borrower or the income and racial composition of the neighborhood. HUD proposed altering the underserved tract definition by reducing the income threshold from 90% to 80% of area median and increasing the minority threshold from 30% to 50% (U.S. Department of Housing and Urban Development, 2000c). Fannie Mae's (2000) extensive comments made it exceedingly clear that the corporation would oppose all attempts to narrow the current underserved definition (which includes almost half of all metropolitan area census tracts). In our view, this dispute provides a reminder that the policy is an attempt to impose a simple dichotomy atop complex, cross-cutting continua of class and neighborhood racial and ethnic composition. At one end of the class continuum are gentrified neighborhoods that qualify for the many policy instruments tied to the definition of underserved tracts.

11. In line with most lending discrimination studies, we exclude observations with data quality flags or with missing values for property location or applicant income. We also exclude observations with extremely high or low applicant income (less than $10,000 or more than $500,000), as well as applications that are incomplete, withdrawn, or approved without subsequent origination of the loan. All 1993 loan and income values are converted to 1999 equivalents with the standard, MSA-level Consumer Price Index deflator for housing expenditures.

12. The extremely high conditional denial probability for subprime lenders is generally consistent with the evidence compiled by the U.S. Department of Housing and Urban Development, which has shown that the growing market share of these lenders explains most of the recent rise in loan rejection rates. But interpretation of the odds ratio is complicated by several factors. First, the variable is coded for applications filed at lenders identified on HUD's list of subprime institutions (Scheeselle, 2000), and is thus a hybrid of loan and lender characteristics. Second, subprime institutions balance the approval decision against loan terms (interest rate, points, fees) in ways that diverge sharply from mainstream lenders; since HMDA provides no information on loan terms,

high denial rates do not mean that subprime lenders are enforcing more stringent lending standards than prime institutions. A third and related factor pertains to market segmentation by loan purpose. Refinance loans account for four-fifths of the subprime market, and thus aggressive marketing and/or predatory practices may yield low denial rates for a subprime institution's refinance and home improvement business while rejection rates remain high in the firm's sideline of home purchase loans (HUD-Treasury Joint Task Force, 2000). For all of these reasons, the parameter estimates for subprime lenders should be interpreted with extreme caution, and are used here primarily to serve as a control to demonstrate that the resurgence of inner-city investment cannot be attributed to the growth of subprime lending.

13. Regulations governing the implementation of HMDA, as amended, require lenders to report racial information on the basis of "visual observation or surname" when an applicant declines to fill out this part of the form. For applications filed by mail, telephone, fax, or electronic communications media without a video component, lenders are not required to make a judgement on the race/ethnicity of the borrower if he or she declines to provide the information; instead, a code is used for "information not provided by applicant in mail or telephone application." (see FFIEC, 1998, Appendix C and D). Applications received by Internet are treated the same way as mail applications, and thus the rapid expansion of online mortgage banking in recent years has propelled a corresponding increase in the proportion of applications without information on borrower race/ethnicity. This regulatory interpretation and technological development creates a substantial, yet ultimately unmeasurable, bias in studies of racial discrimination and credit allocation.

14. These factors can include rental market conditions and state tax policies affecting demand for owner-occupied housing, the market share held by locally-owned institutions, the extent of targeted marketing and outreach programs, the importance of nonprofit lenders or consortia, and the effectiveness of housing counseling programs. See Listokin et al. (1998); Listokin and Wyly (2000).

15. In the vociferous debate over lending discrimination, one of the most serious methodological problems is the paucity of applicant financial characteristics in publicly available mortgage disclosure files. Economists advocating the axiomatic view of discrimination as economically irrational, therefore, argue that racial disparities in loan approval are explained by unobserved differences in credit history. It is possible, however, to derive indirect information on credit characteristics from the HMDA files. When lenders reject an application, they may report up to three reasons for their decision. Options include an excessive debt to income ratio; unstable employment history; poor credit history; insufficient collateral; insufficient cash to meet downpayment or closing costs; unverified information; incomplete application; denial of private mortgage insurance; and an unspecified "other" category. Our instrumental variable is constructed from a logit model predicting the likelihood that any one of the three denial indicators is reported for poor credit history (see Table 7). To avoid circularity, the bad credit model is calibrated with a random sample of half of all applications; the parameters are then used to construct a probability value for each application in the full dataset.

16. It is important to recognize that the limitations of the credit history instrument are biased against any finding of racial discrimination: since lenders are free to report denial codes without any close oversight, there is nothing to prevent them from using the codes to mask discriminatory behavior. Calibrating rejection models with the credit history instrument, therefore, has the effect of controlling for (thereby legitimating) a certain amount of bigotry, disparate treatment, and adverse impact discrimination.

17. Regulatory changes in 1998 complicate the use of the credit history instrument. For most of the 1990s, the denial codes were optional; about two-thirds of all rejections included at least one reason cited by the lender. Lenders regulated by the Office of the Comptroller of the Currency or the Office of Thrift Supervision, however, are no longer given the option of leaving the denial codes blank for rejections (see FFIEC, 1998). Since the codes remain optional for credit unions, independent mortgage companies, and other lenders, there are strong reasons to suspect systematic bias in reasons for denial among different kinds of institutions. For these reasons, our loan denial models for 1999 do not include the credit history instrument.

18. We calculated the likelihood of denial for a "traditional" family couple (white, non-Hispanic male applicant, white, non-Hispanic female coapplicant) seeking a conventional loan for an owner-occupied dwelling from a national (FRB-regulated) lender specializing in the prime market in Chicago. We also assumed that applicants request loans at a standard multiple of 2.5 times their annual income. Adding variables to measure the effects of rising income and loan amounts yields the probability values graphed in Fig. 8.

REFERENCES

Abariotes, A., Ahuja S., Feldman, H., Johnson, C., Subaiya, L., Tiller, N., Urban, J., & Myers, S. L. (1993). Disparities in Mortgage Lending in the Upper Midwest: Summary of Results Using 1992 Home Mortgage Disclosure Act Data. Paper presented at the Fannie Mae University Colloquium on Race, Poverty, and Housing Policy. Minneapolis: University of Minnesota.

Adams, J. S., & VanDrasek, B. J. (1993). *Minneapolis-St. Paul: People, Place, and Public Life*. Minneapolis: University of Minnesota Press.

Bagli, C. V. (2000). As an Industrial Building in Chelsea Goes 'Shabby Chic', Charges Fly. *New York Times*, February 18, p. B8.

Barnes, J. E. (2000). The Trendy Frontier? Eastward Ho in Brooklyn. *New York Times*, July 30, p. 29.

Bennett, L., & Reed, A., Jr. (1999). The New Face of Urban Renewal: The Near North Side Redevelopment Initiative and the Cabrini-Green Neighborhood. In: A. Reed, Jr. (Ed.), *Without Justice for All: The New Liberalism and Our Retreat from Racial Equality* (pp. 175–211). Boulder, CO: Westview Press.

Berry, B. J. L. (1980). Inner City Futures: An American Dilemma Revisited. *Transactions of the Institute of British Geographers*, ns 5: 1–28.

Berry, B. J. L. (1985). Islands of Renewal in Seas of Decay. In: P. E. Peterson (Ed.), *The New Urban Reality* (pp. 69–96). Washington, D.C.: Brookings.

Berry, B. J. L. (1999). Comment on Elvin K. Wyly and Daniel Hammel's 'Islands of Decay in Seas of Renewal: Housing Policy and the Resurgence of Gentrification – Gentrification Resurgent?' *Housing Policy Debate*, 10, 738–788.

Berry, B. J. L., & Kasarda, J. D. (1977). *Contemporary Urban Ecology*. New York: Macmillan.

Boger, J. C., & Welch Wegner, J. (Eds) (1996). *Race, Poverty, and American Cities*. Chapel Hill: University of North Carolina Press.

Bondi, L. (1991). Gender Divisions and Gentrification: A Critique. *Transactions, Institute of British Geographers*, NS 16, 190–198.

Bourne, L. S. (1993a). The Demise of Gentrification? A Commentary and Prospective View. *Urban Geography, 14*(1), 95–107.
Bourne, L. S. (1993b). The Myth and Reality of Gentrification: A Commentary on Emerging Urban Forms. *Urban Studies, 30*(1), 183–189.
Brozan, N. (2000). As Harlem Gentrifies, Brokers Seek Space – For Themselves. *New York Times*, January 28, p. B9.
Canner, G. B., & Gabriel, S. A. (1992). Market Segmentation and Lender Specialization in the Primary and Secondary Mortgage Markets. *Housing Policy Debate, 3*(2), 241–332.
Carliner, M. S. (1998). Development of Federal Homeownership 'Policy.' *Housing Policy Debate, 9*, 299–321.
Carr, J., & Megbolugbe, I. (1993). The Federal Reserve Bank of Boston Study on Mortgage Lending Revisited. *Journal of Housing Research, 4*, 277–314.
Castells, M., & Murphy, K. (1982). Cultural Identity and Urban Structure. In: N. I. Fainstein & S. S. Fainstein (Eds), *Urban Policy Under Capitalism* (pp. 237–259). Beverly Hills, CA: Sage Publications.
Chicago Housing Authority (1997). *HOPE VI Revitalization Plan: Cabrini-Green Extension.* Chicago: CHA.
Chicago Housing Authority (2000a). *CHA Transformation Plan Approved.* Press Release, January 6.
Chicago Housing Authority (2000b). Cabrini-Green. World Wide Web page: http://www.thecha.org/ Cabrini_History1.htm (accessed May 22, 2000).
Claiborne, W. (2000). HUD, Chicago Ink Deal to Reconstruct Public Housing. *Washington Post*, February 6, p. A2.
Clark, E. (1992). On Blindness, Centrepieces and Complementarity in Gentrification Theory. *Transactions, Institute of British Geographers*, NS 17, 358–362.
Clay, P. L. (1979). *Neighborhood Renewal: Trends and Strategies.* Lexington MA: Lexington Books.
Cole, D. B. (1987). Artists and Urban Redevelopment. *Geographical Review, 77*(4), 391–407.
Committee on Government Reform and Oversight, U.S. House of Representatives (1995). *The Federal Takeover of the Chicago Housing Authority.* House Report 104–437. Washington, D.C.: U.S. Government Printing Office.
Dear, M., & Flusty, S. (1998). Postmodern Urbanism. *Annals of the Association of American Geographers, 88*(1), 50–72.
Debord, G. (1994). *The Society of the Spectacle.* Original Publication 1967. Translated by D. Nicholson-Smith. New York: Zone Books.
Dedman, B. (1988). The Color of Money. *Atlanta Journal-Constitution*, May 1–4.
Dixon, J., & Solomon, D. (1997). Motown Rebound: Detroit Sees $5.74 Billion in New Investments. *Detroit Free Press*, April 15, A1.
Downs, A. (2000). Comment on John T. Metzger's 'Planned Abandonment: The Neighborhood Life-Cycle Theory and National Urban Policy.' *Housing Policy Debate, 11*(1), 41–54.
Dunlap, D. W. (2000). For 1930's Behemoth, A New Upscale Life. *New York Times*, February 20, Section 11, p. 1, 6.
Epp, G. (1998). Emerging Strategies for Revitalizing Public Housing Communities. In: D. P. Varady, W. F. E. Preiser & F. P. Russell (Eds), *New Directions in Urban Public Housing* (pp. 121–141). New Brunswick, NJ: Center for Urban Policy Research.
Fannie Mae (2000). *Fannie Mae's Comment Letter on HUD's Proposed Affordable Housing Goals.* Washington D.C.: Fannie Mae.
Fannie Mae Foundation (1997). *Annual Report, 1996.* Washington D.C.: Fannie Mae Foundation.

Federal Financial Institutions Examination Council (1999). *A Guide to HMDA Reporting: Getting it Right*. Washington D.C.: FFIEC.
Federal Financial Institutions Examination Council. Annual. *Home Mortgage Disclosure Act Raw Data*. Machine-readable data file and technical documentation on CD-ROM. Washington D.C.: Federal Financial Institutions Examination Council.
Fishbein, A. (1992). The Ongoing Experiment with 'Regulation from Below': Expanded Reporting Requirements for HMDA and CRA. *Housing Policy Debate*, *3*(2), 601–638.
Fusch, R. (1980). A Case of Too Many Actors? Columbus. In: S. Laska & D. Spain (Eds), *Back to the City: Issues in Neighborhood Renovation* (pp. 156–172). New York: Pergamon.
Giddens, A. (1984). *The Constitution of Society*. London: Polity Press.
Galster, G. C. (2000). Comment on John T. Metzger's 'Planned Abandonment: The Neighborhood Life-Cycle Theory and National Urban Policy.' *Housing Policy Debate*, *11*(1), 61–66.
Giloth, R., & Betanuer, J. (1988). Where Downtown Meets Neighborhood: Industrial Displacement in Chicago, 1978–1988. *Journal of the American Planning Association*, *54*, 279–290.
Glass, R. (1964). *London: Aspects of Change*. London: MacGibbon and Kee.
Goetz, E. (2000). The Politics of Poverty Deconcentration. *Journal of Urban Affairs*, *22*(2), 157–173.
Goldberg, C. (1999). Behind the Curtains of Boston's Best Neighborhood, a New Elite. *New York Times*, February 18, p. A16.
Gotham, K. F. (1998). Blind Faith in the Free Market: Urban Poverty, Residential Segregation, and Federal Housing Retrenchment, 1970–1995. *Sociological Inquiry*, *68*(1), 1–31.
Gotham, K. F., & Wright, J. D. (2000). Housing Policy. In: J. Midgley, M. B. Tracy & M. Livermore (Eds), *The Handbook of Social Policy* (pp. 237–255). Thousand Oaks, CA: Sage Publications.
Gottdiener, M. (1985). *The Social Production of Urban Space*. Austin: University of Texas Press.
Hackworth, J. (2000a). Third Wave Gentrification. Unpublished Ph.D. Thesis. New Brunswick, NJ: Rutgers University, Department of Geography.
Hackworth, J. (2000b). State Devolution, Urban Regimes, and the Production of Geographic Scale: The Case of New Brunswick, NJ. *Urban Geography*, *21*(5), 450–458.
Hackworth, J., & Smith, N. (2001). The Changing State of Gentrification. Accepted and forthcoming at *Tijdschrift voor Economische en Sociale Geografie*.
Hall, T. (2000). Discovering Bed-Stuy's Brownstones. *New York Times*, May 21, p. NJ RE 1.
Hamilton, W. L. (2000). Visions of Greener Pastures Reinvent the Lower East Side. *New York Times*, January 6, p. F1, F7.
Hammel, D. J. (1999). Gentrification and Land Rent: A Historical View of the Rent Gap in Minneapolis. *Urban Geography*, *20*(2), 116–145.
Hammel, D. J., & Wyly, E. K. (1996). A Model for Identifying Gentrified Areas with Census Data. *Urban Geography*, *17*, 248–268.
Hamnett, C. (1984). Gentrification and Residential Location Theory: A Review and Assessment. In: D. Herbert & R. Johnston (Eds), *Geography and the Urban Environment: Progress in Research and Applications* (Vol. 6, pp. 282–319). New York: John Wiley.
Harvey, D. (1973). *Social Justice and the City*. London: Edward Arnold.
Harvey, D. (1985). *The Urbanization of Capital*. Baltimore: Johns Hopkins.
Harvey, D. (2000). *Spaces of Hope*. Berkeley: University of California Press.
Hawley, A. (1981). *Urban Society: An Ecological Approach* (2nd ed.). New York: John Wiley and Sons.
Hevesi, D. (1998). Apartment Prices as Moving Targets. *New York Times*, July 26, Section 11, p. 1.
Hirsch, A. R. (1998). *Making the Second Ghetto: Race and Housing in Chicago, 1940–1960*. (2nd ed.). Chicago: University of Chicago Press.

Historic Landmarks for Living (2000). *Historic Landmarks for Living Web Site.* World Wide Web pages, available at: http://www.historiclandmarks.com, accessed November, 2000.

Holcomb, H. B., & Beauregard, R. A. (1981). *Revitalizing Cities.* Washington D.C.: Association of American Geographers.

Holloway, S. R. (1998). Exploring the Neighborhood Contingency of Race Discrimination in Mortgage Lending in Columbus, Ohio. *Annals of the Association of American Geographers, 88,* 252–276.

Hoover, E. M., & Raymond, V. (1959). *Anatomy of a Metropolis.* Cambridge, MA: Harvard University Press.

HUD-Treasury Joint Task Force on Predatory Lending (2000). *Final Report and Recommendations of the Joint Task Force on Predatory Lending.* Washington D.C.: U.S. Department of Housing and Urban Development.

Housing Research Foundation (1998). *HOPE VI Implementation and Planning Grant Awards, 1993–1998.* Washington D.C.: Housing Research Foundation. World Wide Web Page http://www.housingresearch.org, accessed in September 1998.

Hutchison, R. (Ed.) (1992). *Gentrification and Urban Change.* Research in Urban Sociology, 2. Stamford, CT: JAI Press.

Jackson, K. T. (1985). *Crabgrass Frontier: The Suburbanization of the United States.* New York: Oxford University Press.

Jacobs, A. (1998). A New Spell for Alphabet City. *New York Times,* August 9, Section 14, p. 1, 10.

Jonas, A. E. G., & Wilson, D. (Eds) (1999). *The Urban Growth Machine: Critical Perspectives Two Decades Later.* Albany: SUNY Press.

Kasarda, J. D. (1993). *Urban Underclass Database.* Machine-Readable Data File and Technical Documentation. New York: Social Science Research Council.

Kasarda, J. D. (1999). Comment on Elvin K. Wyly and Daniel J. Hammel's 'Islands of Decay in Seas of Renewal: Housing Policy and the Resurgence of Gentrification' *Housing Policy Debate, 10,* 773–781.

Klages, K. E. (1998). Sweet Home Chicago. *Chicago Tribune Magazine,* February 22, 12–16.

Knox, P. L. (1991). The Restless Urban Landscape: Economic and Sociocultural Change and the Transformation of Metropolitan Washington D.C. *Annals of the Association of American Geographers 81,* 181–209.

Knox, P. (Ed.) (1993). *The Restless Urban Landscape.* Englewood Cliffs, NJ: Prentice Hall.

Lane, V. (1995). Best Management Practices in U.S. Public Housing. *Housing Policy Debate,* 6(4), 867–904.

Laska, S. B., & Spain, D. (Eds) (1980). *Back to the City: Issues in Neighborhood Renovation.* New York: Pergamon.

Lees, L., & Bondi, L. (1995). De-Gentrification and Economic Recession: The Case of New York City. *Urban Geography, 16,* 234–253.

Lefebvre, H. (1991). *The Production of Space.* Original publication 1974. Translated by D. Nicholson-Smith. Oxford: Blackwell.

Leinsdorf, D., and Etra, D. (1973). *Citibank: Ralph Nader's Study Group Report on First National City Bank.* New York: Grossman Publishers.

Ley, D. (1980). Liberal Ideology and the Postindustrial City. *Annals of the Association of American Geographers, 70*(2), 238–258.

Ley, D. (1981). Inner-City Revitalization in Canada: A Vancouver Case Study. *Canadian Geographer, 25,* 124–148.

Lipton, G. S. (1977). Evidence of Central City Revival. *American Institute of Planners Journal*, 45, 136–147.
Listokin, D., & Wyly, E. (2000). Making New Mortgage Markets: Case Studies of Institutions, Homebuyers, and Communities. *Housing Policy Debate*, 11(3), 575–644.
Listokin, D., Wyly, E., Keating, L., Wachter, S. M., Rengert, K. M., & Listokin, B. (1998). *Successful Mortgage Lending Strategies for the Underserved*, Volume I. *Industry Strategies*, Volume II. *Case Studies*. Washington, D.C.: U.S. Department of Housing and Urban Development, Office of Policy Development and Research.
Logan, J., & Molotch, H. (1987). *Urban Fortunes: The Political Economy of Place*. Berkeley: University of California Press.
Long, L. H. (1980). Back to the Countryside and Back to the City in the Same Decade. In: S. Laska & D. Spain (Eds), *Back to the City: Issues in Neighborhood Renovation* (pp. 61–76). New York: Pergamon.
Markusen, A. (1980). City Spatial Structure, Women's Household Work, and National Urban Policy. *Signs*, 5(3), Suppl. 1: S23–S44.
Marcuse, P. (1986). Abandonment, Gentrification, and Displacement: The Linkages in New York City. In: N. Smith & P. Williams (Eds), *Gentrification of the City* (pp. 153–177). Boston: Allen and Unwin.
Marcuse, P. (1998). Mainstreaming Public Housing. In: D. P. Varady, W. F. E. Preiser & F. P. Russell (Eds), *New Directions in Urban Public Housing* (pp. 23–44). New Brunswick, NJ: Center for Urban Policy Research.
Meredith, R. (1999). Detroit, Still Blighted, Puts Hopes in Casinos. *New York Times*, July 30, p. A12.
Metzger, J. T. (2000). Planned Abandonment: The Neighborhood Life-Cycle Theory and National Urban Policy. *Housing Policy Debate*, 11(1), 7–40.
Miller, S. R. (1998). Order and Democracy: Trade-Offs Between Social Control and Civil Liberties at Lake Parc Place. *Housing Policy Debate*, 9, 757–773.
Mollenkopf, J., & Castells M. (Eds) (1991). *Dual City: Restructuring New York*. New York: Russell Sage Foundation.
Munnell, A., Browne, L., McEneany, J., & Tootell, G. (1992). Mortgage Lending in Boston: Interpreting HMDA Data. Working Paper 92–7. Boston: Federal Reserve Bank of Boston.
Munnell, A., Tootell, G., Browne, L., and McEneany, J. (1996). Mortgage Lending in Boston: Interpreting HMDA Data. *American Economic Review*, 86, 25–53.
Nenno, M. (1998). New Directions for Federally Assisted Housing: An Agenda for the Department of Housing and Urban Developmnent. In: D. P. Varady, W. F. E. Preiser & F. P. Russell, (Eds), *New Directions in Urban Public Housing* (pp. 205–225). New Brunswick, NJ: Center for Urban Policy Research.
Newman, O. (1972). *Defensible Space: Crime Prevention Through Urban Design*. New York: Macmillan.
Newman, O. (1980). *Community of Interest*. Garden City, NY: Anchor.
Nieves, Evelyn. (2000). As a Mayor, Jerry Brown is Down to Earth. *New York Times*, February 11, p. A16.
Nyden, P. (1998). Comment on James E. Rosenbaum, Linda K. Stroh, and Cathy A. Flynn's 'Lake Parc Place: A Study of Mixed-Income Housing.' *Housing Policy Debate*, 9, 741–748.
Osborne, D., & Gaebler, T. A. (1992). *Reinventing Government: How the Entrepreneurial Spirit is Transforming the Public Sector*. Reading, MA: Addison-Wesley.
Palen, J., & London, B. (Ed.) (1984). *Gentrification, Displacement, and Neighborhood Revitalization*. Albany: State University of New York Press.

Pataki, G. E. (1998). Governor Pataki Commits $3 Million to Harlem USA. Albany: Office of the Governor, Press Release, July 27.

Pinder, D. (2000). Old Paris is No More: Geographies of Spectacle and Anti-Spectacle. *Antipode*, 32(4), 357–386.

Pristin, T. (2000). Inquiries on Mortgage Deals Crimp Harlem's Realty Boom. *New York Times*, November 26, pp. A1, 51.

Quercia, R., & Galster, G. (1997). The Challenges Facing Public Housing Authorities in a Brave New World. *Housing Policy Debate*, 8, 535–569.

Ricketts, E., & Sawhill, I. (1988). Defining and Measuring the Underclass. *Journal of Policy Analysis and Management*, 7, 316–325.

Rose, D. (1984). Rethinking Gentrification: Beyond the Uneven Development of Marxist Urban Theory. *Society and Space*, 2(1), 47–74.

Rosenbaum, J. E., Stroh, L. K., & Flynn, C. A. (1998). Lake Parc Place: A Study of Mixed-Income Housing. *Housing Policy Debate*, 9, 703–740.

Rozhon, T. (1998). Dreams, and Now Hope, Among the Ruins. *New York Times*, April 16, p. F1.

Rozhon, T. (2000). Battle of the Brokers in Harlem. *New York Times*, April 27, p. F1.

Rusk, D. (1994). Bend or Die: Inflexible State Laws and Policies are Dooming Some of the Country's Central Cities. *State Government News*, February, 6–10.

Salama, J. J. (1999). The Redevelopment of Distressed Public Housing: Early Results from HOPE VI Projects in Atlanta, Chicago, and San Antonio. *Housing Policy Debate*, 10, 95–142.

Sassen, S. (1991). *The Global City*. Princeton: The Princeton University Press.

Schaffer, R., & Smith, N. (1986). The Gentrification of Harlem? *Annals of the Association of American Geographers*, 76(3), 347–365.

Schill, M. H. (1997). Chicago's Mixed Income New Communities Strategy: The Future Face of Public Housing? In: W. van Vliet (Ed.), *Affordable Housing and Urban Development in the United States* (pp. 135–157). Thousand Oaks, CA: Sage Publications.

Schwartz, A. (1998). From Confrontation to Collaboration? Banks, Community Groups, and the Implementation of Community Reinvestment Agreements. *Housing Policy Debate*, 9, 631–662.

Schwartz, A., & Tajbakhsh, K. (1997). Mixed Income Housing: Unanswered Questions. *Cityscape*, 3(2), 71–92.

Shlay, A. B. (1986). *A Tale of Three Cities: The Distribution of Housing Credit from Financial Institutions in the Chicago SMSA from 1980–1983*. Chicago: Woodstock Institute.

Simmons, P. A. (Ed.) (1998). *Housing Statistics of the United States*. Lanham, MD: Bernan Press.

Smith, J. L. (1999). Cabrini Green and the Redevelopment Imperative. Paper presented at the Annual Conference of the Urban Affairs Association, Louisville, KY.

Smith, N. (1979a). Gentrification and Capital: Theory, Practice, and Ideology in Society Hill. *Antipode*, 11(3), 24–35.

Smith, N. (1979b). Toward a Theory of Gentrification: A Movement Back to the City by Capital, Not People. *Journal of the American Planning Association*, 45, 538–548.

Smith, N. (1982). Gentrification and Uneven Development. *Economic Geography*, 58, 139–155.

Smith, N. (1984). *Uneven Development*. Oxford: Basil Blackwell.

Smith, N. (1986). Gentrification, the Frontier, and the Restructuring of Urban Space. In: N. Smith & P. Williams (Eds), *The Gentrification of the City* (pp. 15–34). Boston: Allen and Unwin.

Smith, N. (1996). *The New Urban Frontier: Gentrification and the Revanchist City*. New York: Routledge.

Smith, N. (1997). Social Justice and the New American Urbanism: The Revanchist City. In: A. Merrifield & E. Swyngedouw (Eds), *The Urbanization of Injustice* (pp. 117–136). New York: New York University Press.
Smith, N., & Herod, A. (1991). *Gentrification: A Comprehensive Bibliography*. New Brunswick, NJ: Rutgers University, Department of Geography.
Smith, N., & DeFilippis, J. (1999). The Reassertion of Economics: 1990s Gentrification in the Lower East Side. *International Journal of Urban and Regional Research, 23*, 638–653.
Soja, E. W. (1980). The Socio-Spatial Dialectic. *Annals of the Association of American Geographers, 70*(2), 207–225.
Soja, E. W. (1996). *ThirdSpace: Journeys to Los Angeles and other Real-and-Imagined Places*. London: Blackwell.
Squires, G. D. (Ed.) (1989). *Unequal Partnerships: The Political Economy of Urban Redevelopment in Postwar America*. New Brunswick, NJ: Rutgers University Press.
Squires, G. D. (Ed.) (1992). *From Redlining to Reinvestment*. Philadelphia: Temple University Press.
Staeheli, L. A., Kodras, J. E., & Flint, C. (Eds) (1997). *State Devolution in America: Implications for a Diverse Society*. Urban Affairs Annual Review 48. Thousand Oaks, CA: Sage Publications.
Stegman, M. A. (1996). National Urban Policy Revisited: Policy Options for the Clinton Administration. In: J. C. Boger & J. Welch Wegner (Eds), *Race, Poverty, and American Cities* (pp. 228–269). Chapel Hill: University of North Carolina Press.
Stegman, M. A. (1995). Recent U.S. Urban Change and Policy Initiatives. *Urban Studies, 32*, 1601–1607.
Stegman, M., Quercia, R., McCarthy, G., & Rohe, W. (1991). Using the Panel Survey of Income Dynamics (PSID) to Evaluate the Affordability Characteristics of Alternative Mortgage Instruments and Homeownership Assistance Programs. *Journal of Housing Research 2*, 161–211.
Teaford, J. C. (2000). Urban Renewal and its Aftermath. *Housing Policy Debate, 11*(2), 443–465.
Temkin, K. (2000). Comment on John T. Metzger's 'Planned Abandonment: The Neighborhood Life-Cycle Theory and National Urban Policy.' *Housing Policy Debate, 11*(1), 55–60.
U.S. Department of Housing and Urban Development (2000a). Cuomo, Daley, and Rush Unveil Agreement to Protect Chicago Housing Authority Residents and Support Redevelopment Plan. Press Release, HUD No. 00–25, February 7.
U.S. Department of Housing and Urban Development (2000b). *Fiscal Year 2001 Budget Summary: HUD: Back in Business*. Washington D.C.: U.S. Department of Housing and Urban Development.
U.S. Department of Housing and Urban Development (2000c). HUD's Regulation of the Federal National Mortgage Association (Fannie Mae) and the Federal Home Loan Mortgage Corporation (Freddie Mac); Final Rule. *Federal Register, 65*(211), October 31, 65,044–65,092.
U.S. Department of Housing and Urban Development (2000d). *Rental Housing Assistance: The Worsening Crisis*. Report to Congress on Worst-Case Housing Needs. Washington D.C.: U.S. Department of Housing and Urban Development, Office of Policy Development and Research.
U.S. General Accounting Office (1998). *HOPE VI: Progress and Problems in Revitalizing Distressed Public Housing*. GAO/RCED-98-187. Washington, D.C.: U.S. General Accounting Office.
U.S. House of Representatives (1996). *United States Housing Act of 1996. Report Together with Minority and Additional Views, to Accompany H.R. 2406*. Report 104–461. Washington D.C.: U.S. Government Printing Office.

Vale, L. J. (1998). Comment on James E. Rosenbaum, Linda K. Stroh, and Cathy A. Flynn's 'Lake Parc Place: A Study of Mixed-Income Housing.' *Housing Policy Debate*, 9, 749–756.
van Weesep, J. (1994). Gentrification as a Research Frontier. *Progress in Human Geography*, 18(1), 74–83.
Varady, D. P. (1994). Middle Income Housing Programs in American Cities. *Urban Studies*, 31(8), 1345–1366.
Varady, D. P., Preiser, W. F. E., & Russell, F. P. (Eds) (1998). *New Directions in Urban Public Housing*. New Brunswick, NJ: Center for Urban Policy Research.
Vartanian, T. P., Ledig, R. H., Babitz, A., Browning, W. L., & Pitzer, J. G. (1995). *The Fair Lending Guide*. Little Falls, NJ: Glasser LegalWorks.
Walsh, T. J. (1997). *Developers Converting Board Rooms to Bedrooms*. BizJournals Philadelphia. October 31, p. 1.
Waste, R. J. (1998). *Independent Cities: Rethinking U.S. Urban Policy*. New York: Oxford.
Wicker, T. (1968). Introduction. In: *Report of the National Advisory Commission on Civil Disorders* (pp. iv–xi). New York: Bantam Books.
Wilson, F. H. (1992). Gentrification and Neighborhood Dislocation in Washington, D.C.: The Case of Black Residents in Central Area Neighborhoods. In: R. Hutchison (Ed.), *Research in Urban Sociology: Gentrification and Urban Change, Volume 2* (pp. 113–143). Stamford, CT: JAI Press.
Wilson, W. J. (1987). *The Truly Disadvantaged: The Inner City, the Underclass, and Public Policy*. Chicago: University of Chicago Press.
Wilson, W. J. (1996). *When Work Disappears: The New World of the Urban Poor*. New York: Knopf.
Wyly, E. K., & Hammel, D. J. (1998). Modeling the Context and Contingency of Gentrification. *Journal of Urban Affairs*, 20, 303–326.
Wyly, E. K., & Hammel, D. J. (1999). Islands of Decay in Seas of Renewal: Housing Policy and the Resurgence of Gentrification. *Housing Policy Debate*, 10, 711–771.
Wyly, E. K., & Hammel, D. J. (2000, forthcoming). Capital's Metropolis: Chicago and the Transformation of American Housing Policy. Accepted and forthcoming at *Geografiska Annaler, Series B, Human Geography*.
Yinger, J. (1995). *Closed Doors, Opportunities Lost: The Continuing Costs of Housing Discrimination*. New York: Russell Sage Foundation.
Zorbaugh, H. W. (1929). *The Gold Coast and the Slum*. Chicago: University of Chicago Press.
Zukin, S. (1982). *Loft Living, Culture and Capital in Urban Change*. Baltimore: Johns Hopkins University Press.

HOUSEHOLD SURVIVAL STRATEGIES IN A PUBLIC HOUSING DEVELOPMENT

Joel A. Devine and Petrice Sams-Abiodun

INTRODUCTION

It has been well documented that those living in low-income neighborhoods employ numerous survival strategies to provide for their families and households (see, e.g. Stack, 1974; Edin, 1991; Edin & Lein, 1996, 1997). People who are "welfare reliant" rarely survive solely on financial assistance and benefits from the government. Likewise, many of the working poor cannot depend alone on the meager wages of their unskilled or semi-skilled jobs (Edin & Lein, 1997; Devine & Wright, 1993). The present paper provides a multi-method case study of a single public housing community. While our motivating research question and ultimate focus is on how residents of this impoverished, inner-city community meet their material needs, we first contextualize the investigation by: (1) reconsidering prevailing conceptions of households and families and scrutinizing their applicability vis-à-vis a poor, inner-city, African-American population; (2) tracing the history of this particular housing development and some of the critical policy changes that undermined the viability of this community and much of public housing in the United States; and (3) examining tract-level census data and information gathered in a series of community surveys conducted over the past four years. Limited by the information gleaned from these sources and methods, we subsequently initiated a team ethnography and undertook a series of focus groups in an effort to investigate and more fully appreciate the formal and informal strategies people in a low-income community use to provide for their survival.

THEORETICAL OVERVIEW

In directing our attention to developing an understanding of household survival strategies within a public housing development, we quickly came to the realization that we had to rethink some of our most basic concepts, namely, what constitutes a household, a family, and survival in low-income neighborhoods such as C. J. Peete (formerly known as Magnolia). Much, though by no means all, of the extant literature assumes middle-class conventions and did not appear to fit – or help explain – our early empirical observations. Moreover, the emphasis on the matriarchical family structure of the black poor (see Moynihan, 1965; Wilson, 1987) overlooks the presence and import of black males within the community. We contend that in order to understand household survival strategies in low-income communities such as Peete, the definition of family must be broadened to include kinship systems and especially the residential fluidity of men.

There are three analytically (if not empirically) distinct problems in the limited approach to defining family that hinders our understanding of survival strategies of low-income families. First, there is the problem of distinguishing between family and the household units. Second, is a preoccupation with the nuclear family. Third, is a dichotomous focus on male's presence or absence.

Distinguishing Family From Household

The first problem in the literature is the failure to distinguish family from household. Official (i.e. Bureau of the Census) usage defines family as "a group of two or more persons related by birth, marriage, or adoption who reside together", whereas "households consist of all persons who occupy a housing unit" (Bureau of the Census, 1998: A-1).[1] Not surprisingly, this usage often yields confusion; a situation compounded by the fact that sociologists and demographers typically have been interested in the former, while economists have tended to focus on the latter (e.g. Deaton, 1997). Zollar (1985) claims that using the terms "family" and "household" as though they are interchangeable reinforces the myth of the female-headed black family; asserting that this understandable but unfortunate conflation involves the tendency of scholars to confuse a statistically significant fact – the presence of some female-headed households – with the typical or characteristic domestic arrangement.

Bogue (1969, 368) in his classic, *Principles of Demography*, identified the household as "an economic, ecological, or livelihood unit" consisting of those who, "put their feet under the same table or otherwise join together in an arrangement to provide food, shelter, and other basic residential necessities". However, a number of ethnographic researchers and other demographers suggest that among low-income, urban, black families, this definition based on "feet under the table" might not be sufficient to capture the family unit. Prominent researchers have found that black households have multiple and fluid membership (Stack, 1974; Valentine & Valentine, 1971).

Ryder (1987a, 1992) notes that a central issue with respect to the family concerns the delineation of the time pattern and spatial-interactional association of relationships. Recognizing that families change over time and as a result of "complex intergenerational relationships", Ryder calls for a more dynamic conceptualization that reflects both time and space considerations. In addition, external forces, such as economics and changes in housing and other social policies, influence residency patterns and have a profound impact on male attachment patterns in low-income communities. In *Reconsideration of a Model of Family Demography*, Ryder (1987b) elucidates three divisions of family demography: the marriage market, family history, and adoption market. The latter addresses kinship and residence with family as their intersection and is of particular interest for understanding household survival strategies (p. 114). Ryder argues that the residential group (i.e. household) is much less amenable to a demographic approach than the kinship group. Members who share activities of day-to-day survival, principally the provision of food and shelter, characterize the former. Alternatively, members of kinship groups share in those activities concerned with long-term survival of the group, not only in terms of membership, but also as an institutional structure (p. 115). Ryder concludes that the breakdown into kinship and residence has important implications for family.

We believe that Ryder's adoption market perspective facilitates examination of the empirically dynamic attachments of men across households and over time by relaxing the conceptual constraints imposed by an implicit if not explicit incorporation of a co-residential housing criterion vis-à-vis household and family composition. To understand life and survival in a chronically impoverished minority community and the role and place of black men as an intricate part of the family we must examine not only dynamic residency patterns, but also kinship networks, an issue to which we now turn as these are critical to capturing the socio-emotional and financial contributions that sustain survival.

Preoccupation With the Nuclear Family

While recognizing increasing variations on a theme and the flux and dynamism that the contemporary family is experiencing, much of the extant literature nonetheless continues to exhibit a preoccupation with the nuclear family (Coltrane, 1998; Mann & Grimes, 1997). Arguably, this model may be sufficient for understanding the large majority of (American) families, but the structure of many families in low-income neighborhoods is not always nuclear or best understood as such. This has important implications with respect to survival.

Among the best known and widely cited works of an alternative genre, Stack (1974) describes an extensive network of kin and friends among the poor, supporting and reinforcing one another in a wide variety of ways. Similarly, a number of other researchers have documented the exchange of cooking, housework, childcare, help in finding employment, and financial assistance (see, e.g. Gibson, 1972; Hays & Mindel, 1973; McAdoo, 1978, 1981; Sussman & Bunchinal, 1962; Morgan, 1982; Hofferth, 1984; Kellam et al., 1982; Burton, 1995), with proximity having been identified as a determinant feature of these exchanges (Gibson, 1972; Hogan, Hao & Parish, 1990).[2] Many of these researchers describe shared living arrangements as the most proximate and multifaceted means by which people support their kin, and document how black families, households, and neighborhoods have developed kinship exchange with multiple households, co-residence patterns, fluid household boundaries, and multi-generational membership.

Male-Present/Absent Dichotomy

A third limitation of much of the existing literature is the dichotomous manner in which males are treated as either present or absent in the household. Almost all researchers (and official record keepers and agencies) employ this static all-or-nothing approach[3] In recent decades, the rate of marriage has declined, and divorce, cohabitation, and the proportion of out-of-wedlock births have been on the rise (Murray, 1984; Manning & Smock, 1995; Koball, 1998). Moreover, these patterns are more pronounced among blacks than whites and female-headed households are far more prevalent among black families (Goldscheider, 1995). However, this does not mean that men are necessarily absent or void of roles in families with women and children.

Restricting for the moment the discussion only to fathers, Mott (1990) reports that the manner in which traditional family demographers have examined father's presence and absence substantially misrepresents and under-estimates the reality of father figure contact, particularly for black children. He found that many fathers currently not sharing a residence with their children were likely to see their children at least weekly. This is consistent with much of our ethnographic findings, where male participants reported that they did not live with their children, but saw them quite regularly – even on a daily basis. Mott's (1990) labels this pattern where fathers are residentially in and out of the home as paternal "flux".

In acknowledging the residential fluidity of black men, Mott (1990) suggests that if we want to understand the impact of fathers, we cannot just examine their presence and absence; concluding that part of the overt racial distinction in father presence/absence statistics reflects cultural differences. Because of the manner in which most social scientists define family and measure residency, we fail to count or acknowledge the presence of men due to their "paternal flux" within households.

While higher mortality and incarceration rates do depress the male-female gender ratio to within inner-city communities such as Peete, these empirical trends have been exaggerated by the popular press and mythologized into the view that black men are nearing extinction in many low-income neighborhoods (Majors & Gordon, 1994; Neckerman & Aponte, 1988; Gibbs, 1988; Cordes, 1985), reinforcing the belief that men are no longer a significant part of families and households. Not only is this unfounded and hyperbolic, it is hopelessly myopic. On one hand, it focuses solely on married or cohabitating men. Absent from the current picture are the men who assist in household survival as adult sons, nephews, brothers, friends and boarders. On the other hand, it confounds an externally enforced status of invisibility that largely owes to official policies and economic marginality with the views and understandings of the participants themselves.

Stack (1974) found that it might not be possible to determine the household to which an individual belongs, or to specifically define the membership of each housing unit. She writes, "a resident in the Flats who eats in one household may sleep in another, and contribute resources to yet another. *He* may consider himself a member of all three households" (p. 31, emphasis added). In this brief passage, Stack captures the fluidity of certain members of the family, and men in particular (see also Valentine & Valentine, 1971).

In sum, we are not suggesting that the household is a meaningless unit of analysis. Rather, we are trying to understand residential fluidity, and the social, economic, and political circumstances that influence and determine household composition and survival strategies within an impoverished, inner-city, minority community. Before turning to an examination of these survival strategies and the practices that comprise them, it is useful to gain a sense of the context of our research, to explore the Magnolia/Peete community.

EMPIRICAL CONTEXT

Background[4]

Within six months of passage of the U.S. Housing Act of 1937, the cornerstone legislation of public housing in the United States, funding (90% of which was federal) was authorized for the construction of two public housing complexes in New Orleans: St Thomas and Magnolia. Construction practices, design features, materials, and amenities would be the same in both facilities, but consistent with the prevailing legal segregation then in place throughout the old South, the former was built exclusively for white occupants, the latter for African-Americans.

Located in the Belmont section near Central City in uptown New Orleans, mostly decrepit buildings housing 862 black families were razed just north of Flint-Goodridge Hospital, the city's only black-owned and administered hospital, to make way for the 23.75 acres Magnolia development. Construction began in 1939 and continued through 1941. The first families actually began moving in to the new, well built and appointed two-story units in January 1941 (Mahoney, 1985).

Initial selection criteria employed by the Housing Authority of New Orleans (HANO) established a 20% tenancy ceiling on families on public aid, with the remaining residents comprised of largely intact, working families. Figure 1, photographs from HANO's 1940 annual report, provides visual evidence of the solid, tasteful construction that characterized this early public housing effort. While not flashy, the state-of-the art appliances, molding, and attention to green areas provided much improved housing for the upwardly mobile occupants. Upon completion, Magnolia consisted of 723 units: 252 one-bedroom, 427 two-bedroom, and 44 three-bedroom apartments. All were quickly leased, but the war delayed the project's official opening for several years.

Like many of the nation's then growing central cities, New Orleans faced a serious postwar housing shortage, so a second construction phase to increase Magnolia by another 680 two-bedroom units spread over an additional 17.7 acres and build several other altogether new developments gained approval and funding soon thereafter.[5] Both the second phase of Magnolia and the newly approved projects differed in two critical and ultimately fateful respects. First, design and construction standards were significantly relaxed. While phase I and other projects of that (New Deal) era had been built to last for 60 years and provide for an upwardly mobile working- and aspirant middle-class clientele, the newer (i.e. post-1949) approved construction represented a political compromise between public housing advocates and deeply opposed private housing market interests (see von Hoffman, 2000; Baxandall & Ewen, 2000). In short, the postwar expansion of public housing effectively was constrained to undertake construction at the lowest possible cost to provide cheap housing for that sector of the nation's burgeoning urban population that would have little likelihood of turning to the private housing market to meet their unmet housing needs, i.e. public housing could not compete with the private marketplace – or even pose an indirect threat to that effect. Consequently, cheap and far less durable building and plumbing materials were used. Less attention was devoted to common green areas and such amenities as trees, garden/patio areas, and play equipment. Moreover, by effectively delimiting public housing to a lower and more narrow economic strata, the compromise rendered it less universalistic and, thus, served to further undermine future political and financial support (Radford, 1996).

Residency criteria for the Magnolia Extension (Phase II) were also relaxed and a somewhat younger and poorer population filled the newer units. To this day, the "old" side is populated by older, long-term residents, and is widely understood by the locals as being quieter and safer than the "new" Peete which is known for its "hot" zones (i.e. areas of drug dealing and crime).

In contrast with the practices that informed Phase I, namely, the replacement of aged, often unsafe, and sub-standard facilities, many of which lacked indoor plumbing, Magnolia's second phase partially necessitated the displacement of a more stable and affluent population – including a number of doctors, teachers, shop owners, and clerks – from the more middle-class area stretching from the northern ("lake side" in the local parlance) edge of the development to Claiborne Street (see Figure 2), a major thoroughfare following the entire crescent of the river and the only road running the whole length of the city from its eastern border with St. Bernard Parish to the city's western edge against Jefferson Parish.

Not surprisingly, it did not take long for the former to manifest itself. Within a few years, the new construction completed in only 1955 began to require extensive maintenance and repair. By the mid- to late 1970s, this situation had deteriorated even further and Housing Authority of New Orleans (HANO), landlord to 10% of the city's population, spiraled toward chaos and disarray[6]. Arguably, Magnolia, fared somewhat better than several of the city's other (newer) housing developments, but increasing inattention and continued disrepair, the lack of even basic maintenance, and when undertaken the interminable delays (and escalating collateral damage) surrounding routine repairs such as fixing a broken window or leaking kitchen sink continued to take a direct toll on the physical structure.

The displacement of a sizeable number of middle-class persons possessing numerous much-needed resources foreshadowed by a decade a much larger

Courtyard View

Exterior Views
Above: Rear   Below: Front

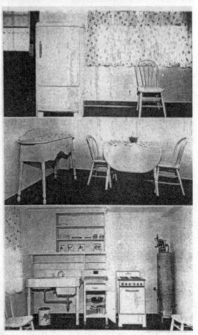

Kitchen Views

Fig. 1. Magnolia Phase I, December 1940.

Bathroom View

Living Room View

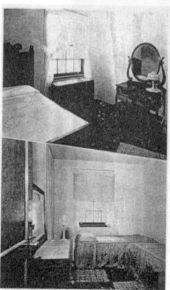

Bedroom Views

Source: HANO (1940)

Fig. 1. Continued.

exodus of middle-class blacks from the central city area to East New Orleans – a dynamic experienced by most if not all older major American cities following the reduction of housing restrictions by the Fair Housing Act of 1965 and the increasing socio-economic and residential mobility of the growing black middle class in the wake of the Civil Rights movement (see Wilson, 1987; Devine & Wright, 1993).

These factors contributed to the diminishing quality of life in Magnolia and the surrounding neighborhood, and had similar sorts of effects in other developments throughout the country, but it wasn't until the early 1960s, when HUD truly took on the mission of "housing of last resort" and explicitly accelerated the departure of intact and working families from public housing in favor of a more deeply impoverished and social capital-challenged clientele that public housing began its fatal descent.[7]

Like other proverbial "roads to hell", this large-scale substitution of those with limited resources (and in retrospect a greater capacity and opportunity to achieve upward mobility) in favor of those with even greater need was presumably well intended. Similarly, the decision to no longer allow and support segregated public housing owed to good intentions, but the immediate consequence of class segregation and attempted racial integration was the very rapid turnover and re-segregation of public housing throughout New Orleans, the rest of the South, and much of the nation. We effectively concentrated the most deeply impoverished, least mobile, most dysfunctional elements of the population (Wilson, 1987; Devine & Wright, 1993). Inasmuch as Magnolia/Peete was defined from the beginning as a "black project", racial desegregation/re-segregation had no direct impact, but the other noted policies helped to concentrate our most chronic and obdurate forms of poverty there, undermining the viability of the community, and establishing it as one among many underclass communities.

Table 1 provides supportive, albeit inferential, decennial census data vis-à-vis much of the above. These data covering the period 1960 to 1990 are highly consistent with the dominant, if not stereotypical, devolution of an already chronically poor neighborhood into an underclass area. Consistent with broader societal-wide trends and the dramatic changes in family and household formation, rates of marriage have fallen precipitously as have the number of employed males.[8] Inversely, the already high rate of poverty has escalated from approximately half to four-fifths of neighborhood persons and families.

As one would expect, data on the occupational distribution of those working – most critically, a declining proportion of the population (as provided by the employment rate) – indicates widespread change. "Private household" and unskilled and semi-skilled manual labor ("handler/helper/laborer") as well as

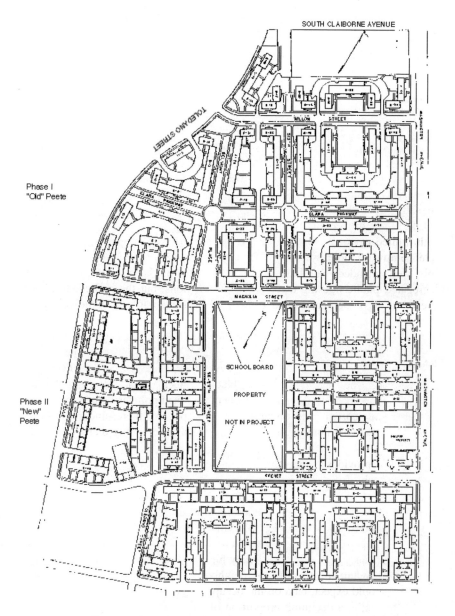

Fig. 2. Magnolia (C. J. Peete) Project (Phase I) and Extension (Phase II).

Table 1. Census Data for Magnolia/C. J. Peete 1960–1990.[1]

	1990	1980	1970	1960
Race				
Black	99.8%	99.6%	99.4%	97.8%
White	–	–	0.4%	2.1%
Unknown/unreported	0.2%	0.4%	0.2%	0.1%
Marital Status[2]				
Persons Ages 15+	2,216	2,452	4,910	5,061
Married	9.1%	15.0%	43.7%	46.4%
Single[3]	90.9%	85.0%	56.3%	53.6%
Males Ages 15+	626	724	1,799	2,033
Married	14.8%	24.0%	39.2%	56.4%
Never Married[4]	71.9%	60.6%	46.0%	30.9%
Divorced/Separated	11.0%	11.6%	10.2%	4.7%
Widowed	2.2%	3.7%	4.6%	8.0%
Females Ages 15+	1,590	1,728	3,111	3,028
Married	6.8%	11.2%	23.9%	39.7%
Never Married[4]	56.4%	38.3%	31.1%	20.4%
Divorced/ Separated	22.5%	32.3%	27.5%	22.0%
Widowed	14.3%	18.2%	17.6%	17.9%
Education, Persons ages 25+	1,625	1,700	3,285	3,852
Some high school	–	31.2%	18.6%	21.9%
H.S. graduates[5]	36.7%	24.4%	13.4%	10.8%
Some college[6]	9.9%	3.7%	3.7%	1.9%
College graduate[7]	3.6%	0.5%	0.5%	0.6%
Employment Status[8]				
Employed persons ages 16+	562	779	1,735	2,170
Rate	14.7%	32.7%	39.2%	42.9%
Distribution by gender				
Male	29.0%	18.1%	45.6%	54.6%
Female	71.0%	81.9%	54.4%	45.4%
Occupation				
Services, excl. protective & hshld.	42.5	38.8	28.9	19.2
Admin support/clerical	14.9	17.7	7.0	4.9
Handler/helper/laborer	7.8	3.1	10.6	16.0
Sales	7.3	4.0	2.9	1.6
Private household	7.1	17.6	20.2	21.6
Professional specialty	6.0	2.3	4.5	2.0
Other	14.2	16.9	22.7	35.0

Table 1. Continued.

Personal Poverty				
Population in poverty	3,126	2,560	4,411	–
Percent in poverty	81.5%	68.3%	61.6%	–
Population	3,836	3,748	7,163	8,431
Poverty Rate by Age				
Ages 0–17	50.8	50.3	52.7	–
Ages 18–64	42.2	42.0	40.4	–
Ages 65+	7.0	7.7	6.9	–
Median income (1990 $s)	$5,268	$7,422	$7,391	$9,313
Gross Rent				
Median rent (current $s)	$99	$56	$54	$38
Median rent (1990 $s)	$99	$88	$182	$167

Notes:
[1] Data for 1990 and 1980 are for census tract 93.02; data for 1970 and 1960 are for tract 93 (which includes tracts 93.01 and 93.02 per later censuses).
[2] In 1970 and 1960 data, age is 14+.
[3] Includes all those persons not currently married.
[4] In 1980, 1970, and 1960 data, category is 'Single'.
[5] In 1980, 1970, and 1960 data, category is '4 years of high school'.
[6] In 1980, 1970, and 1960 data, category is '1 to 3 years of college'.
[7] In 1980, 1970, and 1960 data, category is '4 or more years of college'.
[8] In 1960 data, employment is for respondents ages 14+.
[9] In 1960 data, families in poverty are all those with a family income less than $3,000.
[10] Household income from previous year.

Source: U.S. Bureau of the Census.

the largely uninformative "other" categories exhibit sizeable declines, while "services", "administrative support/clerical", sales, and "professional specialty" show substantial proportional growth. While clearly consistent with larger occupational shifts in American society, the import of these changes is not altogether clear with respect to income and economic well-being. The growth in the "professional specialty" category might appear promising, but it is modest and our survey and focus group data suggest that it may well capture what most of us would consider and classify as low-end service activity. Moreover, it is more than offset by the decline in the (perhaps) low status but relatively well-compensated manual sectors.

Consistent with the above historical chronology of Magnolia/Peete and HANO's escalating problems, the data in Table 1 indicate that between 1960 and 1990, real median income fell by better than 40%. This translated into an escalating rate of poverty characterized by an increasing "income deficit" and all that that implies.[9] As rent in Peete (and all public housing) is keyed to income and could not exceed 25% of cash income, an increasingly impoverished population necessarily paid less (as evidenced in the real median rent figures presented in Table 1). While progressive in principle, HANO, like many other public authorities, was confronted with a situation whereby revenues declined while expenditures and needs increased. Unable to resolve this structural deficit, a situation further exacerbated by the decline in direct federal funding as well as indirect support via other federal aid to cities, the quality of life in public housing typically declined.[10]

The Current Context: Trends in Work, Income, and Government Assistance, 1996-1999

Our research in C. J. Peete did not begin until 1996 but since then and each year thereafter we have hired and trained community residents as interviewers and conducted a simple random sample of households, interviewing 225–260 heads of households each year. With the persistence of our interviewers and the incentive of a $15 stipend, we have achieved an interview response rate of 92% or better in each of the four surveys. All four surveys have included a common core of closed and open-ended questions aimed at gaining demographic, educational and training, employment, income, quality of life, and attitudinal information. In addition, each survey has included one to five distinct one-time modules devoted to such topics as: welfare reform, job training, job search experiences and processes, schools and parental involvement, several health self-assessments, redevelopment and relocation, religiosity, etc.

At the time of our first community survey (Fall, 1996), 1,088 of the 1,403 units built in the two phases were occupied residences and an additional half-dozen units had undergone conversion to administrative use of one sort or another. The remaining 309 units were still standing, but officially vacant.[11] Over the next several years, as relocation, demolition, and de-densification became the operating, albeit still vague plan for the future, the number of occupied units has fallen considerably so that by the end of 1999 a total of only 864 occupied units (852 residences and 12 administrative offices) remained. Parts of the "new" Peete (Phase II) had already been demolished and many of the remaining courtyards have been fenced off and boarded up awaiting the wrecking ball when a final plan eventually gains approval.[12]

The surveys provide continuing information and a useful series of snapshots about the community in a period of rapid transformation inasmuch as HUD efforts to recreate Peete in particular – and public housing in general – into multi-use, mixed income developments have occurred concurrently with the welfare reform initiatives encoded in the "Personal Responsibility and Work Opportunity Reconciliation Act of 1996" and the now longest sustained period of economic growth in the nation's history.

Peete households range from individuals living alone (usually the elderly) to households of ten persons, with a mean of three. Our survey estimates reveal a somewhat larger population than indicated on the lease, especially when we include other persons who "frequently stay over" but are not on the lease.[13] Better than 90% of the official leaseholders are women and 70% of households have children less than eighteen years old. Consistent with the census data presented earlier, almost three-quarters of the respondents report being single (see Panel A, Table 2). In our surveys we find a somewhat higher percentage of high school graduates and lower percentage of college grads than the 1990 census figures (i.e. 53% and < 3% in Panel B, Table 2 vs. 37% and 3.6% in Table 1), but these differences are due to our samples consisting of house-holders only, our inclusion of a GED recipients among high school graduates, and higher rates of completion in more recent years.

Perusal of Panel C, Table 2 indicates that while the poverty rate within C. J. Peete remains extremely high (exceeding 85% by our estimates), income has been rising over the past few years. In 1996, 73% of the households reported receiving less than $6,000 in annual income from all sources (excluding non-cash benefits and the value thereof, i.e. Medicaid, food stamps), while only 3% indicated incomes of $12,000 or more. In our most recent survey (covering 1999), these figures were 39% and 16% respectively.[14]

In the first Peete survey we conducted (Fall, 1996), only 24% of the households reported having someone in the workforce. A year or so later, 30% contained a single labor force participant while 7% reported having more than one worker. Last year, these figures had climbed to 42% and 14% respectively, but according to our latest survey (early 2000), levels of formal employment have reached a plateau over the past year. The proportion of persons working full-time has not appreciably changed over the past few years and continues to hover at just over 60% of those currently employed. Not surprisingly, three-quarters of those working part-time would prefer full-time work, but either cannot find full-time employment or have responsibilities (e.g. child care) that prevent them from taking on full-time work. Per Table 3, mean monthly employment income among those households reporting someone working in the formal economy has increased by $100 (from $599 to $699) over the past three annual surveys.

Table 2. Survey Data.

Panel A: Current Marital Status, % (N = 260)

	Sample 2000
Currently Married	3.9
Cohabitating	1.9
Separated	5.8
Divorced	7.7
Widowed	7.3
Single	73.5

Panel B: Educational Attainment (N = 260)

	Sample 2000
$\bar{X}$ Years School Completed	11.2
S.D. Years School Completed	1.6
Highest Degree,%	
None	42.3%
High School (or GED)	53.5
Associate's Degree	0.8
Bachelor's Degree	1.9

Panel C: Reported Income, 1996, 1997–1998, 1998–1999, 2000, By Percent*

	1996	1997–1998	1998–1999	2000
Less than $6,000	73	64	59	39
$6,000–7,999	15	17	17	27
$8,000–9,999	5	7	10	10
$10,000–11,999	3	3	7	8
$12,000 plus	3	9	7	1
Valid cases (N=	235	242	249	254

* Due to rounding, numbers may not sum to 100%.

Perhaps the most dramatic evidence found in Table 3 concerns the substantial reduction in AFDC/TANF in the wake of welfare reform. Better than half (55%) of Peete households were on AFDC prior to the 1996 welfare reform legislation. Thereafter, as the provisions of the act were implemented and people were moved off the roles, TANF receipt has fallen to 15%. The provision of food stamps has also been reduced, but the decline is substantially less, having dropped from 73% to 57% of households. Continued receipt of food stamps by more than half of the households in Peete clearly indicates that welfare-to-work has been a partial success at best as the vast majority of residents working in the formal economy do so at very low wages.

In sum, our survey evidence suggests that several years into welfare reform occurring within the midst of a rather robust economy, we have clear evidence that more Peete residents are working. However, among those who work, wages average only $700 a month. Assuming steady work (a very liberal assumption), this translates into $8,400 a year. Thus, it is no surprise, that three out of four Peete households still qualify for TANF, Food Stamps, WIC, SSI, Social Security, Disability, or some other government benefit (excluding a housing subsidy, education and job training, child care assistance, energy subsidy, or healthcare) and two out of every five households receives more than one of these benefits. Those fortunate enough to have both work and welfare income may be well off relative to many of their neighbors, but they typically still fall short of escaping their poverty and necessarily continue to engage in daily struggles to survive. It is toward a fuller understanding of these efforts that we now turn.

HOUSEHOLD SURVIVAL

Survey respondents were assured of confidentiality and our interviewers made it quite clear during the pre-interview briefing and consent phase that HANO had nothing to do with the survey and did not have access to any individual or household level data. Despite these assurances and the validity of this claim, we are hard-pressed to believe that substantial under-reporting of the presence (intermittent or otherwise) of adult men, income, and illegal activities did not occur. Though our survey results yielded larger population and income estimates than those revealed to HANO and afforded us extremely useful data about a variety of things, we strongly suspected that the information we were getting regarding residency, income, and related activities, left much to be desired. That the survey evidence did not fully jibe with either the ethnographic evidence we observed hanging out in the community or reveal much detail with respect to many of the activities several of our community contacts reported was not particularly surprising. Many people were reticent to discuss these highly personal and potentially threatening topics – especially within the interaction context of the survey interview. Consequently, in an effort to augment our survey data and gain deeper insight into these activities, it became abundantly clear that we needed to establish longer-term contacts and build greater trust among residents.

For these and other reasons, the CAP research group (headed by the first author) initiated a four-person graduate student ethnography team (which included the second author) in 1998, our third year of research in Peete.[15] Much

Table 3. Trends in Employment Income and Receipt of Government Transfers 1996–2000, By Households, % Receipt and Mean Monthly Dollar Amount.

	1996	1997–1998	1998–1999	2000
Employment				
%	24%	30%	42%	43%
Mean Wages	–	$599	$655	$699
AFDC/TANF				
% receiving	55%	34%	25%	15%
Mean Dollars	$196	$196	$196	$214
Social Security				
% receiving	–	24%	20%	22%
Mean Dollars	–	$434	$401	$411
SSI				
% receiving	–	19%	18%	14%
Mean Dollars	–	$338	$290	$354
Food Stamps				
% receiving	73%	70%	59%	57%
Mean Dollars	–	$218	$209	$227
Govt. Benefit				
Any Benefit*	85%	84%	78%	76%
2 or More Benefits*	62%	56%	50%	38%
Mean Monthly Benefit**	$456	$453	$472	$447
(N=	259	250	254	260

* AFDC/TANF, Food Stamps, WIC, SSI, Social Security, Disability, or other reported government benefits excluding housing subsidy, education and job training, child care assistance, energy subsidy, healthcare (or insurance).
** Value of all benefits among those receiving any governmental benefit per previous note.

of our ethnographic effort was subsequently directed toward developing a fuller understanding of the various dimensions of Peete households' micro-economies. Working with several community residents with whom team members had developed relationships, these residents introduced us to others, vouched for us, and later helped us organize a series of small, intimate conversations and focus groups. The discussion below is based on data gleaned from these sources, specifically a number of in-depth semi-structured interviews undertaken with community residents and a series of 19 focus groups (or community conversations) conducted with another 102 residents (86 females and 16 males) on the subject of "makin" ends meet.

"Makin" Ends Meet

Our research is highly consistent with the seminal work of Edin and Lein (1997), who document how poor women make ends meet via welfare, work, and a combination of welfare and work. At the same time, our research goes beyond their seminal study in several respects. First, we provide further elaboration on the so-called combined or mixed economic strategy, specifically via our expanded investigation of informal economic activity, both in terms of legal and illegal dimensions. We also examine nascent entrepreneurial activity, an often-overlooked aspect of life within poor black communities, and explore the role of men, another often overlooked but critical element of a poor household's survival.

Welfare
Most participants stated that they receive welfare, but said it usually was not enough to survive on. Government assistance in the form of AFDC/TANF, and SSI provide very limited cash assistance and families can rarely make it last an entire month.[16] For those receiving government assistance, food stamps and Medicaid benefits were deemed most essential to survival, with the latter being the most covet aspect of welfare. Without a Medicaid card, people feared being unable to gain needed medical care for themselves, but especially for their children. Over and over again, focus group participants who receive any kind of government assistance told similar, if not the very same, story of how these resources were strategically used to ensure survival. The story always started with food secured by Food Stamps, medical care via Medicaid, and then turning to consideration of available cash (or cash benefits) with rent being paid first followed by necessary household items like tissue paper or cleaning supplies. Finally, if anything was left it, was used to "run them" for the remainder of the month.

Many residents stated that once they got their welfare check, it was gone before they knew it. One participant reported,

> [be] like I said, the stamps took care of the food and out of the $138 the rent and when I had a phone, the phone had to paid, then came household stuff. As far as toilet paper and soap. And if my daughter wanted something, there go the money that's left ... an there ain't nothing but like $25–$30 and what am I suppose to do with that? You know,

to make it for her. I'm not so much worried about me, because I can handle me, but that wasn't no money. I got tired of it. I put myself out there and set myself straight (Focus Group, 4/12/99).

There is continual national debate around welfare benefits, especially Aid to Families with Dependent Children (AFDC), what is now TANF, which stands for Temporary Assistance to Needy Families, since passage of welfare reform in 1996. Historically, AFDC and TANF recipients are portrayed as individuals who do not want to work and who depend on the government to care for them and their children (see Katz, 1989; Devine & Wright, 1993). This persistent message and image have resulted in welfare reform legislation that restricts benefits to those who are not attempting to work. Since 1996, community service may be required in lieu of work to continue to receive welfare benefits. As a direct result of these welfare reform policies, some women report that they volunteer or participate with schools, the Campus Affiliate Program (CAP) office, or other qualifying local institutions in order to ensure that they continue to receive TANF benefits. For example, one woman reported that she volunteers in a job placement office, while also participating in a job readiness program that hopefully will prepare her for the job market. Another participant told of volunteering at her children's school 10 hours per week. Though both of these women are considered volunteers, it is clear that it is a forced activity undertaken to ensure that government assistance continues.

A major dilemma expressed by participants is choosing between welfare and work. One resident expressed it as follows:

> When you do have to work and get money and they take all your money, it's terrible. Yes, it's terrible, when you work. When you work they (Housing Authority of New Orleans) take everything, they try to take your whole check from you. ... If you are working your rent is more than your check. And its terrible when you get on welfare, you get $190.00 and rent [is] $33 and they give you almost $150.00 worth of food stamps, but when you work, they try to take everything. Leave you with $10 left with you. ... If you want to make ends meet, you better stay on welfare (Focus Group, 4/19/99).

The above quote reflects how some participants viewed continuing to receive welfare more as a decision of survival than a strategy to try and avoid work. Welfare is usually not sufficient, but unlike low wage labor it is consistent and dependable.

Work
Work is considered normative in American society. We assume and expect that all able-bodied adults will and must work. Moreover, we believe that if adults work they will not require government assistance. Thus, in American culture, work is necessarily equated with self-sufficiency. While not all Americans

subscribe to the culture of poverty explanation (and its variants), most Americans accept its basic premises (see Wilson, 1987; Katz, 1989; Devine & Wright, 1993), namely, that the poverty that many households experience is strongly linked to a poor work ethic and an unwillingness to accept minimum wage jobs, leading to dependency on welfare (Kaus, 1986; Mead, 1986, 1992; Murray, 1984). Alternatively, it is well documented that many Americans are working and still remain poor (Ellwood, 1988; Devine & Wright, 1993).

The primary thing that people in C. J. Peete report doing to survive is working. Although residents work, many were having a difficult time because they are in low wage, service sector jobs.

> Only thing about working. I'm sure that all of us and that some point in our life [worked]. But the only thing about it, I'm going to give an example, like two years ago, year before last, whenever, I wasn't working, my rent was $29.00. As soon as I start working . . . I wasn't getting nothing but minimum wage and at that time minimum wage was like $4.25, I think. My rent went from $29.00 to $290.00 something, right. . . . I mean when you know they going to go up on your rent, you going to tell them? . . . You already behind in bills because you've been struggling trying to get ends to meet. As soon as you find a little something, they going to [go] up high on your rent. You know, that's ridiculous. And then as soon as you get you a job, you not guaranteed to stay on that job for years and years. You can get on a job and don't like what they're doing or how they're treating you. You might just don't feel comfortable and quit. So why would you as soon as you get a job, go run and tell them, 'Oh I'm working' and boom, your rent three-something (Focus Group, 5/4/99, parenthetical added).

Usually the employment is part-time without benefits, which makes it very difficult for residents to live independent of government assistance. Many participants report working jobs in the formal economy, as cashiers, housekeepers, childcare workers, security guards, and restaurant workers, as well as doing seasonal and temporary work. Several participants report working extra hours or overtime in order to garner a sufficient income to take care of their basic living expenses. More than a few women report working two part-time minimum wage jobs. However, despite such efforts, virtually all of these people report that their work income is insufficient in the face of an emergency, i.e. an illness requiring a clinic visit and/or prescription, an additional school expense, or almost any unanticipated expenditure. The low-wage jobs typically available to individuals in our focus groups rarely if ever allow them to live independently and self-sufficiently, as many of them had hoped. And yet, none of these individuals appeared to lack the work ethic.

Given that most of the jobs are in the service sector, oftentimes in New Orleans' very sizeable hospitality (i.e. tourism) industry, many jobs are part-time and workers often face temporary and seasonal layoffs or further reduced work hours. Working in the hotels, convention facilities, and restaurants downtown often presents transportation costs and barriers and lack of adequate and affordable childcare, especially at night, is highly problematic. As one member of a focus group put it,

> New Orleans is a big tourist spot, but poor people like us don't benefit from this. I work in housekeeping at a downtown hotel, but when tourism gets slow, my work also gets slow, and my money gets real low (Focus Group, 5/11/99).

Or another woman who works full-time in the deli of a large university in the city, talks about how she is laid off for three months during the summer when the kids leave school. She states it's hard for her, because in the summer it is particularly difficult to get work because you are competing with all the school kids for the same jobs.

Mixed Strategies

> A lot of young women don't want to work because it's like this here. When you get out there and try and bust your behind to work, they cut you right off the welfare. They cut your food stamps down to nothing. By the time you get through paying your phone bill, your light bill, your rent, you ain't got nothing to buy your children no clothes to provide what they need . . . That's not helping you, you understand. When you got a welfare check coming and you got the amount food stamps and getting paid under the table, you can make it. You understand what I'm saying? (Focus Group, 5/4/99).

Our focus group conversations make it clear that many residents cannot live on either just government assistance or work in the formal economy and are forced to combine these and other activities to make ends meet and provide for their household's survival. On the basis of our focus group participants' reports, we delineate several analytically, though not necessarily or even typically empirically, distinct strategies that people use to make ends meet include such things as combining welfare and work in either the formal and/or informal economy, household management strategies, income generating activities including hustling and other entrepreneurial behavior, support provided by men, and use of kin and friendship ties.

Combining Work & Government Assistance

Faced with the difficulty of living totally on government assistance or low-wage employment, many residents of C. J. Peete simultaneously receive forms of governmental assistance and are involved in either the formal and/or informal sectors of the economy. The fact that government benefits are contingent on little to no outside income is well understood but commonly circumvented as the poor seek to ensure the survival of their household and family. Edin and Lein (1997),

Harris (1993, 1996), and others have previously documented that many low-income women provide for their households by combining work and welfare, rather telling evidence that welfare does not simply foster dependency, though it necessarily dictates maintaining the appearance thereof. Acknowledging the need and ubiquity of this approach, Jencks (1992, 232–233) goes a step further and argues that welfare should be defined as something that supplements low-wage irregular work.

Women who combine work and welfare are most likely to hold jobs in the service sector paying only minimum or near-minimum wages with little or no benefits. Welfare mothers are aware of the economic consequences of working, a fact previously documented by Edin and Lein (1997), but also echoed by the comments of a young woman who participated in a focus group:

> ... I work part-time, 'til I find some type of, another kind of, you know, full-time job. And then I receive, you know, I still receive some income from the AFDC and some food stamps. And I just take it and manage it and try to work until I find something else (Focus Group, 5/12/99).

Another focus group participant discussed how she actively pursues gainful but "under-the-table" employment in the formal sector so that she can maintain needed food stamp and Medicaid benefits for her small children, while working at a sandwich shop where she is paid cash at the end of each shift. An older resident reported that she receives disability benefits and supplements them by making and selling soup. Still another woman reported that she had learned to combine welfare and work by working in the formal labor force as a childcare worker under a fictional name and social security number while receiving TANF benefits for her children under her real name. Other focus group participants claimed that this strategy was rather commonplace and knew other individuals who used the dual identity approach.

The formal economy can be thought of as officially recognized labor force participation.[17] When the labor market is tight individuals are more likely to have employment that is sufficient for taking care of their family's material needs. However in isolated urban areas where only low wage jobs are readily available and workers are under-skilled and plentiful relative to the demand for workers, labor force participation alone is simply inadequate for survival. Therefore, the urban poor often mix formal employment with other activities, most commonly some form of work in the informal economy.

The informal economy includes both legal and illegal activities. The legal activities involve otherwise legal income generating activities but remuneration is not subject to taxation and other governmental regulations and requirements. Income is "off-the-books" or "under the table" meaning that it is not reported as a source of income to government agencies and officials such as the IRS, state human resources caseworkers, or the housing authority. Off-the-books workers are not afforded any protections (e.g. insurance, workman's compensation) typically afforded legal or above-the board employees. Among our focus group participants, this one job in the formal economy and another job in the informal economy was commonplace. For instance, one woman reported working as a full-time childcare provider during the week and spending Saturdays as the shampoo person at a local hair salon. At the close of business on Saturday the owner pays her cash plus she gets tips. Similarly, other participants reported that they supplement regular and irregular jobs in the formal economy with unreported restaurant and bartending work on a strictly cash basis, while still others regularly provide babysitting and/or housekeeping services.

The informal economy also includes entirely illegal activities such as selling drugs and stolen goods, prostitution, and theft. Participation in these illegal activities obviously could result in legal prosecution and possible incarceration, but most respondents who acknowledge participating in illegal activities report accept the risk as a necessary tradeoff for achieving their means of survival.

Household Management: "Stretching"

> I can only tell you what I do ... Presently, I'm not on AFDC, I been receiving food stamps. I receive child support and it's all right; you know what I'm saying. But I do have to stretcha dollar, you [know] what I mean (Focus Group, 5/12/99, parenthetical added).
>
> ... Well I'm on disability. So my ends meet by my income that I get on the 1st and the 3rd of the month, which is all the income that I receive, you know. There's no other income that I receive but that. Um, so really I have to make it meet by, you know stretching that budget right there. So that's it for me. Once it's gone it's gone, you know (Focus Group, 5/12/99).

Another component of the widely held culture of poverty thesis is the lack of thrift owing to the alleged failure of the poor to internalize the virtue of frugality. The commonly invoked image of the destitute teen wearing a pair of $150 Air Jordan's and other expensive apparel is often used to support this generalization.[18] Whatever the status and basis of this claim, our focus groups uncovered many residents actively engaged in frugality and exercising thrift as a conscious household management strategy. Like the women quoted above, many people refer to this as "stretchin a dollar."

Things participants reported doing to "stretch a dollar" included buying sale items, buying household supplies in bulk, and using coupons – middle class values celebrated in any home economics textbook of the 1950s. One participant reports that she only buys slightly damaged food items at a drastically reduced cost. Others reported constantly being on the lookout for used goods.

Most respondents report that the number one priority in their household is rent and then they take care of whatever else they can. As one respondent captured it:

> I pay my rent and then we can eat beans and rice or grits and butter, because no one else will know what's in my house in my pots. But if I don't pay my rent, everyone knows (Focus Group, 4/12/99).

"Nascent Entrepreneurialism"

The lack of black-owned businesses and the accompanying job opportunities and mentorship presumed to flow from it has figured prominently in policy debates concerning the revitalization of inner-city minority communities (see, e.g. Borjas & Bronars, 1989; Bates, 1993, 1997; Feagin & Imani, 1994). Numerous explanations have been proffered to explain the dearth of black businesses and, as often is the case with theories of poverty, rival claims focus on cultural (i.e. attitudinal and behavioral practices) vs. structural (i.e. access to opportunity, capital, and infrastructure) factors.

Despite widespread lament for the alleged lack of business acumen among poor African-Americans (compared, for instance, against the positive ethnic entrepreneurial stereotype of poor Asians or Jews), it is readily apparent that the majority of the participants know how to "make a dollar." Disregarding tax and regulatory concerns, most often the residents' "off-the books" business activities are otherwise legal means for making ends meet. Because both the jobs C. J. Peete residents hold and the amounts of government assistance they receive typically do not produce sufficient financial resources, people overwhelmingly report that they constantly "hustle." When asked what they mean by "hustling," participants report that it's anything you can do to make a dollar.

The need to generate money is expressed through a wide variety of skills and talents. Among the activities reported to us are numerous forms of gambling, keeping the children of someone that is working, selling cans, doing errands, fixing things, providing rides, or just about anything that people want. Several women indicated that they take in related or unrelated persons as boarders. Another woman states that she buys a case of beer and sells them at "second lines"[19] and other community parties and makes three times her cost. Several focus group participants stated that they give suppers or buy dinners from neighbors sponsoring a supper.[20] Most participants discussed how well you could do on a supper if you are a good cook and can get lunch orders from a local business. And, if you run a card game with your supper, you can make even more money. One participant reports making over $1,000 on a supper and card game she gave.

Other focus group participants either did hair or frequent another resident in the neighborhood that does hair. Some women were well know for their braiding ability and could make anywhere from $50–$65 for braiding one head of hair. Others were skilled in relaxing, cutting and coloring. Getting your hair done by someone in the development was a lot cheaper than going to a licensed cosmetician, $10 opposed to $25 for a wash and style. There are also men in the neighborhood that cut hair on the porches of their homes.

Residents have started businesses that fill the gap of many of the services not available due to urban isolation. For example there is one woman that sells food in her home, Wednesday through Saturday. She posts a sign outside her door listing all the different foods she serves as well as their prices. Her living room is set -up for dining, so guests can eat in if they choose. Another woman has a sweet shop where she sells both homemade and packaged snacks.

In one particular court on the "new" side, entrepreneurial activity is prevalent enough for us to have come to refer to it as the "market". Residents can get just about any type of service they need. There's a young man that does barbering on the porch. In another apartment there's a woman who does hair in her kitchen. In another apartment there is a woman that sells name brand clothing items. Another house gives suppers. Still another neighbor prepares and sells icebergs (i.e. flavored scoops of ice) and snacks for kids. Another woman provides childcare for people in this area. One porch is where an older crowd that drinks all the time hangs out. A porch further down is where the younger people who "smoke" hang out.

The Contribution of Otherwise and Officially Invisible Men
Both the men and women participating in our focus groups and interviews spoke of how men financially contribute to households, even though they are not on the lease. Several men in the group discussed working on jobs and making financial contributions to the household of either their mother, grandmother, or "old lady." Women participants discussed how they use men to make ends meet, with numerous respondents articulating the following expectations: "if you stay, you pay" or "put him out in the court yard," meaning to have him sell drugs. Overwhelmingly, women respondents report that they have a "man friend" who they can depend on for making ends meet. Men in the neighborhood also provide childcare services when their partners work. In fact, several of our discussions with men took place in the presence of the children in their care.

Men have also created businesses that provide necessary services not readily available in the C. J. Peete neighborhood. Men serve as handymen for basic household repairs. They make money by fixing appliances and other broken household items such as washing machines and cars, painting, and by running errands for many of the women in the neighborhood. One man started a unique service for some females in his building. He started bringing their trash down to the dumpster because the women are afraid to go near them because of the rats. Similarly, it is often difficult to get a cab in the housing development, so several men in the neighborhood have filled this void by effectively functioning as taxi drivers by providing transportation for residents for a fee.

Illegal Activities

Drug sales are clearly the most lucrative, but dangerous illegal means for making money. For instance, in one of our early focus groups, one participant acknowledged selling drugs when she was unemployed, because she had to take care of herself and her daughter. However she states that she realized how dangerous this line of work was so she got out as soon as she was able to get on her feet and get a job. Several participants reported knowing people in the development that sold drugs to make ends meet. Many women talked about how other women allowed men, either their sons or boyfriends, to sell drugs out of their homes, because they benefited from the income. A few women in the groups talked about the men in their lives who sold drugs. Most of these women reported that they did not depend on this income, to run their households, but viewed it as something extra for themselves and their children. A smaller number acknowledged their direct and active involvement in the trade.

Other reported illegal activities include mostly petty theft and sales of the stolen goods (auto parts being a favored commodity, but household goods, small electronics, and clothing also figuring in as well), intermittent prostitution, and an extensive gambling system that includes card games, dice, and dog fighting. Gambling is reported as a hobby and a hustle for many focus group participants. Many of the women discussed how they play a game known as "pitty-pat" or run their own card games to make money.

> Card games, card games. You make money on the cut. It all depends on how many people you got playing with you ... And how much you paying ... five dollars is the house cut each game. Some games you must show one hundred dollars. Ya, you have to show $100 to sit at the table ... Some go there with $10 before you know that'll be gone (Focus Group, 4/12/99).

Men in the community frequently shoot dice. Several porches and hallways are known for their almost ubiquitous dice games. While largely a male activity, one young woman also spoke of her shooting dice to make extra money:

> I shoot dice. Well, my brother taught me. When I'm with my brother, I made like $55 ... (Focus Group, 5/25/99).

Another illegal activity dominated exclusively by men in the neighborhood is dog fighting. Men in the neighborhood have big rottweilers and pit bulls that they encourage to fight and bet on. These organized dog-fighting rings have gotten some media attention over the last few years and laws are being enforced to ban dog fighting. Yet, men in this neighborhood still continue to fight their dogs as a means of entertainment and finance.

> ... Like fight dogs sometime. People bet on it. Like a horse race. Or basketball. You might think a brown dog will win and I might think the white dog will win. You might have a side bet going. Its not even related to a fight but it's between the men; like boxing. Instead you're using dogs. There's no number to how much you can make, you know it's just at that time. Whatever, however much you get ... $200–$300. That's small. They got guys that make 5, 6, 7, 10 thousand on dogfights. Just like a bar room fight (Focus Group, 6/2/99).

"Living With Momma" and Kinship and Support Networks

Most participants report that they can depend on family and friends to assist them by babysitting and sharing food and small amounts of money. In case of an emergency there are specific family members or friends that can turn to, but many also acknowledge that their family and friends do not have the financial resources to assist them when and if needed. A few that are working stated they could go to their employer and request an advance.

A common strategy that participants report is "living with momma." Many women and their children are living with their mothers in order to save money. Men in the neighborhood also report that living with their mother is a way that they can assist their mothers financially and save themselves money. For example, one man states that although he is working and making good money, it is better for him to live with his mother and give her a much-needed $250 a month. In so doing, he does not have to get an apartment and pay rent and utilities. Plus, living with momma has other perks like her cooking.

Kinship ties are critical, but the question is more about their generalizability. A primary difference between our findings (and others) and Stack's discussion in the widely cited *All Our Kin* (1975) concerns the extensiveness of the strong ties she captured. Our findings appear more consistent with other research that provides a less sanguine view (see, e.g. Hogan, Hao & Parish, 1990; Hogan, Eggebeen & Clogg, 1993) or suggest that the situation has changed for the worse over the past twenty-five years, with the harshness and isolation of poverty serving to circumscribe the scope and strength of kinship and friendship ties. Our survey evidence pointed to this, but the fact that most of the residents of Peete cannot rely on family nearly as much as Stack suggests is nicely captured in a focus group, where two participants provided the following:

> My people don't have as much as me. Most of the time I have to give them.

> I be out there working for it baby. Ain't nobody give me nothing. I got to go get it. ... I have nobody to lean upon, nobody. (Focus Group, 5/25/99)

This is not to say that ties are non-existent, only that we find them to be far more tenuous and precarious than the seemingly idealized situation presented in Stack's (1974) widely acclaimed ethnography. Interestingly, and perhaps suggestive that a fundamental change has taken place over time, we did observe some distinctions in these patterns based upon what area of the development we were in. The "old side" (see above) had residents that have lived there longer and developed more extensive relationships within their courtyards and our observations and focus groups suggested a greater tendency among neighbors to have looked out for each other's children and share food.[21]

Summary and Conclusion

Our findings indicate that C. J. Peete households survive by using a multitude of strategies to meet their material needs. These include working in the formal and informal sectors of the economy, getting government assistance under legal and extra-legal circumstances, and constantly hustling. Many persons combine a variety of strategies and behaviors as need and circumstance dictate.

The focus group conversations suggest that the majority of the resources received from government assistance were used for food and shelter and usually there was little or no cash remaining for such basic items as clothing, household needs, or uncovered medical expense. Participants in the focus groups report that they combine work with government assistance to address these unmet needs. Some women are able to work part-time and do report this work to their caseworkers and still get food stamps and Medicaid, which they believe are the most important benefits. A greater number of our respondents find this "above-board" work and welfare arrangement untenable – largely because the work restrictions are too constraining and the dollar tradeoff between welfare benefits and wages is punitive. The disincentives are even greater when increasing wages result in loss of health care and higher rent, a situation that fosters participation in the informal sectors of the economy, an illegal but altogether rational response.

Our findings are highly consistent with the recent research of Harris (1993, 1996) and Edin and Lein (1997). While we explore a number of other activities and the contribution made to households by men, who are often invisible to researchers and others, we too find that a combination of work and welfare is the predominant economic survival strategy among Peete households. Women in Peete, like other places analyzed in the work of these other researchers cannot rely on work because they cannot find jobs that are sufficiently steady and pay enough to support their families. Thus, welfare is a necessary supplement. Unfortunately, despite a liberalization of work rules under welfare reform, current welfare guidelines remain punitive regarding work income and, therefore, help to foster widespread participation in informal employment and economic activity. To echo Jencks (1992), the good news is: if you include the informal economy, almost everyone in Peete works.

NOTES

1. The explanation goes on to clarify (at least to some extent) the notion of a housing unit and how multiple units may exist within a single facility and notes that families and households may partially or fully overlap.

2. A substantial body of research, including some of the noted sources, has emerged to challenge the ubiquity, density, and importance of kinship- and friendship-based networks among the inner city poor (see e.g. Benin & Keith, 1995; Hogan, Hao & Parish, 1990; Hogan, Eggebeen & Clogg, 1993). Our own survey evidence raises questions about the pervasiveness and depths of these networks within Peete itself. Some of the apparent difference (i.e. absence-presence, depth, strength, degree of reciprocity, and diversity of exchanges) may owe to idealization, but the differential sensitivity of ethnographic and focus group methods versus survey research may also play a role as may researchers inability to adequately and consistently distinguish between "strong" and "weak" ties (per Granovetter, 1973). Putnam (2000, 317) summarizes this debate and concludes that while such networks that exist are an "important asset" that these "inner-city social networks are not nearly as dense or effective" as those found in previous generations.

3. See Piven and Cloward (1971) for a discussion of the historic role played by public welfare agencies in promulgating the reality and attendant fiction of wholly absent adult males within poor communities. Our research in Peete suggests that while very few adult men are on leases or part of the "official" population, they are extremely visible throughout the development. Knowledgeable residents' estimates of the percentage of households with attached men run as high as 60%. While this exact figure may be substantially inflated, the basic point is that numerous public policies and the economic marginality of these individuals have rendered them officially "invisible." Importantly, this "invisibility" is not limited to husbands and lovers, but extends to all adult males, be they sons, brothers, friends.

4. Much of the material in this section owes to the research of a history student's master's thesis (see Mahoney, 1985) and the senior thesis of an undergraduate sociology major (see Russell, 1995). We are indebted to both students for their painstaking documentation of the history of public housing in New Orleans.

5. In New Orleans as well as many other major cities, the housing shortage was even more acute among African-Americans. During the 1950s, the rate of black in-migration exceeded white in-migration. For blacks this housing supply-demand disjuncture was greatly exacerbated by widespread use of restrictive covenants and other forms of housing discrimination.

6. Even relative to the dismal standards and record characteristic of U.S. housing authorities in general, HANO performed poorly. Several HUD reports identified HANO as among the nation's worst run projects. Numerous experiments, threats, management practices, sub-contractors, etc. were tried over the years, but with little to no positive

results. Finally, in 1996, HUD effectively placed HANO in receivership, appointing a czar-like "Executive Monitor" to oversee the authority and HUD's monetary commitment and ideological reorientation to public housing.

7. Having determined that public housing had come to represent a large part of the problem rather than solution, the Clinton administration's response has been to remove public housing from the mission of "housing of last resort" in favor of a policy of re-development, de-densification, multiple income use, and privatization. Much of this new mission is only in its infancy, so empirical results and evaluation must necessarily await the future. However, current planning for Peete, other HANO developments, and many other low-income housing projects around the country make it painfully obvious that an extremely limited housing solution has been developed for a substantial proportion of current tenants. HUD's rhetoric and plans call for placing these persons in Section 8 housing, but the numbers simply do not add up.

8. Even adjusting for more years of schooling (as suggested by the education data provided in Table 1), later entry into the labor force, and the not unrelated return to marriage at a older age that has occurred throughout American society over the past 40 years, the decline in marriage remains quite substantial.

9. "Income deficit" is the dollar difference between an individual, family, or household's income and the appropriate size-adjusted poverty threshold.

10. Gross mismanagement and corruption often plagued HUD, HANO, and numerous other housing authorities around the country. We do not dismiss the negative impact these factors had on many communities, but these were secondary to the noted structural contradiction.

11. A series of physical inventories by our interviewers revealed that less than 30 of the abandoned units had squatters on more than an occasional basis. While an extremely tempting research target, we nonetheless defined our sample frame as officially occupied units.

12. Since the beginning of calendar year 2000, another 140 or so units have been vacated and plans calling for the demolition of all of "new" Peete (Phase II) have been formulated.

13. We've experimented with different constructions and verbiage of this basic question. No single format has distinguished itself with obvious superiority in terms of reliability and validity.

14. In 1996 the poverty threshold for a 3-person household was $12,516. By 1999, the threshold stood at $13,290.

15. Other faculty and graduate student participants in the team ethnography include the authors of "Kevin and Krista's Report," appearing later in this journal.

16. In Louisiana, the basic monthly TANF benefit for a woman with one dependent child is $138 and increases to $179 for a second child.

17. Our use of the phrase "officially recognized" simply means work in a legal establishment wherein wages are recorded and taxes, workmen's compensation, and other deductions are undertaken in accordance with federal and state requirements.

18. Alternatively, the alleged "hyper-consumerism" of the poor is interpretable as a deeply felt psychological need to appear part of the mainstream, that is to cover the stigma of poverty per se, a form of (over-) compensation that those in the middle class cannot appreciate. Television and other mass media have exacerbated this sense of relative deprivation greatly over the past several decades.

19. A "second-line" is a New Orleans tradition that is a part of the burial process. The first procession is a solemn mournful line to the cemetery. This is the "first-line." After the burial, the journey from the cemetery is the "second-line," which is a joyful

upbeat dance through the streets. Today, "second-line" celebrations are organized throughout the city in lieu of a funeral.

20. The supper is a plate of food that is sold for around $5.00 per plate. It is usually held from lunchtime on Friday and all day Saturday. The plate dinners are usually either fish or chicken, with bake macaroni's, green peas, bread and a slice of cake.

21. While potentially intriguing, this difference may well turn out to be more apparent than real and owe to either the nostalgia of older participants or our focus group participant selection process. The groups were organized by community contacts who invited others to join in. The invitees were hardly random and included people who had some history of shared relationships and trust. Perhaps, some as yet not understood bias occurred across groups organized on the "old" and "new" sides.

ACKNOWLEDGMENTS

A grant from the United States Department of Housing and Urban Renewal to the Tulane-Xavier Campus Affiliates Program supported this research. Any and all views expressed herein are solely the authors and do not reflect the position or knowledge of either HUD or CAP. We wish to thank the residents of C. J. Peete for their support, effort, and willingness to share their time and experiences with us.

REFERENCES

Bane, M. J., & Ellwood, D. T. (1983). The Dynamics of Dependency: The Routes to Self-Sufficiency.

Bane, M. J. (1994). *Welfare Realities: From Rhetoric to Reform.* Cambridge, MA: Harvard University Press.

Bates, T. M. (1993). *Banking On Black Enterprise: the Potential of Emerging Firms For Revitalizing Urban Economies.* Washington, D.C.: Joint Center For Political and Economic Studies.

Bates, T. M. (1997). *Race, Self-Employment and Upward Mobility: An Illusive American Dream.* Washington, D.C.: Woodrow Wilson Center Press.

Baxandall, R., & Ewen, E. (2000). *Picture Windows: How the Suburbs Happened.* New York: Basic Books.

Blank, R. M. (1989). Analyzing the Lengths of Welfare Spells. *Journal of Public Economics, 39,* 245–273.

Blank, R. M. (1995). Outlook for the U.S. Labor Market and Prospects for Low Wage Entry Jobs. In: D. Smith Nightingale & R. H. Haveman (Eds), *The Work Alternative* (pp. 33–69). Washington, D.C.: Urban Institute Press.

Bogue, D. (1969). *Principles of Demography.* New York: John Wiley and Sons, Inc.

Borjas, G. J., & Bronars, S. G. (1989). Consumer Discrimination and Self Employment. *Journal of Political Economy, 97,* 581–605.

Bureau of the Census, U.S. Department of Commerce (1998). *Poverty in the United States: 1997.* Current Population Reports, Consumer Income, Series P60–201 (September).

Burton, L. M., Bengtson, V. L., & Schaie, K. W. (1995). *Adult Intergenerational Relations: Effects of Societal Change.* New York: Springer Publishing Co.

Coltrane, S. (1998). *The Social Construction of Gender and Families*. Thousand Oaks, CA:Pine Forge Press.
Cordes, C. (1985). Black Males at Risk in America. *APA Monitor*, (January), 9–10, 27–28.
Deaton, A. (1997). *The Analysis of Household Surveys*. Baltimore: The John Hopkins University Press.
Devine, J. A., & Wright, J. D. (1993). *The Greatest of Evils: Urban Poverty and the American Underclass*. New York: Aldine deGruyter.
Edin, K. (1991). Surviving the Welfare System: How AFDC Recipients Make Ends Meet in Chicago. *Social Problems*, 38(4), 462–474.
Edin, K., & Lein, L. (1996). Work, Welfare and Single Mothers Economic Survival Strategies. *American Sociological Review*, 61, 253–266.
Edin, K. (1997). *Making Ends Meet: How Single Mothers Survive Welfare and Low-Wage Work*. New York: Russell Sage.
Ellwood, D. T. (1988). *Poor Support*. New York: Basic Books.
Feagin, J. R., & Imani, N. (1994). Racial Barriers To African-American Entrepreneurship: An Exploratory Study. *Social Problems*, 41, 562–584.
Fitzgerald, J. (1991). Welfare Durations and the Marriage Market: Evidence from the Survey of Income and Program Participation. *Journal of Human Resources*, 26, 545–561.
Gibbs, J. T. (1988). *Young, Black, and Male in America: An Endangered Species*. Dover, MA: Auburn Publishing House.
Gibson, G. (1972). Kin Family Network: Overheralded Structure in Past Conceptualizations of Family Functioning. *Journal of Marriage and the Family*, 34, 13–23.
Goldscheider, F. K. (1995). Interpolating Demography with Families and Households. *Demography*, 32(3), 471.
Granovetter, M. (1973). The Strength of Weak Ties. *American Journal of Sociology*, 78, 1360–1380.
Gutman, H. (1976). *The Black Family in Slavery and Freedom 1750–1925*. New York: Vintage.
Hao, L., & Brinton, M. C. (1997). Productive Activities and Support Systems of Single Mothers. *American Journal of Sociology*, 102(5), 1305–1344.
Harris, K. M. (1991). Teenage Mothers and Welfare Dependency. *Journal of Family Issues*, 12(4), 492–518.
Harris, K. M. (1993). Work and Welfare among Single Mothers. *American Journal of Sociology*, 99(2), 317–352.
Harris, K. M. (1996). Life After Welfare: Women, Work and Repeat Dependency. *American Sociological Review*, 61, 407–426.
Hays, W. C., & Mindel, C. H. (1973). Extended Kinship Relations in Black and White Families. *Journal of Marriage and the Family*, 35, 51–57.
Hofferth, S. L. (1984). Kin Networks, Race and Family Structure. *Journal of Marriage and the Family*, 46, 791–806.
Hofferth, S. L. (1985). Updating Children's Life Course. *Journal of Marriage and the Family*, 47, 93–115.
Hogan D., Hao, L.-X., & Parish, W. L. (1990). Race, Kin Support and Mother-Headed Families. *Social Forces*, 68(3), 797–812.
Housing Authority of New Orleans. (1940). *Report of the Housing Authority of New Orleans for the Year Ending December 31, 1940*. New Orleans: HANO.
Jencks, C. (1992). *Rethinking Social Policy*. New York: Harvard University Press.
Katz, M. (1989). *The Undeserving Poor: From the War On Poverty To the War On Welfare*. New York: Pantheon.

Kellam, S., Adams, R., Brown, C. H., & Ensminger, M. (1982). The Long-Term Evolution of the Family Structure of Teenage and Older Mothers. *Journal of Marriage and the Family, 44*, 537–554.

Koball, H. (1998). Have African-American Men Become Less Committed to Marriage? Explaining the Twentieth Century Racial Crossover in Men's Marriage Timing. *Demography, 35*(2), 251.

Mahoney, M. R. (1985). The Changing Nature of Public Housing in New Orleans, 1930–1974. Unpublished master's thesis, Department of History, Tulane University, New Orleans.

Majors, R. G., & Gordon, J. U. (1994). *The American Black Male: His Present Status and His Future.* Nelson-Hall Publishers: Chicago.

Mann, S. A., & Grimes, M. D. (1997). Paradigm Shifts in Family Sociology? Evidence From Three Decades of Family Textbooks. *Journal of Family Issues, 18*(3), 315–349.

Manning, W. D., & Smock, P. (1995). Why Marry? Race and the Transition to Marriage Among Cohabitors. *Demography, 32*(4), 509–520.

McAdoo, H. P. (1978). Factors Related to Stability in Upwardly Mobile Black Families. *Journal of Marriage and the Family, 40*, 761–776.

McAdoo, H. P. (1981). *Black Families.* Beverly Hills, California: Sage Publications.

McLanahan, S., & Garfinkel, I. (1989). Single Mothers, the Underclass, and Social Policy. *Annals of the American Academy of Political and Social Science, 501*, 92–104.

Mead, L. M. (1986). *Beyond Entitlement: The Social Obligations of Citizenship.* New York: Free Press.

Mead, L. M. (1992). *The New Politics of Poverty: The Nonworking Poor in America.* New York: Basic Books.

Morgan, J. N. (1982). The Redistribution of Income by Families and Institutions and Emergency Help Patterns. In: M. S. Hill et al. (Eds), *Five Thousand American Families, Vol. 10.* (pp. 1–59). Ann Arbor, MI: Institute for Social Research.

Mott, F. L. (1990). When Is a Father Really Gone? Paternal-Child Contact in Father-absent Homes. *Demography, 27*(4), 499–517.

Moynihan, D. P. (1965). *The Negro Family: The Case for National Action.* Washington, D.C.: Department of Labor.

Murray, C. (1984). *Losing Ground: American Social Policy.* New York: Basic Books.

Neckerman, K. M., & Aponte, R. (1988). Family Stricture, Black Unemployment, and American Social Policy. In: M. Weir, A. S. Orloff & T. Skocpol (Eds), *The Politics of Social Policy in the United States* (pp. 397–420). Princeton, N. J.: Princeton University Press.

Parish, W. L., Lingxin, H., & Hogan, D. P. (1991). Family Support Networks, Welfare and Work among Young Mothers. *Journal of Marriage and the Family, 53*, 203–215.

Piven, F. F., & Cloward, R. (1971). *Regulating the Poor.* New York: Vintage.

Putnam, R. (2000). *Bowling Alone: The Collapse and Revival of American Community.* New York: Simon and Schuster.

Radford, G. (1996). *Modern Housing For America: Policy Studies in the New Era.* Chicago: University of Chicago Press.

Russell, J. L. (1995). The Story of Public Housing in New Orleans, 1935–1990. Unpublished senior thesis, Department of Sociology, Tulane University, New Orleans.

Ryder, N. B. (1987a). Discussion. In: J. Bongaarts & T. K. Burch (Eds), *Family Demography* (pp. 345–356). Oxford: Clarendon Press.

Ryder, N. B. (1978b). Reconsideration of a Model of Family Demograpjy. In: J. Bongaarts & T. K. Burch (Eds), *Family Demography* (pp. 102–122). Oxford: Clarendon Press.

Ryder, N. B. (1992). The Centrality of Time in the Study of the Family. In: E. Berquo & P. Xenos (Eds), *Family Systems and Cultural Change* (pp. 161–175). Oxford: Clarendon Press.

Stack, C. (1974). *All Our Kin*. New York: Harper.
Sussman, M., & Bunchinal, L. (1962). Kin Family Network: Unheralded Structure in Current Conceptualizations of Family Functioning. *Marriage and Family Living*, 24, 231–240.
Valentine, C., & Valentine, B. (1971). *Missing Men: A Comparative Methodological Study of Under-enumeration and Related Problems*. Bureau of the Census, Contract 1510. Washington DC: U.S. Bureau of the Census.
von Hoffman, A. (2000). A Study in Contradictions: The Origins and Legacy of the Housing Act of 1949. *Housing Policy Debate*, 11(2), 299–326.
Wilson, W. J. (1987). *The Truly Disadvantage: The Inner City, the Underclass and Public Policy*. Chicago: University of Chicago Press.
Zollar, A. C. (1985). *A Member of the Family*. Chicago: Nelson-Hall Publishers.

ABSTRACT SPACE, SOCIAL SPACE, AND THE REDEVELOPMENT OF PUBLIC HOUSING

Kevin Fox Gotham, Jon Shefner and Krista Brumley

INTRODUCTION

This chapter examines conflicts and struggles between public housing residents and housing authority officials over the redevelopment of public housing in New Orleans. In recent years, the federal government has pressured local housing authorities to adopt policies to promote tenant self-management, mixed-use and mixed-income communities, market-based leasing, scattered site low-income housing, and decentralize management by adopting site-by-site needs analysis and site-based budgets. More important, the federal government has called for the demolition of existing public housing units and converting larger projects into small-scale communities as a vehicle for promoting "self-sufficiency" among tenants (U.S. Department of Housing and Urban Development, 1996). Indeed, the centerpiece of all federal housing policy since the 1970s has been retrenchment, privatization, and devolution of authority, responsibility, and funds from the federal government to state governments and, then, to local municipalities (see Gotham & Wright, 2000 for an overview). As Quercia and Galster (1997) have recently argued, public housing authorities are now being forced to change their management practices, the types of tenants they house, the kinds of developments they operate, and to attract private capital for the development and operation of public-private public housing venues. Public housing authorities and residents now face a situation of government-imposed redevelopment, demolition, and displacement – a situation exacerbated by uneven urban development, lack of affordable housing, and the general anti-poor sentiment shared by the nation.

In this chapter, we argue that the redevelopment of public housing, in New Orleans and elsewhere, is an attempt by federal and local officials to recommodify space, to enhance its exchange-value through privatization while reducing its use-value for low-income people. We use Henri Lefebvre's (1991) conceptual tools of "abstract space" and "social space" to understand the conflicts between public housing residents and housing authority officials over the redesign of public housing, broad policy shifts involving public housing space, and the role of the state in using space for social control. "Abstract space," according to Lefebvre, is the space of instrumental rationality, fragmentation, homogenization, and, most important, commodification. It is the use of space by capitalists and state actors who are interested in the abstract qualities of space, including size, width, area, location, and profit. In contrast, "social space" is the space of everyday lived experience, an environment as a place to live and to call home. For Lefebvre, the uses proposed by government and business for abstract space, such as planning a new highway or redeveloping older areas of the city, may conflict with existing social space, the way residents think about and use space. This conflict between abstract and social space is a basic one in modern society, according to Lefebvre, and involves spatial practices (spatial patterns of everyday life), representations of space (conceptual models used to direct social practice and land-use planning), and spaces of representation (the lived social relation of users to the built environment) (1991, pp. 33, 38–9).[1] The inseparability of spatial and social relations, interconnectedness of power and space in the city, and the association of space with social stratification and inequality are the central themes of this chapter.

We begin by discussing our theoretical orientation. The recent "spatial turn" in urban sociology has directed attention to the ways in which spatial arrangements operate as constitutive dimensions of social phenomena (Foucault, 1977; Harvey, 1993; 1989; Bourdieu, 1990; Giddens, 1990). Examining topics as diverse as racialized spaces (Gotham, 1998; Haymes, 1995), the social organization of gang activity (Venkatesh, 1997), the spatial attributes of corporate interlocking directorates (Kono, Palmer, Friedland & Zafonte, 1998), the emergence of social movements (Shefner, 2000; Zhao, 1998; Wright, 1997), space and community identity (Gotham, 1999; Swearington & Orellana-Rojas, 2000; Kasinitz, 2000), the militarization of urban space (Davis, 1992), and the political economy of art and entertainment (Scott, 2000), scholars have delineated why space is important and how the consideration of socio-spatial relations and land-use conflicts can illuminate our understanding of social change. As a component of social organization, spatial patterns and conflicts permeate social relations at the personal, collective, regional, and global levels (Lefebvre, 1979, p. 290; Gottdiener, 1994, pp. 126–127). Although urban scholars disagree about how space influences social relations, they agree by viewing space as a means of production (i.e. land and real estate), an object of consumption, and a geographical site of social action.

In the third section of the paper we discuss our data sources and method, and then introduce our empirical case. Next, in our findings section, we identify the themes, symbols, and motifs used by public housing residents to interpret the causes of the disinvestment and deterioration of their community. We also examine the rhetoric employed by public agencies such as the Housing Authority of New Orleans (HANO) and HUD to legitimate their "definitions of the situation" and discredit alternative views of this redevelopment process. In conclusion, we discuss the policy implications of our analysis. Rather than viewing the problem of urban disinvestment and poverty as only "structural" problems, we encourage scholars to consider them also as "land-use" issues directly connected to urban redevelopment, displacement, and cultural meanings of what the city "should look like" by those who have the power to shape its development.

Many empirical studies illustrate the powerful influence of government agencies and capitalist firms in shaping and transforming urban space (Feagin, & Parker, 1990; Jonas & Wilson, 1999; Logan & Molotch, 1987), the impact of "globalization" (Wilson, D, 1997; Van Kempen & Marcuse, 1997; Smith & Feagin, 1987), and macro-level economic "restructuring," especially the shift from mass industrial production to high technology and information processing (Sassen, 1990, 2000; Castells 1996, 1997, 1998). Yet little sociological research has investigated how cultural meanings and the interpretive and symbolic aspects of socio-spatial struggles are central to the social construction of urban space. The work of Sharon Zukin and colleagues (1998) on the symbolic economy and visual consumption, Robert Beauregard (1993) on the discourse of urban decline, and Mark Gottdiener (1997) on theming, represent exceptions to this tendency. Yet, overall, questions concerning how spatial arrangements influence social relations and reinforce social inequalities has been a secondary topic in sociological research. In this chapter, we connect power with its spatial context to highlight socio-spatial inequality, concentrated disadvantage, and conflicts over the use of space. Our goal is to supplement structural analysis and open new avenues to exploring the mechanisms that reinforce and perpetuate urban poverty and uneven development.

URBAN SPACE, POWER, AND CONFLICT

> The paradigmatic (or 'significant') opposition between exchange and use, between global networks and the determinant locations of production and consumption, is transformed here into a dialectical contradiction, and in the process becomes spatial. Space thus understood is both abstract and concrete in character: abstract inasmuch as it has no existence save by virtue of the exchange-ability of all its component parts, and concrete inasmuch as it is socially real and as such localized. This is a space, therefore, that is homogeneous yet at the same time broken into fragments (Henri Lefebvre 1991, pp. 341–342).

This chapter builds upon several areas of sociological research and theory. Ecological studies have long been interested in the forces behind the spatial ordering of urban communities, land-use conflicts, and social movement networking (Park, Burgess, & McKenzie, 1925; Warner, 1963; Lofland, 1973; Suttles, 1968, 1972). Moreover, Marxian scholars have devoted considerable attention to specifying the role of different user groups, land developers and real estate interests, and other economic and political elites in continually refashioning land-uses to assure the reproduction of profit and capitalist social relations (Harvey, 1985; Castells 1978; Molotch, 1979). Despite their competing assumptions and modes of analysis, both perspectives have treated spatial relations as expressions of social relations and have downplayed the spatial character of social action and structure. Whereas for urban ecologists urban space is an outcome of a noncontentious process of human adaptation and functional integration, for Marxian scholars it is a consequence of the logic of capital accumulation. In both accounts, urban space becomes a "residual phenomenon" (Logan & Molotch, 1987, p. 100), a container of social action, or a product of social conflict rather than a constitutive component of human agency and identity.

In response to the limitations of urban ecology and Marxian theory, a growing number of contemporary theorists view urban space as a medium of social relations and a material product (e.g. the "built environment") that can affect social relations (Lefebvre, 1991; Gottdiener, 1994, 1993; Gotham, 1998; Milligan, 1998; Friedland & Boden, 1994; Liggett & Perry, 1995; Wright, 1997; 2000; Zukin, 1991). In this critical sociological literature on cities and urban life, social relations exist to the extent that they posses a spatial component: they project themselves into space, becoming inscribed there, and in the process producing that space itself (Gottdiener, 1994; Swearingen & Orellan-Rojas, 2000; Wright, 2000; Gotham, 2000, 1999, 1998). According to Gottdiener (1985, p. 123), "Space cannot be reduced merely to a location or to the social relations of property ownership – it represents a multiplicity of sociomaterial concerns. Space is a physical location, a piece of real estate, and simultaneously an existential freedom and a mental expression. Space is both the geographical site of action and the social possibility for engaging in action."

One important outcome of the emphasis upon space has been to challenge the notion of space as merely a container of social action, or a derivative of the logic of capital accumulation. Indeed, the assumption that space is a "reflection" of exogenous social processes is a form of spatial fetishism. In his recent trilogy on the rise of the global network society, Manual Castells argues that space is "not a reflection of society, it is its expression. In other words: space is not a photocopy of society, it is society" (1996, p. 410). Castells defines space as having a form, function, and social meaning that shapes, and is shaped by, individuals and groups engaged in historically determined social relationships (1978, p. 152). According to Lefebvre, the production of different spaces shapes individual behavior, social action, and group formation. Moreover, the consequent layers of space – individual, local, metropolitan, national, global, and so on - interpenetrate and superimpose on one another, connecting global and local socio-economic processes with the production of fragmented and yet homogeneous spaces. As a "brutal condensation of social relationships," (Lefebvre, 1991, p. 227), space reflects power relations while also being a "site" for contesting relations of domination and subordination.

The emphasis within urban scholarship on delineating the links between space and social structure has led analysts to rethink basic social categories such as time, class, ethnicity and race, and gender through the prism of space (DeSena 2000; Spain, 1993; Urry, 1996; Miranne & Young, 2000; Friedland & Boden, 1994; Castoriadis, 1987; Hanson & Pratt, 1995; Simonsen & Vaiou, 1996; Wekerle, 1980; Stimpson, Dixler, Nelson, & Yatrakis, 1981; Hayden, 1986; Wilson, E., 1992). Urban geographer Doreen Massey (1984) argues in her book, *Spatial Divisions of Labor*, that changing social relations of production or social structure alone does not determine the spatial distribution of labor activities. For Massey, social structures and forms of conflict – including class-based divisions of labor and capital-labor struggles – take on a number of different spatial forms and there is no particular historical ordering in the emergence of new "regimes of accumulation" or "modes of political regulation." The fact that "space matters," according to Massey, requires us to consider that "processes take place over space, the facts of distances, closeness, of geographic variation between areas, of individual character and meaning of specific places and repair – all these are essential in the operation of social processes themselves" (1984, p. 14). As a set of productive forces, space plays an active part in the constitution of social reality, a point highlighted, in various ways, by Henri Lefebvre (1991, 1979, p. 290), Manuel Castells (1996, pp. 410–428, 1998, pp. 130–145), David Harvey (1989, p. 204), Edward Soja (1989), and sociologists such as Mark

Gottdiener (1994, pp. 110–156) and Talmadge Wright (2000; 1997). As John Urry (1996, pp. 378–379) puts it, "space makes a clear difference to the degree to which ... the causal powers of social entities (such as class, the state, capitalist relations, patriarchy) are realized."

The theoretical work of Pierre Bourdieu, Michel Foucault, and Anthony Giddens represent attempts to refashion social theory to take into account the reflexive relationship between space and social action. Specifically, Bourdieu (1988, pp. 774–777, 1993, p. 45) develops two concepts: "habitus" – the "ensemble of dispositions" that orient action and perception; and "field," the structured space where social struggles emerge (Bourdieu, 1989; Bourdieu & Wacquant, 1992, pp. 16–19) to highlight and explain the spatial and temporal attributes of agency and structure (for an overview see Camic & Gross, 1998). For Bourdieu, space helps to generate the "habitus" of everyday life for local residents and the factor that can produce place-specific forms of identity, consciousness, and knowledge (Bourdieu, 1989). Similarly, Foucault uses the concept "heterotopias" to refer to those oppositional spaces that form with relations of domination and subordination and serve as the birthing place for political mobilization and revolution. Just as "space is fundamental in any form of communal life; space is fundamental in any exercise of power" (1984, p. 253). Space, Foucault (1977, p. 70) suggests, can no longer be treated as the "the dead, the fixed, the undialectical, the immobile" (quoted in Liggett & Perry, 1995, p. 3).

Giddens (1973, pp. 108–10) draws attention to the importance of spatial segregation as a "proximate factor of class structuration ... an aspect of consumption rather than production which acts to reinforce separations" (quoted in Logan & Molotch, 1987, p. 19). According to his "structuration" approach, agency and structure are expressed through micro relations, routinized activities, and the "repetitive nature of day-to-day life" (Giddens, 1984, p. xxiv). Structuration means that "structure," as Giddens puts it, "is always both constraining and enabling" (1984, pp. 25, 163) and embedded in and reproduced through specific rules, procedures, and social relationships in time and space. In Giddens' terms, agency and structure are a duality that manifest themselves in historical, processual, and dynamic ways. Underscoring the importance of space in Giddens' work, Saunders (1989, p. 218) asserts that "any sociological analysis of *why* and *how* things happen will need to take account of *where* (and when) they happen" (emphasis in original).

In the rest of the paper, we apply Lefebvre's insights on the conflict between social space and abstract space to understand the redevelopment of public housing. Specifically, we examine the efforts of residents to resist displacement and demolition of their social space, and housing authority strategies to recommodify public housing, transform it into a privatized space, and make it subject to exchange-value pressures. Our goal is to specify why space is important and how the consideration of socio-spatial relations can illuminate our understanding of the recent policy shifts involving public housing. First, we investigate how housing authority officials and other local elites use various rhetorical devices, imagery, and motifs to define public housing as an undesirable area of overcrowding and social pathology and convince the public and housing residents of the necessity of redevelopment. The dominant discourse of urban redevelopment views public housing as a symbol of urban "decline" (Beauregard, 1993), a "failure" of federal policy, and a rampant ghetto of crime, drugs, and violence. This vision is a key to understanding the cultural and racial politics that supports the redevelopment effort and, at the same time, discourages alternative views, a point we explore further in the paper. Following Lefebvre's insights, we ground our analysis of the processes of place construction in material practices but recognize that these processes cannot be separated from representational and symbolic activities. Exploring the relationship between materiality and the symbolic interpretations people assign to space is crucial to understanding the inseparability of power and space in the city.

DATA AND METHOD

Our research is part of a multi-year team ethnography composed of graduate students and sociology faculty working on a number of different research projects within the C. J. Peete (CJP) public housing development in New Orleans (see chapter by Sams-Abiodun & Devine for an overview). The three authors of this ethnographic study use long-term field observations of daily life in the housing development, including field notes from numerous public meetings, and informal and semi-structured interviews with housing authority officials and residents from September 1, 1998 through August 31, 2000. The third author took on the role of primary data gatherer and field researcher while the first and second authors monitored and directed the activities of the field researcher. Sociology faculty and graduate student field researchers shared data collection strategies, data, and analyses while working on different agendas, questions, and issues within public housing.

We divided our ethnographic research and data collection into two major phases. During the first phase, from August 1998 through May 1999, our overall research strategy was to "hang out" with residents, spending time with them in varied settings (at their place of residence, at the community center, in surrounding neighborhood stores). The basic task was to acquire an appreciation for the nature of life in public housing and the ways in which residents managed project life both experientially and cognitively. The field researcher followed the residents she encountered through their daily routines and listened not only to what they told her but also to what they told one another. She asked questions and probed from time to time and recorded her observations and conversational dialogues in a detailed field narrative.

In the second phase, starting in May 1999, the field researcher began to supplement her field observations with semi-structured interviews. These respondents included ten African-American females ranging in age from thirty-two to seventy-eight who have lived in CJP for ten years or more. She gathered some of these interviews through a snowball sample while others were with residents who worked or volunteered within the CJP office. In August 1999, the field researcher expanded her interview sample to include six housing authority officials, including the on-site redevelopment coordinator, the assistant director of modernization, the director of management, the home ownership specialist, the development manager, and program coordinator.

Finally, in December 1999, the field researcher conducted interviews with six persons working for the Tulane-Xavier Campus Affiliates Program (CAP) at CJP. Established in 1996, CAP is a collaboration between Tulane and Xavier universities, HANO, and HUD that seeks to address the problems of poverty in public housing, and improve the living conditions and economic social well-being of residents in CJP. HUD helped establish CAP as part of the agreement under which Tulane University became the executive monitor over HANO (Shefner & Cobb, 2000). The field researcher interviewed CAP's office manager, the social work supervisor, three full-time social workers, and the "relocation liaison." The purpose of these interviews was to uncover information about the status of the redevelopment, CAP's role in this process, and finally, identify the concerns CAP staff had regarding the effects of the redevelopment on residents. We use pseudonyms to protect the confidentiality of all interviewees quoted in the paper.

In recent years, urban scholars have attempted to integrate structural analyses of poverty with ethnographic accounts of the everyday life of the poor using multiple data sources and by triangulating methods. This integrative framework has emerged as a result of criticisms of "structural" explanations that ignore the voices of the poor themselves (Jarrett, 1994), and ethnographic reports that often suppress the mention of structural features in understanding the conditions of poverty and inequality. Attempts to integrate ethnographic analyses with structural explanations are evident in the work by David Wagner (1993), Talmadge Wright (1997) and David Snow and Leon Anderson (1993) on homelessness, Jon Shefner (1999) on Mexican community organizations, Philippe Bourgois (1996) on crack dealers and David Wilson (1993) on inner city disinvestment. In this paper, we integrate our ethnographic analysis with larger structural concerns of macroeconomic change, transformations in federal housing programs, and recent trends in urban redevelopment. Structural accounts argue the poor are unable to overcome poverty, resist disinvestment, and mobilize to fight displacement because of intractable "barriers" they face (Wilson, 1987, 1996; Massey & Denton, 1993). These barriers include institutional discrimination; social marginalization and lack of access to quality education, health care, and employment opportunities; political exclusion from the policy-making and policy-formulation process; denial of political representation through electoral channels, and so on. We contend that this structural account has neglected a key issue – the interconnectedness of power and space – in attempts to explain the causes and consequences of urban poverty and disinvestment. In its explicit appreciation of space, our integrative framework supplements structural explanations by investigating how spatial arrangements reinforce social inequalities.

ABSTRACT SPACE AND "EXCLUSIVE" REDEVELOPMENT

> In connections with the city and its extensions (outskirts, suburbs), one occasionally hears talk of a 'pathology of space,' of 'ailing neighborhoods,' and so on. This kind of phraseology makes it easy for people who use it – architects, urbanists, or planners – to suggest the idea that they are, in effect, 'doctors of space'. This is to promote the spread of some particularly mystifying notions, and especially the idea that the modern city is a product not of the capitalist or neo-capitalist system but rather some putative 'sickness' of society. Such formulations serve to divert attention from the criticism of space and to replace critical analysis by schemata that are at once not very rational and very reactionary (Henri Lefebvre. 1991, p. 99).

In recent years, housing authorities around the country have begun to demolish units, convert existing units to mixed income occupancy, and privatize substantial portions of their public housing stock under the Urban Revitalization Demonstration Program, or HOPE VI. Currently, CJP is undergoing a massive $90.7 million "revitalization program" aimed at razing 881 units, constructing ninety-four senior housing units, and renovating 526 rental units and 150 for-sale properties in the adjacent community outside the CJP development. HOPE VI program goals for distressed public housing in New Orleans include: (1) changing the shape of public housing, through selective demolition, dedensifying sites, and creating safe and secure housing developments; (2) lessening the isolation of residents, creating mixed-income communities, and integrating public housing sites into surrounding neighborhoods; (3) establishing positive incentives for residents' self-sufficiency. The rhetoric employed here casts revitalization as "innovative" and "trend-setting" and its supposed effect is to transform a deteriorating and poorly managed public housing project into a vibrant living space that reflects the diversity and cultural vitality of New Orleans (Housing Authority of New Orleans, 1998, 2000). The multimillion dollar redevelopment effort attempts to remedy the problems of public housing space in a holistic fashion, to change the very structure – physical, social, and economic – of public housing.

On the national level, a recent series of reports from HUD maintains that HOPE VI public housing revitalization is reducing isolation by providing opportunities for employment and education and engaging residents in the life and prospects of the community (Department of Housing and Urban Development (HUD), 2000, 1999, 1996). The evidence from New Orleans suggests, however, that the revitalization process is generating substantial conflict between CJP residents and HANO over when demolition and redevelopment will be completed, who will be forced to move, how HANO will reimburse residents for the cost of relocation, and where residents will live once demolition and conversion to mixed income takes place. As we see it, rather than remedying the social inequalities and problems the poor face, the redevelopment of public housing reinforces such inequalities by being "exclusive" rather than "inclusive" in its design and intent. In the case of C. J. Peete, HANO has designed urban redevelopment to "exclude" the poor, from both the redevelopment process itself and from living in the renovated complex once it is complete.

Indeed, the very policy defined by the HOPE VI program in CJP is exclusive. HANO documents count 1403 resident units in the housing development, 881 of which the agency has scheduled for demolition and 517 for conversion to rental units. In addition, HANO will build 128 "for-sale" units, and a private developer will construct 94 units. Following the policy of creating a mixed-income community, only 40% of the 517 rental units will be available for "extremely" low income households, and an additional 35% of these units will be reserved for low-income households. This division of housing resources does not match the needs of the immediate community, where the majority of residents live below the poverty line, and more than 40% are unemployed (see chapter from Devine & Sams-Abiodun). According to HANO's own figures, they will reduce the number of units available for the existing residents within the housing development from 1088 to 482 (Brumley, 2000). The policy, in effect, legitimizes exclusion.

As of November 2000, the housing authority had demolished several buildings within CJP, rehabilitated others, and allowed others to stand vacant and fall into disrepair. In many of our conversations and interviews, residents repeatedly condemn the housing authority for being insensitive to their concerns, accuse officials of disseminating inaccurate and conflicting information, and charge that the housing authority is refusing to work with them to find affordable housing. As a number of residents told us:

> Sadie (age 53): I don't think HANO is being straight up with everybody. You know, one time they'll tell you one thing then when you look again they're telling you something different. And at each meeting something different comes up. The first contractors did some set of details and the second contractor come in with something altogether different and turn it around now so, you really don't know where you stand. All you know is that you got to get out.

> Tangie (age 44): All the top guys, they're not being honest because they are telling the people 'hey, why don't you go to another development and if you decide you don't like it you can come back here in C. J. Peete.' You can't. How in the world can these people come back in C. J. Peete when you have so many units now and it's going to be downsized to four hundred and something?

> Gillian (age 32): Well my opinion is that they're moving all these people and they don't have nowhere to put them – first of all ... They don't give them all the information that they really need to tell them about, you know you have to pay the water, lights in some houses, plus your phone bill. The other thing I think they're lying about when they say they're going to bring them back. They're not going to bring those people back. None of them, not even the older people.

As these excerpts suggest, residents believe that the housing authority is being dishonest and that the objective of the redevelopment effort is to displace them. Conflicting information about how much low-income housing will be available after the revitalization is completed exacerbates this fear of displacement. In other conversations, some residents doubt that revitalization will actually occur because of the time lag between information given and action taken by HANO and the private revitalization firm. For some, the fact that the housing authority undertook some demolition in 1997 and then left the land vacant fuels resident skepticism. Neither have CJP families been forced to move in anticipation of further construction. The private developer's intentions to demolish all the buildings on one side the development has yet to be fulfilled due to many delays in signing the contract with the housing authority. HANO's expressed commitment to offer 90-day notification of relocation contrasts with housing authority employee interviews, who suggest HANO will not forcibly remove residents. The many extensions granted to residents similarly creates ambiguity over the redevelopment schedule and further reinforces resident uncertainty and doubt over the future of the project (Brumley, 2000).

Moreover, HANO's policy of giving residents relocation money well in advance of their move has also caused much discontent and frustration, due to the realistic fear that the money will be spent before the move. Housing Authority representatives have alerted residents that they will receive only one check and if they spend it before their move they will not receive anymore money when it is time to move. This policy has come under attack by residents who charge that HANO is erecting barriers to successful relocation and is unsympathetic to their housing needs. According to the third author's field notes of a conversation she had with one resident:

> I then asked her what she thought about the revitalization in general. She said that she was particularly concerned about the elderly. "They don't want to leave because they feel safe and secure here. How can you tell a 64 year old woman that she has to relocate." She then went on to say that it doesn't make sense how the housing authority is handling the redevelopment. She says that HANO has already started handing out Section 8 vouchers. She raises her voice and argues that HANO or UNIDEV (the private firm in charge of redevelopment) need to give classes on Section 8. "You don't give vouchers to clients without explaining and without training. It won't work."
>
> I then replied that I thought at the town meeting in December 1998 that HANO announced they were going to have workshops and that in order to get your certificate you must complete the training. She shook her head and said they had already started giving them out. "What are they planning on doing? Building a homeless shelter? You know not everyone is gonna get a job. It just makes no sense" (field notes 1/25/99).

Most Housing Authority officials around the nation assume that pubic housing residents are dissatisfied (or that they should be) with their living arrangements, and that most would be interested in leaving public housing, or would enthusiastically support redevelopment (Vale, 1997). Yet no one has confirmed this view using systematic surveys or in-depth ethnographic research. More often than not, press accounts and conservative poverty researchers have conveyed empirically unsubstantiated views that totalize public housing residents by emphasizing the most extreme and pathological dimensions of life in public housing environments – e.g. violence, drugs, joblessness, hopelessness, and despair. From the perspective of the housing authority, the relocation of residents and redevelopment of public housing is a solution to the problems of poverty and deterioration. Thus, the challenge is to recommodify public housing, to create an economic mix of residents, and redesign projects so that renovated area attracts new residents. The cost of this policy, however, is the exclusion of large numbers of needy people. Demolishing units (without a one-to-one replacement) and reconfiguring public housing for mixed-income occupancy eschews responsibility for housing the traditional clientele of public housing, those who cannot afford private market housing.

According to housing authority officials, if there is a problem in the relocation policy and process, that problem is with the residents themselves. For example, one HANO official we interviewed maintains that the "residents need a new 'mind-set' in order to get through this process easily" (Interview with Mark Burton). Another HANO official defended the redevelopment process as a way to achieve "a mix of different folks both socially and economically" in order to repair the "incredible dysfunctional community" currently in place. That official further stigmatized public housing, calling the developments "a microcosm of every social and economic problem at an exacerbated high [level]." When pressed about the redevelopment process, this same official stressed he would advocate no changes, saying "more or less, it works. It is strictly business redeveloping a site – it is not about people, it is about business."

According to Lefebvre (1976, p. 28), the domination of abstract space does not result from some autonomous logic of capital but emanates from socio-spatial strategies employed by powerful groups and actors – socio-spatial strategies that shape public views, images, and understandings of particular groups of people and spaces. One strategy used by powerful land-use actors and organized interests to commodify space is to assign certain negative imagery, metaphors, and symbols to that space thereby stigmatizing inhabitants, their culture, social relations, and so on. In New Orleans, city leaders view public housing as undesirable areas of overcrowding and social pathology, as "obstacles to regeneration," according to Cook and Lauria (1995). The effect of this stigmatization enhances the abstract qualities of space, fragments social relations and action, and, in the process, represents an attempt to homogenize social relations under the aegis of commodification. The denigration of public housing fits well with Lefebvre's analysis. Here the socio-spatial strategy is to reduce the use-value of housing for low-income residents and, at the same time, legitimize the demolition of dwellings and the displacement of residents through the redevelopment effort. Thus, the redevelopment of public housing is the mechanism to expedite the commodification of public housing.

SOCIAL SPACE AND RESIDENT VIEWS OF REDEVELOPMENT

> Let us now turn our attention to the space of those who are referred by means of such clumsy and pejorative labels as 'users' and 'inhabitants.' No well-defined terms with clear connotations have been found to designate these groups. Their marginalization by spatial practices thus extends even to language. The word 'user' (*usager*), for example, has something vague – and vaguely suspect - about it. 'User of what?' one tends to wonder. Clothes and cars are used (and wear out), just as houses are. But what is use value when set alongside exchange and its corollaries? As for 'inhabitants,' the word designates everyone - and no one. The fact is that the most basic demands of 'users' (suggesting 'underprivileged') and 'inhabitants' (suggesting 'marginal') find *expression* only with great difficulty, whereas the *signs* of their situation are constantly increasing and often stare us in the face.
>
> Henri Lefebvre. 1991, p. 362 (emphasis in original).

In the *Production of Space*, Lefebvre points out that the use of space by "inhabitants" or "users" is shaped by the social relations of capitalism and state action that fragment and homogenize social space (p. 129). The juggernaut of commodification transforms the use value of social space into the exchange value of abstract space, while the modern bureaucratic state imposes hierarchy and instrumental control over space. Concurrently, spatial interchangeability and compartmentalization destroy the ability of people to forge meaningful social contacts and relationships (pp. 49–50, 287, 341–342). In a commodified world, abstractions come to dominate reality such that "everyday life" becomes degraded, privatized, and divided among competing private interests (pp. 308–320). The imposition of technical planning schemes and redevelopment plans, for example, restricts the multidimensional character of everyday life and fragments life activities. At the same time, such plans impose homogeneity by integrating transportation and communication technologies with schools, work, housing, and differentiated land-uses. The instrumental value of space eclipses its intrinsic value: People no longer value space for its own sake but come to view space for the advantages it may bring, especially as a means to the single end of financial gain. As far as possible, the abstract space of commodification, quantification, calculability, and control seeks to crush the social space of lived experience (pp. 59–60). The incompatibility of use and exchange value causes space itself to become an object of social conflict (pp. 52, 356).

Although many residents express fondness for their past and present in CJP, many public housing residents with whom we talked with dislike public housing and feel disdain for its social fabric. Many readily complain of the poor living conditions they are forced to live with (e.g. leaky ceilings, drafty apartments, poor public services, etc.), in addition to the ominous threat of crime, violence, and drug trafficking. From these observations, it seems likely that many would prefer to leave public housing, even in the face of extensive redevelopment and modernization of units. Yet many residents we talked with fear leaving the relative security of public housing and being forced to face new financial burdens (i.e. utility bills) and potential homelessness. These observations corroborate findings from other studies of public housing around the country (Feldman, Stall & Wright, 1998; Feldman & Stall, 1994; Vale, 1997). For the CJP residents we talked with, public housing is what Lawrence Vale (1997) calls an "empathological place," where "profound ambivalence is the ruling emotional response" (p. 159) where fear of remaining is counterbalanced by the fear of departure. As the following residents told us,

> [HANO] don't give them [the residents] all the information that they really need to tell them about, you know you have to pay the water, lights in some houses plus your phone bill. A lot of people not used to this, used to doing them things. I know a lot of people personally that have given back their Section 8 vouchers because they could not afford them.

> Farrah (age 38): There's only so many vacancies in public housing. For those who do have to relocate, that means they would have to go outside of public housing – then you deal with bills that they're not accustomed to dealing with. Or having added adequate funds to pay those bills. And that's my concern for the redevelopment and relocation for the community.
>
> Margerie (age 47): I know I've heard a lot of people say that when they go in the (HANO) office they don't have any place for them to go and look at. You know. So really what can you do if they don't have any place. I've been there several times.

As the above excerpts show, residents feel that alternative housing outside public housing is too expensive and therefore unavailable. Here the desire to remain in public housing accompanies feelings of entrapment and lack of real choice, given the dearth of affordable housing in the private market. Not surprisingly, a number of residents we interviewed disagree with the Housing Authority's optimistic view that redevelopment and relocation will benefit them, surrounding neighborhoods, and the city. Some residents do indeed welcome redevelopment as an opportunity to escape distressed living conditions. Others express a firm desire to stay in CJP, despite their negative evaluations of life in public housing. Moving out of public housing, as residents recognize, will disrupt connections to schools, church, friends, work, shops, and other organizations and resources. As Rhoda and Vivian put it,

> Rhoda (age 78): at this age I am not thinking about [moving] . . . No indeed, no part. Because I'm living comfortable there's no sense in me moving now. Been here fifty some odd years or longer, as long as they've been up. There's no sense in me moving.
>
> Vivian (age 33): This is where I grew up at. I would like to stay in this development . . . Really and truly I like this project, I really do, cause I grew up around here from a baby. I grew up around this area. And, you know, it would be strange for me to move in another development.

Rhoda, Vivian, and other residents with whom we interacted with express a desire to stay in CJP because redevelopment plans as they stand today will destroy their immediate homes, the spaces they congregate and regularly and frequently interact with others, and their community ties and friendship networks – connections so many poor people rely on for resource and survival (Stack, 1974). These residents view public housing as a community and oppose the housing authority efforts to redefine and recommodify the social space they call home. These voices may not represent the CJP population as a whole, and they may represent more of a "pocket of resistance" or important sub-group rather than a norm. Nevertheless, the antipathy toward HANO, the deep sentimental attachment to public housing, and the firm resistance to the redevelopment effort clearly express the basic conflict between abstract space and social space. The redevelopment represents an attack on their place of residence, an assault on their "home," and a disruption of the "habitus" of social activities and everyday life.

A number of employees from the Campus Affiliates Program (CAP) have observed that many residents have a common stake in the future of public housing not only because it is the only form of housing that they can afford but because of the long-term community ties many have forged over the years. The stakes involved in the relationship to place are high, reflecting material and psychological connections to space, people, and organizations. The CAP employees express similar notions as the residents about the significance of place. As these employees told the third author:

> I see a lot of community ties and affiliations and the way that the community copes with lower paying jobs and lack of childcare and things like that. This is a tragedy, and I know that is a strong word, but that is reality, families being separated. I see a loss of relationships (CAP interview 12/15/99).

> There are a lot of systems set up here to help them make it that they won't have outside the community. Like most they are placed somewhere else, and I don't think that they will have the same support, even a grandmother's help (CAP interview 12/16/99).

> This is all they know. This is where many people spend most of their days and it is hard for people to separate themselves from this place (CAP interview 12/9/99).

As the above excerpts show, residents' understandings and interpretations of public housing are undergoing a major transformation as they find themselves forced to ask what kind of secure and meaningful place they can make in the face of impending demolition and displacement. The tenuousness of their ties to public housing does not keep residents from drawing on a distinctly spatial discourse, which they use as a political tool to contest the efforts of the housing authority to transform public housing space. As we see it, the construction of a meaningful place can and does have political meaning, even under circumstances where the daily practices of people in that place show little commonality. For Lefebvre (1976, p. 31) "space is not a scientific object removed from ideology or politics; it has always been political and strategic." From the point of view of the housing authority, the problems of public housing are "technical" and "administrative" problems amenable to policy recommendations and redevelopment "strategies" emanating from their particular constructions of urban reality. Urban geographers Helen Liggett and David C. Perry (1995, p. 18) argue that various elites, including urban planners, policy analysts, and state actors "are more adept at representing or discussing reality in precisely those terms that conceptually structure their intervention of which they are agents . . ." Thus, how various groups define and "represent" space are "goal-oriented activities that aim at furthering particular interests" (p. 16), including elite interests in recommodifying space.

Yet even as housing authority officials seek to convert public housing into an abstract space of commodity exchange, residents assign their own meanings to refute the premises imposed by abstract space. Philip Kasinitz (2000, pp. 254–255) argues that "the social construction of urban spaces is highly contingent on the local construction of meanings, often in response to idiosyncratic local circumstances. The ways in which people, including poor people, think about their spaces plays an important, though by no means exclusive, role in how those spaces are shaped." The meaning or symbolic significance a particular place takes on is in part the outcome of a struggle among different groups that compete to control that space – e.g. capitalist and state "utilizers" versus community "users." It is these socio-spatial conflicts that structure the "city trenches" (Katznelson, 1981) in which government officials and planners debate the policy issues and people defend their sentimental attachments to place. In short, the fierce contest over images and counter-images of places is an arena in which the cultural politics of places and the political economy of their development frequently interlock in indistinguishable ways.

CONCLUSION

In this chapter, we have used Henri Lefebvre's theoretical insights on abstract space and social space as an empirical basis for understanding the redevelopment of public housing in New Orleans. Lefebvre's task is to bring together objective and subjective understandings of space by tracing them both back to the process in which individuals and groups produce space. He questions the validity of any understanding of space that does not locate spatial conflicts and struggles within the political economy of capitalism and the production of space. The strength of Lefebvre's work, therefore, is that it refuses to see materiality and representation as separate spheres and it denies the particular privileging of one realm – e.g. class, race, gender, and so on – over another. For Lefebvre, the crushing of lived social spaces by the imposition of abstract space results in the ghettoization of all sectors of society, from the ghettos of the elite, of the bourgeoisie, of the intellectuals, of immigrants, and the poor. These fragmented and segregated spatial forms lead to the creation of new socio-spatial conflicts that transcend class divisions. For example, the emergence in the 1960s and later of powerful and diverse social movements based on civil rights, gay rights, women's issues, housing and poverty, NIMBY conflicts, nuclear weapons, and the environment derived from the "explosion of spaces," from the proliferation of new spatial boundaries in response to the chaotic expansion of abstract spaces. As social inequality expands and land-uses constantly change, new mobilizations of the very poor, the homeless, the unemployed, home owners fearing a decline in property values and crime, and other community organizations appear at the borders of the newly created social and physical spaces. Today, retrenchment in federal housing policy combined with the commodification of public housing is leading to the creation of new spatial forms and therefore new conflicts and antagonisms, including struggles for fair housing, tenants' rights, and homeless mobilizations, among others.

Rather than viewing urban space as a container of social action, this analysis suggests that space has a strategic and political character that is an active component in producing, reinforcing, and contesting social inequalities in society. In his award winning book, *Out of Place*, Talmadge Wright (1997) suggests that one of the key areas of struggle over urban space in contemporary urban America is "the allocation of land-use according to 'higher and better' uses" (p.10). "What these uses are is established by constructing a hierarchy of worthy spaces, people and things related to the labeled moral and physical comportment of a given population." For example, the myths of self-reliance and self-sufficiency, combined with possessive individualism, work to maintain distinctions between the "deserving" and "undeserving" populations, establishing objective and subjective distinctions and hierarchies of worth. These meanings specify how deviant populations, especially the urban poor, should be treated, where they should live, and what should be done to keep them in their "place." In turn, stigmatizing public housing residents with pejorative labels – e.g. lazy welfare queens, drug abusers, violent criminals, and so on –works to legitimize individualistic interpretations of social inequality, divert attention away from the structural causes of urban poverty and inner city deterioration, and justify the commodification of public housing. HOPE VI efforts to ensure that public housing developments are mixed-income does little to solve the critical housing problems of the poor. What the program does is exacerbate the problems of housing and poverty by displacing residents, making fewer units available, and forcing local housing authorities to compete with the private sector for tenants.

Understanding how space is a component of social organization and political struggle also has implications for understanding public policy and urban redevelopment. To date there has been little empirical work that examines the dialectical interplay between public policy, changes in the built environment, and the socially "constructed" aspects of urban space – e.g. the ways different actors and user groups produce, negotiate, and assign different meanings, imagery, and discourages to spaces in the city. Yet Robert Beauregard (1995, p. 77) suggests that to "contemplate public policy for our cities or to consider acting collectively requires not merely an analysis of the conditions available for success but also a reflective understanding of the language with which we represent these conditions." This is a call "to direct our attention to the simultaneity of discourse and history, the importance of negotiated imaginations, and the power of material conditions." Rather than viewing the social problems of poverty and lack of affordable housing as social welfare issues, this ethnographic study suggests we should view them as land-use issues connected to urban imagery, typifications, and cultural and symbolic representations of the city. Various motifs and themes, including rhetoric such as "growth," "development," "revitalization," and the like, are not objective, fixed, and stable categories but constitute authoritative and yet contested representations of space that shape peoples' perceptions of different living spaces and the people that live in them. In turn, these meanings of space shape peoples' understandings of the causes of urban problems that bias and select against some urban planning and policy choices rather than others. The assignment of meaning to places in the city, by urban planners, policy makers, poor people, and others, is not arbitrary but established through social practices that connect with the material conditions of uneven development, political-economic power, and public policy.

NOTE

1. Lefebvre's work is the seminal source of critical American and European scholarship on the city and urban life. David Harvey, Manual Castells, Mark Gottdiener, Joe Feagin, and Talmadge Wright have elaborated on Lefebvre's political economy of space. Many other urban geographers and postmodern scholars have incorporated Lefebvre's ideas into their perspectives and analyses, including Fredric Jameson, Michael Dear, Neil Smith and Edward Soja, among others (for reviews see Gottdiener, 1993; Stewart, 1995; Dear, 1997; Benko & Strohmeyer, 1997; Liggett & Perry, 1995).

REFERENCES

Beauregard, R. A. (1995). If Only the City Could Speak: The Politics of Representation. In: H. Liggett & D. C. Perry (Eds), *Spatial Practices: Critical Explorations Social/Spatial Theory* (pp. 59–80). Thousands Oaks, CA: Sage Publications.
Beauregard, R. A. (1993). *Voices of Decline: The Postwar Fate of U.S. Cities*. Blackwell.
Benko, G., & Strohmeyer U. (Eds) (1997). *Space and Social Theory: Interpreting Modernity and Postmodernity*. Malden, MA: Blackwell Publishers.
Bourdieu, P., & Wacquant, L. (1992). *An Invitation to Reflexive Sociology*. Chicago: University of Chicago Press.
Bourdieu, P. (1988). Vive le crise! *Theory Sociology, 17,* 773–787.
Bourdieu, P. (1990). *The Logic of Practice*. Translated by R. Nice. Stanford, CA: Stanford University Press.
Bourdieu, P. (1993). *The Field of Cultural Practice*. New York: Columbia University Press.
Bourdieu, P. (1989). Social Space and Symbolic Power. *Sociological Theory, 7,* 14–25.
Bourgois, P. (1995). *In Search of Respect: Selling Crack in El Barrio*. Cambridge: Cambridge University Press.

Brumley, K. (2000). Final Report: Redevelopment and Relocation in: C. J. Peete, *Unpublished Manuscript*. Department of Sociology, Tulane University.
Camic, C., & Gross. N, (1998). Contemporary Developments in Sociological Theory: Current Projects and Conditions of Possibility. *Annual Review of Sociology, 24*, 453–476.
Castells, M. (1996). *The Rise of the Network Society*. New York: Blackwell.
Castells, M. (1998). *The End of Millennium*. New York: Blackwell.
Castells, M. (1997). *The Power of Identity*. Oxford, U.K.: Blackwell.
Castells, M. (1978). *The Urban Question*. Cambridge, MA: MIT Press.
Castoriadis, C. (1987). *The Imaginary Institution of Society*. Cambridge, MA: MIT Press.
Cook, C., & Mickey. L. (1995). Urban Regeneration and Public Housing in New Orleans. *Urban Affairs Review, 30*, 538–557.
Davis, M. (1992). Fortress Los Angeles: The Militarization of Public Space. In: M. Sorkin (Ed.), *Variations on a Theme Park: The New American City and the End of Public Space* (pp. 154–180). New York: Hill and Wang.
Dear, M. (1997). Postmodern Bloodlines. In: G. Benko & U. Strohmeyer (Eds), *Space and Social Theory: Interpreting Modernity and Postmodernity* (pp. 49–71). Malden, MA: Blackwell Publishers.
Department of Housing and Urban Development (2000). *HOPE VI: Community Building Makes a Difference*. Washington, DC: Government Printing Office.
Department of Housing and Urban Development (1999). *HOPE VI: Building Communities, Transforming Lives*. Washington, DC: Government Printing Office.
Department of Housing and Urban Development (1996). *An Historical and Baseline Assessment of HOPE VI*. Washington, DC: Government Printing Office.
DeSena, J. N. (2000). Gendered Space and Women's Community Work. In: R. Hutchison (Ed.), *Constructions of Urban Space. Research in Urban Sociology. Vol. 5* (pp. 275–297). Stamford: CT: JAI Press.
Devine, J. (1999). *Peete Community-Wide Survey 4 (Winter 1998–99)*. Department of Sociology. Tulane University, New Orleans, LA.
Emirbayer, M., & Mische. A. (1998). What is Agency? *American Journal of Sociology, 103*, 962–1023.
Feagin, J. R., & Parker, R. (1990). *Building American Cities: The Urban Real Estate Game*. (2nd Ed.) Englewood Cliffs, NJ: Prentice Hall.
Feldman, R., & Stall. S. (1994). The Politics of Space Appropriation: A Case Study of Women's Struggles for Homeplace in Chicago Public Housing. In: I. Altman & A. Churchman (Eds), *Women and the Environment* (pp. 167–199). New York: Plenum Press.
Feldman, R. M., Stall. S., & Wright, P. A. (1998). 'The Community Needs to Built By Us': Women Organizing in Chicago Public Housing. In: A. Nancy (Ed.), *Community Activism and Feminist Politics: Organizing Across Race, Class, and Gender* (pp. 257–274). Naples. New York: Routlege.
Foucault, M. (1977). *Discipline and Punish: The Birth of the Prison*. Translated from French by Alan Sheridan. New York: Vintage books.
Friedland, R., & Boden, D. (Eds) (1994). *Nowhere: Space, Time, and Modernity*. Berkeley: University of California Press.
Giddens, A. (1990). *Consequences of Modernity*. Stanford, CA: Stanford University Press.
Giddens, A. (1984). *The Constitution of Society*. Berkeley: University of California Press.
Giddens, A. (1973). *The Class Structure of Advanced Societies*. New York: Harper and Row.
Gotham, K. (1998). Suburbia Under Siege: Low-Income Housing and Racial Conflict in Metropolitan in Kansas City, 1970–1990. *Sociological Spectrum, 18*, 449–483.

Gotham, K. F., & Wright, J. D. (2000). Housing Policy. In: J. Midgley, M. Livermore & M. B. Tracy (Eds), *Handbook of Social Policy* (pp. 237–255). Sage Publications.
Gotham, K. F. (2000). Representations of Space and Urban Planning in a Post-World War II U.S. City. In: R. Hutchison (Ed.), *Constructions of Urban Space. Research in Urban Sociology. Volume Five* (pp. 155–180). Stamford, CT: JAI Press.
Gotham, K. F. (1999). Political Opportunity, Community Identity, and the Emergence of a Local Anti-Expressway Movement. *Social Problems, Vol. 46*, No. 3. August 1999.
Gottdiener, M. (1994). *The Social Production of Urban Space*. (2nd Ed). Austin: University of Texas Press.
Gottdiener, M. (1985). *The Social Production of Urban Space*. (1st Ed.). Austin: University of Texas Press.
Gottdiener, M. (1993). Marxian Urban Sociology and the New Approaches to Space. In: *Urban Sociology in Transition. Research in Urban Sociology. Vol. 3* (pp. 209–230). JAI Press.
Gottdiener, M. (1997). *The Theming of America*. New York: Westview Press.
Government Accounting Office (GAO) (1996). *Public Housing: HUD Takes Over the Housing Authority of New Orleans*. Washington, DC: Government Printing Office.
Hanson, S., & Pratt, G. (1995). *Gender, Work, and Space*. New York: Routledge.
Harvey, D. (1985). *Consciousness and the Urban Experience: Studies in the History and Theory of Capitalist Urbanization*. Johns Hopkins Press.
Harvey, D. (1993). From Space to Place and Back Again: Reflections on the Condition of Postmodernity. In: J. Bird, B. Curtis, T. Putnam, G. Robertson & L. Tickner (Eds), *Mapping the Futures: Local Cultures, Global Change* (pp. 3–39). London: Routledge.
Harvey, D. (1989). *The Condition of Postmodernity*. New York: Free Press.
Hayden, D. (1986). What Would a Non-Sexist City Be Like? Speculations on Housing, Urban Design, and Human Work. In: R. G. Bratt, C. Hartman & A. Meyerson (Eds), *Critical Perspectives on Housing* (pp. 230–247). Philadelphia: Temple University Press.
Haymes, S. N. (1995). *Race, Culture, and the City: A Pedagogy for Black Urban Struggle*. Albany, NY: State University of New York (SUNY) Press.
Housing Authority of New Orleans (HANO) (1998). *Continuous Improvement Plan: A Strategy for Moving the Housing Authority of New Orleans Into the Next Millennium*. New Orleans: The Housing Authority of New Orleans.
Housing Authority of New Orleans (HANO) (2000). *Building the Bridge: A 12th Month Transition Plan for Fiscal Year 2000*. New Orleans: The Housing Authority of New Orleans.
Hutchison, R. (Ed.) (2000). *Constructions of Urban Space. Research in Urban Sociology. Vol. 5*. Stamford, CT: JAI Press.
Jarrett, R. (1994). Living Poor: Family Life Among Single Parent, African-American Women. *Social Problems, 41*, 30–49.
Jonas, A. E. G., & Wilson, D. (Ed.) (1999). *The Urban Growth Machine: Critical Perspectives Two Decades Later*. Albany: State University of New York Press.
Kasinitz, P. (2000). Red Hook: Paradoxes of Poverty and Place in Brooklyn. In: R. Hutchison (Ed.), *Constructions of Urban Space. Research in Urban Sociology. Vol. 5* (pp. 253–274, 275–297). Stamford: CT: JAI Press.
Katznelson, I. (1981). *City Trenches: Urban Politics and the Patterning of Class in the United States*. New York: Pantheon Books.
Kono, C., Palmer, D., Friedland, R., & Zafonte, M. (1998). Lost in Space: The Geography of Corporate Interlocking Directorates. *American Journal of Sociology, 103*, 863–911.
Lefebvre, H. (1991). *The Production of Space*. Blackwell.
Lefebvre, H. (1976). *The Survival of Capitalism*. London: Allison and Busby.

Lefebvre, H. (1979). Space: Social Product and Use Value. In: J. Freiberg (Ed.), *Critical Sociology: European Perspective*. New York: Irvington Publishers.

Liggett, H., & Perry D. C. (Eds) (1995). *Spatial Practices: Critical Explorations in Social/Spatial Theory*. Thousands Oaks, CA: Sage Publications.

Lofland, L. (1973). *A World of Strangers: Order and Action in Urban Public Space*. Prospect Heights, IL: Waveland Press.

Logan, J., & Molotch, H. (1987). *Urban Fortunes: The Political Economy of Place*. Berkeley: University of California Press.

Massey, D. S., & Denton. N. A. (1993). *American Apartheid: Segregation and the Making of the Underclass*. Cambridge, MA: Harvard University Press.

Massey, D. (1984). *Spatial Divisions of Labor*. London: MacMillan.

Milligan, M. (1998). Interactional Past and Potential: The Social Construction of Place Attachment. *Symbolic Interaction*, 21, 1–33.

Miranne, C. B., & Young, A. H. (Eds) (2000). *Gendering the City: Women, Boundaries, and Visions of Urban Life*. Lanham, MD: Rowman and Littlefield.

Molotch, H. (1979). Capital and Neighborhood in the United States: Some Conceptual Links. *Urban Affairs Quarterly*, 14, 289–312.

Park, R. E., Burgess, E., & McKenzie, R. D. (Eds) (1925). *The City*. Chicago: University of Chicago Press.

Quercia, R. G., & Galster, G. (1997). The Challenges Facing Public Housing Authorities in a Brave New World. *Housing Policy Debate*, 8(3), 535–69.

Sassen, S. (2000). *Cities in a World Economy*. New York: Prentice Hall.

Sassen, S. (1990). Economic Restructuring and the American City. *Annual Review of Sociology*, Vol. 16. (pp. 465–490).

Saunders, P. (1989). Space, Urbanism, and the Created Environment. In: D. Held & J. Thompson (Eds), *Social Theory of Modern Societies: Anthony Giddens and His Critics* (pp. 215–234). New York: Cambridge University Press.

Scott, A. J. (2000). French Cinema: Economy, Policy, and Place in the Making of a Cultural-Products Industry. *Theory, Culture, and Society*, Vol 17(1), 1–38.

Shefner, J. (2000). Austerity and Neighborhood Politics in Guadalajara Mexico. *Sociological Inquiry*, 70, 4.

Shefner, J. (1999). Sponsors and the Urban Poor: Resources or Restrictions. *Social Problems*, 46, 3.

Shefner, J., & Cobb, D. (2000). *Crisis and University Response in New Orleans*. Unpublished Manuscript.

Simonsen, K., & Vaiou, D. (1996). Women's Lives and the Making of the City: Experiences from 'North' and 'South' of Europe. *International Journal of Urban and Regional Research*, Vol. 20, No. 3.

Smith, M. P., & Feagin, J. R. (Eds) (1987). *The Capitalist City: Global Restructuring and Community Politics*. New York: Basil Blackwell.

Snow, D., & Anderson, L. (1993). *Down on Their Luck: A Study of Homeless Street People*. Berkeley, CA: University of California Press.

Soja, E. W. (1989). *Postmodern Geographies: The Reassertion of Space in Critical Social Theory*. London: Verso.

Spain, D. (1993). *Gendered Spaces*. Chapel Hill: University of North Carolina Press.

Stack, C. (1974). *All Our Kin: Strategies for Survival in a Black Community*. New York: Harper and Row.

Stewart, L. (1995). Bodies, Visions, and Spatial Politics: A Review Essay on Henri Lefebvre's The Production of Space. Environment and Planning D: *Society and Space*, Vol. 13 (pp. 609–618).

Stimpson, C. R., Dixler, E., Nelson, M. J., & Yatrakis, K. B. (Eds) (1981). *Women and the American City*. Chicago: University of Chicago Press.
Suttles, G. D. (1968). *The Social Order of the Slum: Ethnicity and Territory in the Inner City*. Chicago: University of Chicago Press.
Suttles, G. D. (1972). *The Social Construction of Communities*. Chicago: University of Chicago Press.
Swearington, S., & Orellana-Rojas, C. (2000). Conflict, Space, and Identity: Two Cases, One Process. In: R. Hutchison (Ed.), *Constructions of Urban Space. Research in Urban Sociology. Vol. 5* (pp. 59–80). Stamford: CT: JAI Press.
Urry, J. (1996). The Sociology of Space and Time. In: B. S. Turner (Ed.), *Blackwell Companion to Social Theory.* (pp. 369–396). Malden, MA: Blackwell Publishers.
Vale, L. J. (1997). Empathological Places: Residents' Ambivalence Toward Remaining in Public Housing. *Journal of Planning Education and Research, 16*, 159–175.
Van Kempen, R., & Marcuse, P. (Eds) (1997). The Changing Spatial Order in Cities. *American Behavior Scientist, Vol. 41*, No. 3, November/December.
Venkatesh, S. A. (1997). The Social Organization of Street Gang Activity in an Urban Ghetto. American. *Journal of Sociology, 103*, 82–111.
Wagner, D. (1993). *Checkerboard Square: Culture and Resistance in a Homeless Community*. Boulder, CO: Westview Press.
Warner, W. L. (1963). *Yankee City*. New Haven, CT: Yale University Press.
Wekerle, G. (1980). Women in the Urban Environment. Signs: Journal of Women in Culture and Society. Vol. 5, No. 3.
Wilson, W. (1996). *When Work Disappears: The World of the New Urban Poor*. New York: Knopf.
Wilson, W. J. (1987). *The Truly Disadvantaged: The Inner City, the Underclass, and Public Policy*. Chicago: University of Chicago Press.
Wilson, D. (1993). Everyday Life, Spatiality, and Inner City Disinvestment in a U.S. City *International Journal of Urban and Regional Research, Vol. 17*, No. 4. December, 578–594).
Wilson, D. (Ed.) (1997). *Globalization and the Changing U.S. City. Annals of the American Academy of Political and Social Science, Vol. 551.* May. Sage Publications.
Wilson, E. (1992). *The Sphinx and the City: Urban Life, the Control of Disorder, and Women*. Berkeley: University of California Press.
Wright, T. (1997). *Out of Place: Homeless Mobilizations, Subcities, and Contested Landscapes*. Albany, NY: State University of New Press.
Wright, T. (2000). New Urban Spaces and Cultural Representations: Social Imaginaries, Social-physical Space, and Homelessness. In: R. Hutchison (Ed.), *Constructions of Urban Space. Research in Urban Sociology. Vol. 5.* (pp. 23–58). Stamford: CT: JAI Press.
Zhao, D. (1998). Ecologies of Social Movements: Student Mobilization During the 1989 Prodemocracy Movement in Bejing. *American Journal of Sociology, 103*, 1493–1529.
Zukin, S. (1991). *Landscapes of Power: From Detroit to Disney World*. Berkeley: University of California Press.
Zukin, S. et al. (1998). From Coney Island to Las Vegas in the Urban Imaginary: Discursive Practices of Growth and Decline. *Urban Affairs Review, Vol 33*, No. 5, May (pp. 627–654).

WORLD CITY THEORY: THE CASE OF SEOUL

Jamie Paquin

INTRODUCTION

Over the past 30 years, a novel phase of economic globalization has given rise to new global production networks, a new international division of labour, and new urban social and spatial conditions (Castells, 1996; Harrison, 1994; Harvey, 1989; Sassen, 1991, 1998). In addition to the serious implications of these changes for nation-states and workers, cities are dramatically transformed, particularly in relation to the phenomenon of deindustrialization and the de-territorialization of a large portion of economic activity.

Those working on the issue of 'world cities' generally paint a troubling picture of the globalization process in relation to the city, stressing that the global economy is intensifying tensions between elite and common interests, and is placing new demands on cities to attend to the demands of global capital. Such conclusions have however been drawn largely from research based on major cities in the United States and England (i.e. Chicago, New York, London) – cities with histories as pre-eminent centres of global capital (cf. King, 1990; Knox & Taylor, 1995; Sassen, 1991), although research on other cities in this context has begun to emerge in recent years (cf. Isin, 2000; Oncu & Weyland,

1997). This limitation suggests that cities in other regions need to be studied before 'world cities' theory can be considered an accurate depiction of the contemporary global/urban dynamic, since it is reasonable to assume that integration into, or responses to, the global capitalist system may vary by cities due to differences in developmental stages, histories, power relations and cultural contexts.

This paper looks at the recent past and projected future of urban development and restructuring of Seoul, South Korea in an effort to bring to light the way(s) in which economic globalization is experienced in and engaged with outside of the United States and Western Europe, with the intention of contributing to a more comprehensive understanding of the social consequences of urban restructuring in an era of globalization. I begin with an examination of Seoul's recent and planned urban restructuring through an engagement with the 'world cities hypothesis' as described by Friedmann (1986) and others (cf. King, 1990; Knox & Taylor, 1995; Sassen, 1991, 1998) in an attempt to gauge the validity of their arguments and observations. Secondly, I consider the social consequences of Seoul's recent urban restructuring, for a major premise of the world cities discussion is the position that social polarization is a central part of the changing urban landscape in major cities around the globe. Sociologically it is this question that is most relevant within the larger line of enquiry related to contemporary urban restructuring.

Why should we consider Seoul? Its location within a newly-industrialised, semi-peripheral nation, the magnitude of the urban restructuring currently underway there, its rapid insertion into the global economic system, and the fact that political and business leaders (at both the municipal and national levels) have enthusiastically adopted the notion of the 'world city', makes it a prime site for the consideration of the relationship between economic globalization and urban processes outside the West.

'WORLD CITIES' THEORY AND THE NEW SIGNIFICANCE OF CITIES

In the ever-increasing 'informational' economy, a city's built environment and infrastructure (in concert with certain types of human capital, and 'pro-business' economic policies), is challenging, and even superseding, geographical location as the main consideration to corporations and investors. As Harvey (2000 [1989]) explains:

> The decline in the significance of natural spatial barriers has made the 'distance from the market or from raw materials ... less relevant to locational decisions' - strengthening capital's choice over location – while at the same time 'small differences' in labour supply

(quantities and qualities), in infrastructures and resources, in government regulation and taxation, assume much greater significance than was the case when high transport costs created 'natural' monopolies for local production in local markets (55).

In other words, there has been 'a historical shift in the meaning of comparative advantage away from natural resource endowments and toward created assets' (Douglass, 1998: 112) with the shrinking of transportation costs, the decline of industrial manufacturing and resource extraction and processing as portions of total economic activity, and the growth of services as the largest segment of economic activity in advanced capitalist economies.

It should also be noted that in conjunction with changes in trade regulations and various advances in communications and transportation technologies, these spatial changes include a 'new logic to the organization of industrial space' that is:

Characterized by the technological and organizational ability to separate the production process in different locations while reintegrating its unity through telecommunications linkages, and microelectronics-based precision and flexibility in the fabrication of components. Furthermore, geographic specificity of each phase of the production process is made advisable by the singularity of the labour force required at each stage, and by the different social and environmental features involved in the living conditions of highly distinct segments of this labour force (Castells, 1996: 386–387).

What this discussion suggests is that to a significant degree, older geographical constraints have diminished with new technologies, allowing for a new set of criteria to determine the location of various economic activities. The ability to separate production processes across space thus allows for high-wage, high-skilled employment to concentrate in a relatively small number of cities, mainly in the West (and some areas in Asia), with more labour-intensive and assembly based functions being located mainly in Asia and Latin America.[1] This economic restructuring has created problems in many cities in the developed world over the past few decades as they have seen their manufacturing industries largely disappear. For cities in developing regions, rapid urbanization, urban poverty and pollution have been some of the effects. Yet the consequences of a shift towards a less geographically-determined economy are complex, since for cities like Seoul which lay in relatively resource-poor countries, the de-territorialisation of production may level the playing field somewhat; that is, if the city can provide the types of 'created assets' attractive to investors and multi-national corporations.

John Friedmann's (1986; see also Wolff & Friedmann, 1982) 'world cities hypothesis', as well as the work of others on the issue of 'global' or 'world' cities (King, 1990; Knox & Taylor, 1995; Sassen, 1991; Soja, 2000), provides a theoretical base for understanding the interaction between the global economy and cities. In Friedmann's words, 'the world cities hypothesis is about the spatial organization of the new international division of labour' and how the city is affected by this new economic system' (1986: 317). This approach is original and important in suggesting that contemporary urban processes are responses to the global economy rather than only locally or nationally determined as urban research emphasised in the past.

A central argument of world cities theory is that a hierarchy of cities performing different functions within the global economy is emerging. At the top are a relatively small number of world cities that are distinguishable by their 'relatively high concentrations of corporate headquarters (major TNCs), international finance, global transport and communications, high-level business services and the dissemination of information, news, entertainment and other cultural artefacts' (Friedmann, 1986: 322). 'Truly' world cities are those that perform these above-mentioned functions simultaneously, while cities that specialize in only a portion of these or other functions are cities of a lesser order (Lyons & Salmon, 1995: 99). Due to the high amount of capital in such cities, the world city is likely to attract migrant labour both domestic and foreign.[2] Friedmann stresses that the world cities hypothesis is not meant to be deterministic, as history, policy, degree of global integration and culture are all mitigating factors, however, he also states that 'the economic variable... is likely to be decisive for all attempts at explanation' (317), making this perspective one which takes the range of possibilities for urban development outside the global economic system to be extremely limited.

On the issue of winners and losers, the argument is that those cities that come to be the control hubs of the global capitalist system 'win' in that they become spaces of intense capital and infrastructural investment and home to some of the most profitable sectors of the global economy. Losers on the other hand, include both those cities that can only attract the less capital-intensive elements of production, and also, those people in world cities who become peripheral to the new types of production underway in the cities at the top of the production chain. Thus, a major point of the world cities research is to explore the possibility that inequality is increasing both within and across cities in concert with an increasingly global economy.

In sum, according to the world cities paradigm, the globalization of production results in some cities becoming the centres of advanced financial, research, and various professional services that manage the global production system as well as produce new products and services. Such cities, including London, Tokyo and New York, which have secured a major role in the command and control of the global economy, can be considered world cities. Other cities struggle to find their particular niche whether it is in manufacturing of high-technology goods, low-technology assembly, tourism, entertainment or other specialities.

Cities unable to find their function face stagnation or decay. Most importantly, the claim is that in concert with these structural changes, socio-economic inequality is growing both across and within cities. The overall picture provided by world cities theory and research is thus somewhat bleak, captured in the words of Riccardo Petrella, who states that world cities are 'a high-tech archipelago of affluent, hyperdeveloped city regions ... amid a sea of impoverished humanity' (in Friedmann, 1995: 41). Yet this is not a static hierarchy or state of affairs, for many cities are actively seeking to change their position in the global urban hierarchy – i.e. Seoul – making any categorization of cities temporally contingent.

DATA AND METHODS

As stated above, the aim of this investigation is to 'test' world city theory as developed by Friedmann, Sassen and others in a non-Western setting. Specifically, I am interested in two main questions: (1) to what extent is Seoul working towards world city status as predicted by Friedmann; and (2) to what extent is the social polarization thesis that Friedmann, Sassen and others argue is characteristic of the world city valid in this case? These questions are important ones for if cities are indeed engaged in extensive urban restructuring in response to global economic transformations, it stands to reason that major resources are being devoted to these efforts which may or may not come at the expense of social needs. Moreover, if social polarization is taking place in globalizing cities, we can expect this experience to be a new source of conflict in the coming century, and there would be a strong rationale for rejecting the current global economic agenda. But as Lippman Abu-Lughod (1995) points out, while one of the most important contributions of Friedmann and Wolff's work is their consideration of the *consequences* of world city-ness, not simply its characteristics, 'it is in this section (the consequences) of their article that premature generalizations are most prevalent' (183). Therefore, while the consequences of world city-ness should be of the utmost concern to social scientists, it is important to make sure that we have thoroughly investigated the implications of globalization for a variety of cities before declaring the world cities system inherently problematic.

In order to assess the urban restructuring of Seoul in the context of globalization I reviewed business and government publications, academic journals, and conference papers pertaining to urban restructuring, globalization and Seoul between 1999 and 2000 (four months of that time was spent conducting research in Seoul where I visited various research centres and libraries). In addition, I have tracked four Korean English-language daily newspapers over the past two years. Many sources, especially municipal government documents, were located on the Internet. My experience of living in Seoul for a total of twenty months intermittently between 1996 and 1999 also gave me an opportunity to observe and experience Seoul's globalization.

While finding government and business literature related to major restructuring initiatives was relatively easy, it was a formidable challenge to find local critiques of the major urban restructuring initiatives underway. This is largely because I do not speak Korean and was thus limited to English materials (although this limitation was mitigated the growing number of publications about Seoul being produced in English). The analysis here is also partially limited by the fact that many of the 'mega' urban projects planned for the Seoul region are in the early stages and hence the consequences of these projects are not fully realized at this time. These shortcomings aside, this paper has the advantage of being timely. I have been able to assemble very recent data on the urban restructuring underway in the Seoul region. This data contributes to the world city literature by confirming that there is indeed – in Seoul at least – a globally driven urban restructuring process underway, while at the same time (as will be shown below) Seoul appears to present a challenge to the 'social polarization' thesis that has thus far been a central tenant of world cities theorizing.

GLOBALIZATION, URBAN RESTRUCTURING AND SEOUL

Despite the fact that world city formation is considered to have a number of problematic features by those engaged in academic research on the phenomenon, many city leaders consider 'world city status' to be highly desirable. There does indeed appear to be great rewards for those who can elevate their cities to the top of the hierarchy: 'those gaining the top positions would be on the cutting edge of high technologically driven production and producer services and would enjoy a position of power unprecedented in their history' (Douglass, 1998: 111). They would also enjoy a disproportionately high amount of capital investment and considerable international prestige, and greater prestige would likely feedback into more capital investment.[3] Indeed, Seoul may have little other choice than to pursue the sort of economy characteristic of the world city, as the success of labour in the region in achieving higher wages and benefits over the past decade means that it is no longer competitive in terms of labour-intensive manufacturing (Hong, 1996).

Seoul is one of the ten largest cities in the world with a population near 10 million, or 25 million when including the Seoul Metropolitan Region (SMR); a figure greater than half of Korea's 46 million people (Business Korea June 1999: 34). Along with being South Korea's national capital, it is clearly the economic, educational and cultural centre of the country, hosting nearly all of its major corporate headquarters, most international conferences, 42 universities and 16 colleges, and nearly all research centres, major television networks and cultural facilities (Seoul Metropolitan Government, 1999). Seoul is also a wealthy city by any standards, as its GDP per capita (in PPP terms) is U.S.$14,886 (Asiaweek, 2000), generated from an economy that is over 80% services based (Cho, 1997).

Yet, despite some notable attributes (and the fact that Seoul aspires to world city status), some observers do not perceive Seoul as a major hub in the global network of cities, either now or in the near future. In 1995, Friedmann had this to say about Seoul:

> The South Korean capital appears to be losing some of its appeal to outside, especially Japanese, capital. There are several reasons for this, including rising labour costs and a complex bureaucracy. Although South Korea's economy is doing well it remains cut off from North Korea and so provides no access to markets other than its own. Korean capital is outward-bound, and its markets are increasingly abroad, especially on the west coast of the United States. But Seoul is not so much a world city, rather it is the capital city of South Korea. If the two Koreas should ever reunite this picture may change, especially if north china markets can be accessed from Seoul (37).

Lee (1996: 70) similarly sees Seoul's potential to be a world city as unfavourable:

> Seoul has too many shortcomings to be regarded as a genuinely world-class city of the type defined by John Friedmann. Specifically it lacks the transportation and communications infrastructure necessary for 'decisions made by transnational corporations to be conveyed to all parts of the world (in Douglass, 1998: 263).

If we consider Seoul's position only within East Asia, it appears that Seoul still has some distance to go before achieving world city status. In a recent survey of the quality of life in Asian cities done by Asiaweek (Dec 1999), Seoul ranked a mediocre thirteenth[4] (see Table 1). This poses a serious problem for Seoul because living conditions can attract or repel capital investment; in choosing the location of headquarters or regional offices, corporations give considerable weight to the quality of life offered in a given city since attracting skilled employees/professionals is more costly and difficult when the environment is considered undesirable. And while Seoul is making progress on a number of fronts, its breakneck pace of economic development over the past four decades has created an urban landscape with many problems, as indicated by its ranking.

Table 1. Asian City Rankings.

Rank	City	Points
1.	Fukuoka	73
2.	Osaka	72
3.	Taipei	72
4.	Tokyo	68
5.	Singapore	66
6.	Bander Beri Begawan	65
7.	Georgetown	64
8.	Kuala Lampur	63
9.	Hong Kong	62
10.	Shanghai	62
11.	Chiang Mai	60
12.	Kaohsiung	60
13.	Bangkok	59
14.	Seoul	59
15.	Kuching	58

Source: Asiaweek, Dec 1999.

According to the sources mentioned above, Seoul is not a world city, nor is it clearly moving in that direction. Moreover, the opinions of Lee and Friedmann were expressed before the 1997 financial crisis hit Korea, an event which would likely be considered by these authors as a further obstacle to Seoul's world city aspirations. However, we must keep in mind that a major element of the new global economy is that intense competition makes location within the world urban system capricious, or as Friedmann explains, 'the world economy is too volatile to allow us to fix a stable hierarchy for any but relatively short stretches of time' (1995: 23). Aware of this contestability, leaders at both the local and national levels are making serious efforts to improve the city's position within the global hierarchy of cities, in part by addressing some of the concerns raised above, and also, by becoming far more 'entrepreneurial' in terms of urban governance.

'ENTREPRENEURIAL' URBAN GOVERNANCE IN RESPONSE TO ECONOMIC GLOBALIZATION

In the increasingly global economy, cities are engaged in an intense competition with one another. This instigates a more pro-active orientation to economic development on the part of urban governments than was the case in the past. This is not to say that municipalities have only recently taken an interest in economic development, for there is a long history of the city operating as a 'growth machine' (Logan & Molotch, 1997). However, what is different today is that social and spatial planning have been subordinated to the pursuit of financial investment (Hall, 1988: 343). Harvey's (2000 [1989]) statement summarizes well the shift from 'managerial' to 'entrepreneurial' urban governance while additionally drawing attention to the global element of this shift:

> There is general agreement ... that the shift has something to do with the difficulties that have beset capitalist economies since the recession of 1973. Deindustrialization, widespread seemingly 'structural' unemployment, fiscal austerity at both the national and local levels, all coupled with a rising tide of neoconservatism and much stronger appeal (though often more in theory than in practice) to market rationality and privatization, provide a backdrop to understanding why so many urban governments, often of quite different political persuasions and armed with very different legal and political powers, have all taken a broadly similar direction. The greater emphasis on local action to combat these ills also seems to have something to do with the declining powers of the nation state to control multinational money flows, so that investment increasingly takes the form of a negotiation between international finance capital and local powers doing the best they can to maximise the attractiveness of the local site as a lure for capitalist development (51).

According to this depiction, urban governments are today dealing more directly with global finance, and are more dependent on it for their prosperity. Perhaps not surprisingly then, Seoul's municipal government has become far more entrepreneurial in the past decade. Especially since hosting the 1988 summer Olympics, the city has been very proactive in its efforts to increase Seoul's international character and stature, making frequent use of phrases like 'Seoul's rise to a world city' and 'Seoul is going global' in official documents and media releases (in Kim, 1998: 213). City officials consistently employ the concept of the world city to signify 'the highest level of successful urban development' – a usage considerably different than the meaning intended by Friedmann's concept of the world city.

To encourage economic development, Seoul is actively marketing and promoting itself to prospective investors, and like many other cities, competing for 'global spectacles such as major sporting events, cultural festivals and trade fairs, which generate considerable multiplier effects' (Lever, 1993, in Kim, 1998: 49). Such events are believed to not only generate income for cities, but also, they are seen as promotional vehicles that serve to raise a city's global status and thus attract investor and corporate attention. Seoul has been particularly successful in attracting such 'global spectacles' including the 1986 Asian games, the 1988 summer Olympics, the 2002 World Cup (to be co-hosted with Japan), and an increasing number of international trade-shows, conventions and conferences. The city has also established a number of 'Sister City' agreements with other major cities around the world (16 cities in 15 countries) in an effort to expand cultural, knowledge and trade relations, but especially, to raise the city's profile abroad (Seoul Metropolitan Government, 1999).

The governments of the SMR – with support from the national government – are now heavily engaged in the pursuit of foreign direct investment (FDI), as well as in the production of numerous infrastructure projects that require significant financial capital, as expected by world cities theorists. The pursuit of FDI by the city government marks a paradigmatic shift as the economic development of Korea was accomplished with insignificant levels of FDI (Castells, 1998: 257). Notable is the fact that the municipal and national governments have not just reacted to the city's internationalization but have been actively 'leading and promoting the whole process' (Kim, 1998: 211). In other words, the Seoul government is deeply involved in bringing to fruition a globalized city, a fact that complicates attributing recent urban restructuring solely to external factors.

Other indications of the city's entrepreneurial and pro-global approach are plentiful. In 1998, Seoul City Hall s main floor was converted into the *Seoul Information Centre* which offers, free consultation services' concerning 'travel information, relocation problems, and investment/trade ... one can also surf the Web, read government publications, purchase books and souvenirs'. In February of this year the city launched its own computer game titled *Virtual Seoul*, which in the government's own words is designed for promoting Seoul city world-wide'. Incorporated into game is the city's current agenda. Player's are expected to build a future Seoul, which will have become the capital of the Unified Korea and hub of the global economy in the 21st century' which involves hosting the 2002 World Cup, recovering from an economic crisis, and building a new airport (Seoul Focus, 1999).

MEGA PROJECTS

The entrepreneurial orientation of Seoul's leadership is materially and spatially manifested in a large number of infrastructural initiatives. In fact, looking at Seoul's urban development and restructuring over the past two decades (and considering what is planned for the next two), it is easy to think that city leaders have followed Friedmann's description of a world city as a blueprint for urban development. In order to contextualize these projects, each will be discussed here briefly.

The number of major projects underway or planned in the SMR towards the goal of regional primacy is quite striking and illustrates – perhaps epitomises – the type of urban restructuring that cities are undertaking in order to secure a premiere position in the global economy. Table 2 provides a synopsis of the main infrastructure projects underway in the Seoul region. Many of the projects reveal the degree to which government resources are currently mobilized towards attracting global investment in the context of hyper-mobile capital and labour.

Especially implicated in these plans is the satellite-city of Inchon, where the *Triport* Project (which includes the construction of an airport, seaport, and - teleport) is currently underway (B. K., July, 1999: 18). By developing its transportation, distribution and information infrastructures simultaneously, Inchon is attempting to outdo Malaysia, Singapore and Hong Kong who have focused on one or some, but not all of these services (19); a pointed example of the sort of regional and global urban competition that is occurring at present. Arguably, this project goes a long way to answering Lee's earlier quoted critical assessment of Seoul's transportation infrastructure.

The centrepiece of the *Triport project* is *Inchon International Airport* (IIA). Promoted as the 'Winged City' to signify its mobile and comprehensive character, it will include the world's largest ever passenger terminal (B. K. July, 1999: 29). A virtual 'airport city-state', the *IIA* and the surrounding islands are to become a 'twenty-four hour operational city where any nationality and cargo can freely pass without taxes and visas' (B. K., May 1998: 52). It is the 'completeness' of the planned development project that is expected to distinguish Seoul as a pre-eminent urban centre in East Asia – *Songdo International City* is being built nearby as a part of a larger free trade zone which will allow for international business people to enter and exit the zone with little interruption. Kang Dong-Suk, Chairman and CEO of the *Korean Airport Construction Authority*, explains the intent of the airports design: 'in contrast to those 20th century models that placed their chief aim on transporting passengers and cargos, the 21st century (airport) will become a final destination where one can engage in his or her business or enjoy leisure and shopping and, at the same time, will play a role as a center of international politics, economics, and cultural activities where information and culture is gathered and dispersed' (51).

Songdo International City, envisioned as a 'New World City', will be a 'visa-free', supra-national zone which will rely heavily on foreign investment and foreign high-tech workers and business people. Central to the new 'city' is the *Songdo Media Valley* – Korea's answer to California's Silicon Valley – a complex that intended include six zones: 'Soft Park' (for software and content developers), 'Techno Park' (R&D and exhibition halls), 'Media Valley Academia' (for educational institutions), 'Intelligent Venture Building' (support centre for venture enterprises), and 'Ecological Village' (for residential, educational, and cultural facilities). Media Valley Academia will include a university campus: 'providing universities around the world with classrooms

Table 2. Major Project Underway or Planned in the Seoul Metropolitan Region.

Project	Cost/Funding Source	Project Schedule	Details	
*Seoul World Cup Stadium	200 billion won/ 30% Municipal, 30% Federal, 40% Private	1998/2002	**Accompanied by a four-year project to plant 10 million trees in Seoul in an effort to 'foliate' the capital	
Inchon International Airport	U.S.$8.45 billion	1992/2020	**Phase I(2001)** Annually: 170,000 flights 27m passengers 1.7 m/tons cargo	**Phase 2 (2020)** Annually: 530,000 flights 100m passengers 7 m/tons cargo
Songdo 'International City'		-/2012	43 square metres 80,000 pop limit for low density Songdo Media Valley will be the city's core	
Yongyu-Muwi Tourism Complex	3.2 trillion won foreign capital M.C.T. to invest 45 billion won	1999/2012	Marine World, Dragon City, Alice Land, casinos, condominiums, parks, shopping centres, beaches, convention centre, hotels, art galleries, golf course Goal: 16 million tourists and income of 4 trillion won annually	
Sea Port	1.46 trillion won, Private Ivestment	-/2011	Handling capacity goal: 61.44 million tons	
***Songdo 'Media Valley' Teleport	Private Investment	2001-2008	To include: media park, media academy, multimedia information centre, International Info-communications University 1/3 of media valley development earmarked for foreign companies	
Yongsan Station Development Project	U.S.$550m Private and public investment	1999/2019 2 Phases	'New downtown', high-tech, multinational base. A 'city within a city' which will include: 5-star hotels, department stores, shopping centres, movie theatres, theme plazas and speciality shops Transportation hub for high-speed and other trains linking other regions of Korea and Inchon airport, subways, highways	

Sources: Business Korea, February 1999; * Korea Herald 1998a; ** Korea Herald 1998b, ***www.metro.inchon.kr

within the campus building *free of charge*, the university will provide not only regular curriculum but also internship training courses through which students will receive two years of courses in addition to a four-year master's program: *the government will also pay their salaries* (B. K., May, 1998: 58, italics added).

Striking – but not surprising when from a world city theory perspective – is both the scope of this project and the degree of government support. According to Kim Tae-Yon, President and CEO of Media Valley Inc., 'the success of the Media Valley project depends largely upon participation of major U.S. corporations ... in this connection, we need to make huge investments to build necessary infrastructure facilities and offer land at low prices ... so the government is required to actively support this project' (B. K., June, 1999: 54). The governments 'active support' consists of providing discounted land to foreign companies and 'sweeping tax benefits' to entrants into the Media Valley. Incentives are to include 'full exemption of the taxes such as corporate tax, acquisition tax, property tax, and real estate tax for the first seven years and a 50% reduction for three years thereafter. 'Further, no restriction on capital flow or on remittances of profit will be put on the resident companies' (B. K., October 1998: 26). Planners estimate that the project could attract up to U.S.$4 billion of foreign capital and provide 300,000 jobs in the process of site development and building, but it is clear that the competitive global environment is pressuring governments to make significant financial contributions/concessions to investors and multinationals in order to attract investment.

In order to service the leisure demands of Songdo's residents as well as to draw tourists to the region, a major theme park and entertainment development is also planned. The *Yongyu-Muwi Tourism Complex* is designed to capitalize on the increased flow of visitors to Inchon with the new airport. To be situated west of Inchon Airport on Yongyu and Muwi islands, the development will consist of *Marine World* and *Dragon City* theme parks (which will include hotels, condominiums, parks, shopping centres and beaches on Yongyu island), and Aliceland (on Muwi and which will included five casino hotels, a shopping mall, convention centre, art galleries, and an 18-hole golf course). Despite the fact that these projects will consume a substantial amount of currently undeveloped land, the project is being touted – as are virtually all the above projects – as 'green developments' in reference to the declared intent to develop these sites in an environmentally conscientious manner.

The city of Seoul itself is also undertaking a number of initiatives at this time as part of the 'Urban Restructuring of Seoul in the Era of Internationalization' (Kim, 1998: 215). In preparation for the 2002 World Cup, the *Sangam-dong Millennium City Project*, which includes the building of Seoul World Cup Stadium, is underway. Previously the site of a garbage dump the SMCP will involve 'on a budget of about 896.9 billion won, construction of a digital media industrial complex, an environment-friendly housing complex accommodating 7,000 households, and an ecological park bigger than Youido[5] on a 6.6 million square meter area. Upon completion in 2010, the 'Millennium City' is expected to be inhabited by 30 thousand residents, and host a large number of daily commuters, amounting to a population of approximately 400 thousand' (Seoul News & Events, 2000).

The *Yongsan Station Development Project* (YSDP) is a major undertaking with a number of intended functions. Like Songdo City, the *YSDP* emphasises the business and residential needs of foreigners. Currently, the area is rather run-down and neglected but will soon be dramatically transformed. Linked to the new Inchon airport, the Seoul subway system, and the new high speed Seoul-Pusan railway, *YSDP* is to serve as a new downtown, and will become the major transportation hub in Seoul. The project is also designed to nurture strategic industries, especially knowledge oriented business services – and will include an information centre, semiconductor plants, an international financial centre, and office buildings for multinational corporations (B. K., February, 1999: 50).

The various projects discussed above reveal some key elements of contemporary urban restructuring. There is a strong emphasis on the non-local, both in terms of attracting foreign investment to finance various mega-projects and also to attract non-local businesses and people. *Songdo International City*, designed to accommodate 300,000 residents (almost entirely foreign), is to be largely self-contained. Simply put, the city will 'fully reflect the preferences of foreigners since it is designed to attract foreign companies into the city' and will essentially be a 'global space' on Korean soil: 'multinational companies will be able to utilize the city as their regional headquarters in East Asia' (43). A Key consideration of many of these developments is the necessity of providing a pleasant living environment. The projects underway in Seoul and Inchon all emphasize environmental quality as not only desirable, but also elemental to their competitive advantage over such places as Malaysia, China, and Taiwan. In the words of the former CEO of Media Valley Inc, 'there needs to be a pleasant environment that can raise creativity and that enables one to relax in a comfortable environment' and 'therefore, we will be building half of the complex as green space' (B. K., May, 1998: 59).

The global orientation of these initiatives obviously raises the question of to what extent local residents are served by these projects. Put differently, will the new economy of Seoul be for Seoulites or will it merely take place in their proximity? It is too early to assess the degree to which these economic spaces will contribute to or detract from the economic conditions of the broader urban landscape and citizenry. What is clear at this time is that the mega projects underway in the SMR are attuned to contemporary global processes that pressure cities to compete against one another in an attempt to attract the higher-order economic functions of the global economy (or at least capitalizing on the rhetoric of economic globalization), and this competition is extremely costly as it is based on providing an urban infrastructure to facilitate capital-intensive economic activity.

THE SOCIAL CONSEQUENCES OF SEOUL'S GLOBALIZATION

Seoul is currently engaged in a dramatic degree of urban restructuring, lending support to the world city theory claim that economic globalization induces new urbanization processes. However, the social conditions expected by some world city theorists to accompany this process are not clearly evident in this case. For while a number problematic events and processes have coincided with Seoul's globalization (see below), it is also the case that in recent years (especially if a full recovery from the 1997 financial crisis can be accomplished), social conditions have been improving on a number of levels, and many types of social expenditure are on the rise. Seoul is by no means a city without social problems, but the various socially problematic conditions and events in Seoul's recent history should not for the most part be blamed on the forces of globalization for they are the result of the policies of various authoritarian regimes that undertook an undemocratic and overly hasty program of modernization (cf. Pai, 1997). This is not to dispute that globalization has the capacity to induce new forms of social inequality but rather to argue that distinctions need to be made between local and global sources of social injustice. When this is done, we can see that blaming globalization for too much can result in distracting us from identifying and challenging more local sources of social inequality.

As just mentioned, globalization does indeed have the capacity to inflict social suffering: the Asian financial Crisis that hit Korea in 1997 illustrates that integration into the global economy comes with risks and problems. In the aftermath of the crisis, the lives of many people were tragically affected: eight months after Asian markets were thrown into turmoil, Korea's unemployment rate rose four-fold, the number of homeless tripled and in just three months after the crisis hit, 2,300 people committed suicide (BBC, September 30, 1998). However, many of Seoul's urban problems predate the IMF era and are less clearly related to global financial or other forces. The city's history of human displacement may be unparalleled, as church activists claim that 'the 230 redevelopment projects put forward by the city between 1985 and the early 1990s displaced more than 3 million people, or nearly 30% of the population of the metropolis' (Douglass, 1998: 264). During this time, the city was preparing for the 1986 Asian games and the

1988 Summer Olympics, precisely the types of projects that ambitious cities seek to enhance their global standing, but which also create great controversy in relation to their social consequences. In these cases, efforts to 'clean up' the city resulted in the removal of squatters and the poor who resided either along routes visible to foreign visitors and television crews, or who occupied the cheap land required for developing the facilities required for hosting these events.

Other recent urban development projects have lacked the same degree of displacement, but the costs associated with these projects have nonetheless been felt by the more disadvantaged in the city. As we have seen, urban redevelopment in recent years involves a high degree of government involvement (in terms of facilitation rather than regulation), a fact that strains the resources of cities like Seoul. A Korean government official from the Ministry of Finance acknowledged as much when speaking about the city's financial contribution to the *YSDP*, stating that 'education and public housing programs would be the major victims in terms of government spending' (in Kim, 1998: 215).

Yet at the same time, unlike many cities in the West that have suffered a significant decline in welfare expenditures over the last decade, social expenditures are not declining in the city (although nor are they at levels comparable to Western welfare states). In fact, they are generally on the rise despite global competition and a general adherence to a neo-liberal economic agenda (see Seoul City Budget, Seoul Focus, 1999).[6]

It is also true that the city has become more liveable on many levels during the past two decades due to a number of infrastructural investments. There have been improvements in water quality, noise reduction measures and sidewalks. The housing supply ratio is expected to meet 85% of demand by 2011 as compared to 71.1% in 1998, while park areas are forecasted to increase from 12.48m sq. per person in 1998 to 14.43m sq. per person in 2011 (Ibid.).[7]

Major parks have been constructed in recent years toward this end, most notably, Youido Park, completed in 1997, which lies in the middle of Seoul's key financial district. The 'Han River Development Plan' is supposed to develop three new riverside parks in the city as well as refurbish existing parks along the river (Korea Herald, July 9,1999).

Seoul has also built a remarkable mass transportation system, which makes the city's extreme density far more manageable. In addition to a fleet of approximately 8,655 buses and 69,720 taxis, the city is projected to have 400 kilometres of subway track consisting of 12 different subway lines by 2005 (Seoul Metropolitan Government, 1999).[8] The subway and bus system receive public funding and are quite reasonably priced and the extensiveness of the system means there are virtually no neglected areas of the city. Thus, although Seoul is plagued by serious traffic related problems, the problem is not a lack of public transportation but that of powerful allure of automobility, a problem faced in virtually every city in the world to some level.[9] And while the degree of resources devoted to providing an infrastructure for automobiles seems excessive considering the limited degree of social welfare provision in the city,[10] there is nonetheless a commitment to the transportation needs of all city residents to a far greater extent than is found in many other major cities of the world either in developed or developing city regions.

Unlike many developing cities of the world today that are heavily engaged in the global economy, Seoul has a *relatively* equitable income distribution. The 1997 Asian financial crisis initially increased income disparities, although wealth distribution appears to be moving in the direction of pre-crisis levels. According to World Bank figures for 1998, income distribution in Korea as represented by the GINI coefficient stood at 0.325, which represents a more equitable distribution of wealth than is found in many economically advanced nations (Canada, Italy, the Netherlands, Spain and Britain were all between 0.31 and 0.33, 0.337 for Australia, 0.34 for Japan, and 0.401 for the United States).

An equally important though non-quantifiable consideration is the city's aesthetic. Few would characterize Seoul as an aesthetically pleasing urban landscape, but newer buildings and facades are more visually appealing and better quality than those of the previous few decades. Importantly, one does not generally see in Seoul the stark socio-economic inequalities characteristic of many North American or other Asian cities. That is, while Seoul's built environment may not sparkle and dazzle, neither does it tend to display gross disparities of wealth and living conditions the way many other cities do.[11]

On a number of fronts then, efforts to integrate more deeply into the global economic system appear to coincide with a variety of improvements to the quality of life in Seoul.

CONCLUSION

This examination of Seoul lends support to claims by world cities theorists that cities are undergoing tremendous process of restructuring in relation to globalization. Virtually all of the city's urban development efforts over the past two decades have been motivated by global considerations, and the shift of urban governance towards 'entrepreneurialism' further demonstrates that the global economy is intensely competitive. However, despite the fact that there have been incredible numbers of household displacements, limited social spending, and political repression in Seoul's history, more recent initiatives such as quantitative and qualitative improvements in housing, mass transit, parks and green spaces point to the possibility that the quality of life for most citizens may actually be improving simultaneously with the city's globalization efforts, and hence, the claim by the world cities literature that social polarization accompanies globalization-induced urban restructuring was not established. To recognize this fact is not to valorize the project of neoliberal globalization, but rather it is to point out that there appears to be a greater degree of manoeuvrability within the global system than has sometimes been assumed, and that many of the social injustices of recent decades are not the consequences of globalization but are instead the result of local power relations and decision making. It is thus important when considering the implications of globalization that we do not unduly attribute social injustices to global phenomena.

In closing, I would like to suggest that an important element of globalization studies should be an effort to distinguish between the contributions of local and global forces to the social conditions of cities, and studies of the urban/globalization dynamic should investigate the sources of social injustice carefully before unduly attributing them to globalization processes.

NOTES

1. Although it is also the case that pockets of affluence exists within some of the poorest cities of the world while the 'third world' is increasingly visible in many major cities in the affluent west.
2. Catherine Kim estimates that there are as many as 500, 000 migrant workers in Korea (University of Toronto, April 24, 2000). The vast majority are manual labourers from South and South-East Asia while a small portion are professionals, technicians and teachers, mainly from North America and Europe.
3. Yeong-Hyun Kim alerted me to the issue of prestige. For many people in many cities, international recognition is a strong desire and source of pride.
4. 24 statitical indicators were included that carried different wightings; life expectancy, hospital beds per 1,000 people, per-capita state expenditure for education, average clas size (primary), univeristy-educated people as a percentage of total population, vaious environmetnal quality measures, housing price/income ratio, park space, vehicles per kn of city raods, mass transit.
5. Youido (park) is a large new park in the heart of Seoul's financial district which was completed in 1998.
6. One could argue however, that the social welfare spending might otherwise be growing if not for global competition.
7. It is admittedly problematic to take forecasts as likely outcomes, however such targets have generally been met in recent years.
8. By comparison, the current length of other major cities subway systems is as follows: Tokyo 467, London 408, New York 368, and Paris 199.
9. Ben-Huat Chua's (1998b) article on Singapore shows that the desire to own a car is not directly related to an absence of alternatives or 'need'.
10. Nearly 40% of the city's 'finance has and continues to be funnelled into the transportation sector to accommodate constantly increasing cars, although (the city's)

population has decreased since the early 1990s' (Seoul Metropolitan Government, 1996, in Kim, 1998: 217),

11. Beijing is an interesting comparison. While in Beijing in the summer of 1999 I was amazed at the visibility of new wealth alongside clearly third world conditions – a Mercedes parked in front of a Starbuck's Coffee shop while across the street, unkempt children begged for money. During my 20 months in Seoul I am not sure I saw such a dramatic contrast of living conditions.

REFERENCES

Asiaweek (1997). Special Report on Cities: The Asiaweek Quality of Life Index, (http://www.pathfinder.com/asiaweek/97/1205/cs2.html).
Asiaweek (2000). Asian Cities 2000. (http://www.asiaweek.com/asiaweek/features/asiacities2000/index.html).
Business Korea (1998). Cover Story: Inchon City, May, 40–59.
Business Korea (1999). Special Report: The City of Seoul, February, 46–58.
Business Korea (1999). Korea's Capital is Ideal Destination for Foreign Capital, June, 34–35.
Business Korea (1999). Korea is Building Silicon Valley of the East, June, 54.
Business Korea (1999). Hub of Northeast Asia's Global Village: The City of Inchon Aims to become the New York of its Hemisphere, July 16, 7.
Castells, M. (1996). *The Information Age: Economy, Society and Culture Volume 1: The Rise of the Network Society*. Oxford: Blackwell.
Castells, M. (1998). *The Information Age: Economy, Society and Culture Volume 3: End of Millennium*. Oxford: Blackwell.
Cho, M-R. (1997). Flexibilization Through Metropolis: The Case of Postfordist Seoul, Korea. *International Journal of Urban and Regional Research*, 21, 2, June 180–201.
Choe, S-C. (1996). *The Evolving Urban System in North-East Asia, Lo, Fu-Chen & Yeung Yue-Man, Emerging World Cities in Asia Pacific*. New York: United Nations Press.
Chua, B. H. (1998b) World Cities: Globalization and the Spread of Consumerism: A View From Singapore. *Urban Studies*, 35, 981–1000.
Dalton, B., & Cotton, J. (1996). New Social Movements and the Changing Nature of Political Opposition In: G. Rodan (Ed.), *South Korea, Political Oppositions in Industrializing Asia*. London: Routledge.
Douglass, M. (1998). World City Formation on the Asia Pacific Rim: Poverty, 'Everyday' Forms of Civil Society and Environmental Management In: M. Douglass & J. Friedman (Eds) *Cities for Citizens: Planning and the Rise of Civil Society in a Global Age*. Rexdale: John Wiley & Sons Ltd.
Fouser, R. J. (2000). Income Distribution. *Korea Herald*, 08/30.
Friedmann, J. (1995). Where We Stand: A Decade of World City Research, In: P. L. Knox & P. J. Taylor (Eds), *World Cities in a World-System*. Cambridge: Cambridge University Press.
Friedmann, J. (1995 [1986]). The World City Hypothesis In: P. L. Knox & P. J. Taylor (Eds), *World Cities in a World-System*. Cambridge: Cambridge University Press.
Friedmann, J., & Wolff, G. (1982). World City Formation: An Agenda for Research and Action, *International Journal of Urban and Regional Research*, 6, 3, 309–344.
Hall, P. (1988). *Cities of Tomorrow*. Oxford: Blackwell.
Harrison, B. (1994). *Lean & Mean: Why Large Corporations Will Continue to Dominate the Global Economy*. New York: Guilford Press.

Harvey, D. (1989). *The Condition of Postmodernity*. Oxford: Blackwell.
Harvey, D. (2000 [1989]). From Managerialism to Entrepreneurialism: The Transformation in Urban Governance in Late Capitalism. In: Miles, Hall & Borden (Eds), *The City Cultures Reader*. London: Routledge.
Hong, S. W. (1996). Seoul: A Global City in a Nation of Rapid Growth In: F-C Lo. & Y-M. Yeung (Eds), *Emerging World Cities in Asia Pacific*. New York: United Nations Press.
Isin, E. (2000). Istanbul's Conflicting Paths to Citizenship: Islamization and Globalization In: A. J. Scott (Ed.), *Global City-Regions: Trends and Prospects, Policy Questions, Theoretical Debates*. Oxford: Oxford University Press.
Kim, Y-H (1998). *Globalization, Urban Changes and Seoul's Dreams: A Global Perspective on Contemporary Seoul*. Unpublished dissertation: Syracuse University.
King, A. D. (1990). *Global Cities: Post-Imperialism and the Internationalization of London*. London: Routledge and Kegan Paul.
Knox, P. L., & Taylor, P. J. (1995). *World Cities in a World-System*. Cambridge: Cambridge University Press.
Korea Herald (1998a). Seoul to Build New Stadium for 2002 World Cup, May 7. (http://www.korea-herald.co.kr).
Korea Herald (1998b). Seoul City to Plant 10 Trees by 2002 World Cup, Oct 21. (http://www.korea-herald.co.kr).
Lippman A. J. (1995). Comparing Chicago, New York, and Los Angeles: Testing Some World Cities Hypothesis In: P. L Knox & P. J. Taylor (Eds), *World Cities in a World-System*. Cambridge: Cambridge University Press.
Logan, J. R., & Molotch, H. L. (1996). The City as a Growth Machine, In: S. Fainstein & S. Campbell (Eds), *Readings in Urban Theory*. Oxford: Blackwell.
Lyons, D. A., & Salmon, S. (1995). World Cities, Multinational Corporations, and Urban Hierarchy: The Case of the United States. In: P. L. Knox & P. J. Taylor (Eds), *World Cities in a World-System*. Cambridge: Cambridge University Press.
Oncu, A., & Weyland, P. (Eds) (1997). *Space, Culture, and Power: New Identities in Globalizing Cities*. London: Zed Books.
Pai, H. M. (1997). Modernism, Development, and the Transformation of Seoul: A Study of the Development of Sae'oon Sang'ga and Yoido. In: W. B. Kim, M. Douglass, S. C. Choe & K. C. Ho (Eds), *Culture and the City in East Asia*. Oxford: Clarendon Press.
Sassen, S. (1995). On Concentration and Centrality in the Global City. In: P. L. Knox & P. J. Taylor (Eds), *World Cities in a World-System*. Cambridge: Cambridge University Press.
Sassen, S. (1991). *The Global City: New York, London and Tokyo*. Princeton NJ: Princeton University Press.
Sassen, S. (1998). *Globalization and its Discontents: Essays on the New Mobility of People and Money*. New York: The New Press.
Seoul Focus (1999). http://www.metro.seoul.kr/eng/index.html
Soja, E. (2000). *Postmetropolis: Critical Studies of Cities and Regions*. Oxford: Blackwell.
World's Seoul, The (1998). Media Valley Project: Toward Rivalling Silicon Valley in the 21st Century, October 26–27.

THE CITY AS AN ENTERTAINMENT MACHINE

Richard Lloyd and Terry Nichols Clark

INTRODUCTION

Traditionally urban theory presumes a division between the economy of cities and their culture, with culture subordinate in explanatory power to the "work" of the city. However, categories of production and labor in the urban context have been severely impacted by post-industrial and globalizing trends; cultural activities are increasingly crucial to urban economic vitality. Rather than opposing culture to economic analysis, we show how culture is intertwined with the contemporary urban economy as new patterns of production and consumption emerge. Models used to explain the growth of cities during industrial, "Fordist" capitalism are outmoded. Loss of heavy industry impacts the dynamics of urban growth, increasing the relative importance of the city both as a space of consumption and as a site for "production" which is distinctly symbolic/expressive. Even in a former industrial power like Chicago, the number one industry has become entertainment, which city officials define as including tourism, conventions, restaurants, hotels, and related economic activities (Clark, 2000). New urban growth sectors like information technology and FIRE (Finance, Insurance and Real Estate) change the occupational structure of cities with crucial consequences for our thesis. Workers in the elite sectors of the postindustrial city make "quality of life" demands, and in their consumption practices can experience their own urban location *as if tourists*, emphasizing aesthetic concerns. These practices impact considerations about the proper nature of amenities to provide in contemporary cities. The city becomes an Entertainment Machine leveraging culture to enhance its economic well being. The entertainment components of cities are actively and strategically produced through political and economic activity. Entertainment is the work of many urban actors. We will elaborate on the production of the Entertainment Machine, showing how it is linked to the growth sectors of the global economy, and the new competition to capture and enhance expert labor.

Harvey Molotch's well known metaphor suggests that the city is a machine geared to creating "growth," with growth loosely defined as the intensification of land use and thus higher rent collections, associated professional fees and locally based profits (1976). However, a quarter center later, in the contemporary competition among U.S. cities, the growth machine model requires revision. The "new economy" has not spelled the demise of older central city areas, but it has changed the basis for urban economic viability. Traditional forms of capital give way to the primacy of human capital in the form of an educated and mobile workforce. Large developers are still important players, but they compete with the citizen preferences advanced by this new class of urban residents. An ideology of growth at any cost, in the form of land use intensification, is no longer a given. For a number of cities that have evinced considerable success in the competition for knowledge workers, "smart growth strategies" replace the growth machine as the driving civic ideology. Portland, Oregon, has gone so far as to implement "a program of financial penalties designed to discourage excessive growth by one of its largest employers, Intel, Inc." (Florida, 2000 p. 24). Likewise, Nevarez documents the increasing importance of "quality of life districts" around the Los Angeles metropolitan area which concentrate expert labor and make growth limitations an explicit part of their amenity profile (1999). Such policies run counter to political strategies in which the provision of manufacturing jobs and corresponding patterns of capital intensification is taken as quasi-automatically desirable.

Molotch's use of the growth concept is too imprecise, more appropriately replaced with a concept of centrality. Size and density are not in themselves essential to new rankings of cities in, for example, global city theory (Friedman & Wolff, 1982; Sassen, 1991). What matters is the continuing influence which the city, as a socially structured space, exerts in the operation of the current economy and of human life. Maintenance of economic relevance increasingly responds to new imperatives, as key urban enterprises involve attracting mobile, well-educated workers who in addition to professional training are well-trained participants in the mature consumer economy. They are so much in demand that they can make location decisions where the consumption elements often overshadow (traditional) production elements. The ongoing centrality of older cities, along with the emergence of newer ones in continuous competition, requires attention to the city as an Entertainment Machine, producing consumption opportunities and leveraging cultural advantages. The features of the Entertainment Machine are not altogether new, as cities have long been sites for consumption and aesthetic innovation. What is new is the degree to which these "cultural" activities have become crucial to urban fortunes.

Explosive urban growth in Europe and the United States occurred during the 19th and early 20th centuries under conditions of rapid industrialization. Some European theorists recognized the importance of the city as a cultural space, stressing consumption, especially Benjamin in his fragmentary analysis of the Paris Arcades (1939; [1999]). This suggestive account demonstrates not only the implication of the city in new modes of production, but also its importance as a platform for emerging consumption practices. However, following the morphology of the Chicago sociologists, American theorists have predominantly emphasized transformations from agriculture to industrial manufacturing as core to urban spatial patterns. To focus on culture may have seemed too elitist in populist America. But by 2000, much has changed. We need to update the conceptual tools that are now too linked to our urban past. The high Fordist period of the 1950s and early 1960s saw immigrant slums solidify into relatively stable "blue collar communities" such as those documented by Kornblum (1974), with unionized blue collar workers a significant feature of the urban polity and economy. For the New Deal's national politics, in cities like Chicago these neighborhoods provided the base for political organizations such as Daley's machine, manufacturing votes in exchange for public works agendas geared toward blue-collar job creation. Politicians reflected their concerns, and political leaders in U.S. industrial cities rarely emphasized aesthetic features of the city in their agendas.

Much has changed in the last half century. In prototypical industrial cities like Chicago, we have seen a steady decline in employment in manufacturing and a growth of the service sector, followed in turn by more subtle high tech and globalizing processes. Displacement of manufacturing from central city space changes the class structure of large cities, with political and cultural consequence. Workers whose social location renders them less adaptable to structural change suffer from chronic unemployment or are forced into subsistence occupations in the service sector (Wilson, 1987); they occupy "spaces of devastation" (Zukin, 1991) within most large cities. Nightmare landscapes of poverty are a feature of all former industrial cities in the U.S., and they have been studied extensively, ethnographically and demographically.

At the same time, however, a new elite economy has emerged in large cities featuring educated workers employed in finance, producer services, information technology, and media production. Castells has referred to this economy as "informational" (1989). Participants tend to be educated, and fluent at manipulating diverse symbolic systems, leading Robert Reich to term them "symbolic analysts" (1991). Their activities are embedded in a global economic system; they bring a cosmopolitan sensibility and new demands on the "quality of life" of the cities in which they live and work. But this leap from post-industrial production as stressed by Castells and Reich, for example, is still not clearly joined to entertainment by them or most theorists who have long used production-based interpretations to explain general urban processes. We thus stress the critical epistemological implications of the next conceptual step in our analysis: the "informational city" described by Castells and others implies the "city of leisure." While many have described the increased emphasis on entertainment in urban policy and entrepreneurial activity, they struggle to interpret the broader implications of this shift toward consumption. For instance Bennett has documented the importance of new consumption-oriented strategies for U.S. urban (re)development (1999), but does not locate them in a broader conceptual framework. The fine book by Judd and Fainstein (1999) documents the huge and critical role of tourism in the modern world economy, linking tourist practice to new spatial organization within cities.[1] However, these important trends are not explicitly linked to the new growth sectors of the urban economy, and the consumption habits of local residents.

To help systematize these changes in urban growth dynamics, we postulate several new components of change: (1) There is a rise of the individual citizen/consumer in explanatory power, which follows from increases in citizen income, education, and political empowerment. This translates into more individualization and volatility of tastes, creating more numerous and complex "niche markets." The growth of this "new class" however coexists with substantial numbers of structurally disadvantaged within the city, and the development of the Entertainment Machine is structurally uneven; (2) Conversely, we note a decline in large bureaucratic decisions-makers in both the public and private sector. In the past they could produce large quantities of basic products inexpensively. But as tastes and sub markets differentiate, they are less nimble than small firms and individuals in adapting to rapid change. The public sector acts in a more "entrepreneurial" fashion to promote development (Clarke & Gaile, 1998); (3) There is a relative decline in explanatory power of classical variables affecting the economic base, like distance, transportation costs, local labor costs, and proximity to natural resources and markets – since air travel, fax, the Internet, and associated changes have drastically facilitated contacts among physically distant persons globally. This shifts the mix of inputs for location of households and firms, increasing the importance of more subtle distinctions in taste, quality of life concerns, and related considerations; (4) There is a rise of leisure pursuits compared to "work," increasing the relative importance of new or more refined occupations like tour guide or restaurant critic, and creating an increasing differentiation among providers of personal services; (5) There is a rise of the arts and aesthetic considerations alongside more traditional considerations, in people's lives as well as in modeling the dynamics of cities. This corresponds to increases in education among key sectors of the urban workforce; (6) These create a new role for government and public officials, as they seek to implement these new concerns, many of which are for "public goods" (clean air, attractive views, pedestrian responsiveness) contrasting with more "private goods" (jobs, contracts, tax breaks to separate persons and firms) in the past. There is a rise of zoning, construction of new public spaces, support for public art in many forms, and the introduction of a host of new considerations into urban political decision-making, since judging the demand for competing public goods is far more complex than private goods. These last elements are elaborated elsewhere (Clark & Inglehart, 1998; Clark, 2000); our focus here is on dynamics of urban growth (and decline). These changes are more profound for heavily industrial cities, but the fact that they emerge there too shows their pervasiveness and power.

1. DECLINE AND RENEWAL: POSTINDUSTRIAL TRENDS IN U.S. CITIES

Disinvestment and fiscal crises in large U.S. cities during the 1970s led to a bleak prognosis concerning urban fortunes. The growth of telematics and globalization, which appears to undermine the place boundedness of economic activity, suggested that the dense, central investment of capital in urban cores was no longer desirable. New information technologies are an advance with extreme potential impact on spatial organization: "they represent the opportunity to conduct many more economic transactions at a distance – from an employee at home to a central office, from a consumer to a store, from one company to another" (Atkinson, 1998 p. 134). The changes in the technological foundation of economic activity have been consequential for spatial organization. Our theories need corresponding updating. The urban morphology suggested by the Chicago School, encapsulated in Burgess' famous map and predicated on the centralized locational tendencies of manufacturing is no longer adequate. Edge cities and deconcentration are now catchwords in what Gottdiener terms the "new urban sociology" (1994). As unionized manufacturing jobs declined in the old center cities, structural mismatch occurred between workers and jobs, and between the built environment and new economic activities, producing patterns of extreme poverty and blight.

However empirical results have not matched predictions of the center city's demise. Contrary to expectations, the 1980s and 1990s experienced a growth in the density of economic activity in many of the world's leading central business districts, even as the importance of globalization and telematics increased. Some urban researchers documented elements of these processes, but almost no theorists have seriously sought to address the deeper implications for urban modeling of this major turn-around. One partial exception is Saskia Sassen who has noted that "This explosion in the number of firms locating in the downtowns of major cities during that decade goes against what should have been expected according to models emphasizing territorial dispersal; this is especially true given the high cost of locating in a major downtown area" (Sassen, 1994, p. 2). Sassen has pointed out that central cities have enjoyed renewed vitality as postindustrial production sites (1991, 1998). In Castells' terms, they are important "milieux of innovation" in the information economy (1989). Postindustrial production differs from industrial production in key ways. In particular, it is design intensive, and highly flexible vs. the "long run" durable assemblage of Fordist production (Lash & Urry 1994). The proliferation of media provides the content for one such postindustrial activity, since the production of media images is an activity significantly concentrated in urban cores, along with finance and elite producer services.[2] The analyses of Reich, Castells, Lash and Urry, and others highlight the highly symbolic and expressive content of these activities, and the distinct competencies of their most valuable workers. The question of why some such activities continue to cohere in what were industrially-based city spaces is one of the most crucial puzzles of contemporary urban sociology. The Entertainment Machine provides a key piece of the puzzle.

Contemporary consumption practice extends to the consumption of space. The lifestyle concerns of social participants are increasingly important in defining the overall rationale for, and in turn driving, other urban social processes. Quality of life is not a mere byproduct of production; it defines and drives the new processes of production. It has been advanced to explain the population shift from the frost belt into the more (consumption- friendly) climates of the Southern and Western U.S. Castells questions this order of causality, foregrounding other priorities in economic location: "so, the 'quality of life' of high technology areas is a result of the industry (its newness, its highly educated labor force) rather than the determinant of its location pattern" (1989 p. 52). His interpretation seems to reflect an earlier reality. It is important is that in many urban locales, migration patterns of residents, especially elite participants in postindustrial growth sectors, are driven by new quality of life demands. In *City Money* (1983), Clark and Ferguson argued that urban job growth increasingly turned on citizen's consumption patterns and tastes, not on production, and showed that certain past migration and job growth studies could be productively reinterpreted in these terms. Evidence of such patterns has mounted in the subsequent decade and a half, evidenced in the suggestive studies of Bennett (1999), Nevarez (1999), and Florida (2000) among others. Increasing importance of tourism and convention dollars to central city coffers, both public and private, raise the stakes in the lifestyle game. Talented high tech staff who can locate where they choose drive cities to compete for them with public amenities.

Residential patterns since the 1980s have run counter to bleak expectations for some older industrial cities. The concentration of poverty documented by Wilson as a near-automatic response to de-industrialization coexists with re-valorization of some former slums by black, brown and white residents. Gentrification trends indicate that affluent workers, particularly the young, are finding the city not simply a clear destination for work, but also a desirable place to live and play. These changes in the residential profiles of urban neighborhoods are treated by some as indicators of postmodern consumption trends (Harvey, 1989); however as Neil Smith points out "systematic gentrification. . . is simultaneously a response and a contributor to a series of wider global transformations: global economic expansion in the 1980s: the restructuring of national and urban economies in advanced capitalist countries toward services, recreation and consumption: and the emergence of a global hierarchy of world, national, and regional cities" (1996 p. 8). The gentrified neighborhood as a distinct type of urban community differs from the neighborhoods studied in past classics of urban sociology such as Gans' *The Urban Villagers* (1962). The important local amenities are no longer schools and churches, as in the ethnic enclaves of the urban mosaic described by the old Chicago school. A residential population of young professionals with high levels of education and lower incidence of children creates a social profile geared toward recreation and consumption concerns. They value the city over other forms of settlement space because of its responsiveness to a wide array of aesthetic concerns, because it can become a cultural center offering diverse, sophisticated and cosmopolitan entertainment lacking elsewhere.

In the "new economic geography" of Entertainment, cities like Seattle and Portland have become central locations for the development of information technologies. A common "explanation" for location of a firm like Microsoft in firm location discussions is often presented with a perhaps disgruntled sigh or laugh: the "personal choice of the top executive," like Bill Gates. The conceptual fallacy here is in implying that the top executives are merely idiosyncratic, simply wrong, or personally selfish – since they did not select a lower-cost or more production-driven location. But this may just be conceptual tunnel vision by the interpreters. Behind it lies a key to reinterpretation: the top executives may have had in mind not merely themselves in locating in attractive places, but a concern to attract top talent globally to work with them. Provision of lifestyle amenities has become a key feature of urban development that we must recognize conceptually; these two cities are extremes ("deviant cases") in being leaders in "smart growth" strategies, and in their recent dramatic growth. Both have aggressively included cultural initiatives in their public agendas. Seattle, home to Microsoft, has been a site of cultural as well as technological innovation, especially in youth culture. The celebrated "Seattle scene" produced several of the most popular musical acts of the early 1990s, and the appreciation for such popular culture on the part of tech workers extends to Microsoft co-founder Paul Allen's recent opening of a lavish, technologically sophisticated rock and roll museum in the city. Both Portland and Austin, Texas have seen thriving youth cultures match the growth of their technology sectors, correspondences that deserve further investigation (extending Florida's fine study).

Meanwhile, older cities pursue new strategies to attract talented workers in the "new entertainment economy." Chicago's current Mayor Daley faces an economic reality unfamiliar to his father. He was said to be in a "red rage" when the *Wall Street Journal* omitted Chicago from a list of "high tech hot spots" in 1999, just as he was described as "beaming with pride" when Chicago was included among *Wired Magazine's* 46 locations across the globe that "matter most in the new digital economy" (Cruze, 2000). Daley's strategies for "building post-industrial Chicago" (Clark, 1999) are quite different from his father's public works agenda, with a new emphasis on aesthetic improvements, and encouragement of neighborhood re-development (i.e. gentrification) through liberal use of Tax Increment Financing (TIF).[3] These strategies aim to make Chicago more attractive as an Entertainment Machine, producing leisure and consumption opportunities for talented knowledge workers. But equity concerns have led Chicago to pursue related strategies that benefit the entire population.

As we will see, even the derelict spaces of the industrial production economy are selectively re-valorized in former industrial powers as sites for consumption, or for knowledge industry production. In San Francisco, for example, loft spaces, whose aesthetic rehabilitation was initiated by artists (Simpson, 1981; Zukin, 1982), are also popular office locations for the technological artistry of Internet site and software producers. In Chicago, warehouse spaces in the old manufacturing zone along the Central Business District's western fringe now house the nightclubs and restaurants of the postindustrial glamour circuit, where fashion models and options traders share sushi and chocolate martinis. The mix of industrial grit, high tech, and exotic consumption is a distinctive urban experience of the Entertainment Machine.

The resurgence of central cities encourages reconceptualizing urban space as embedded in the global economic system. Global city theories have recognized that urban economies are transcending old Fordist economic practice, and associated spatial categories like region and nation state. Select first world cities they term "command centers," coordinating a globally dispersed system of production and distribution. But we stress that they are also production sites for economic processes more grounded in symbolic content, i.e. inputs for Castell's posited informational economy. Media images and cultural content circulate around the globe as objects of production and profit; they are increasingly central to the new "work" of cities, and such global images are realized in material space in local contexts. Globalization encourages an aesthetic cosmopolitanism (Lash & Urry, 1994) impacting consumer demands and motivating the direction of economic strategies toward quality of life and entertainment in large cities. Urban leisure is no passive project; practical political and economic actions produce it, which in turn changes the occupational structure. Consumers must no longer travel vast distances to experience a magnificent diversity of consumption opportunities. For their convenience, flourishing districts of urban entertainment concentrate objects, or at least their facsimiles, from the world over. In a few square blocks of Chicago's Gold Coast, one encounters Thai, Japanese, Mexican, Indian, French, Cajun and Italian cuisine, or one may settle at Gibson's for the old Chicago standby, a Midwestern prime. In the desert creation of Las Vegas, unencumbered by indigenous cuisine or culture, casino entrepreneurs entice tourists with their distinctive versions of Paris, Monte Carlo, New York, and ancient Rome. Peter Jencks suggests that it is natural to satisfy cosmopolitan impulses of affluent consumers this way. "Why, if one can afford to live in different ages and cultures, restrict oneself to the present, the local? Eclecticism is the natural evolution of a culture with choice" (1984; in Harvey 1990 p. 87).[4] Residents increasingly act like tourists in their own cities.

2. THE NEW CLASS

The dual city formulation of Mollenkopf and Castells (1991) stresses that inequality still reigns in the postindustrial city. Yet despite the continued linguistic momentum of terms like class and exploitation, the cleavages of the new division of resources do not map well onto old industrial occupational categories. In an essay asking "Are Social Classes Dying?" Clark and Lipset pointed to the decline of class cleavages in the traditional Marxist sense (1991), sparking an exchange which continues. One approach is to posit a "new class" of workers, with more education and new political and lifestyle concerns. Clark and Inglehart suggest that a "new political culture" maps onto the structural changes emerging from the old class system. A new generation of citizens emerges for which material scarcity was not core to the shaping of values: "The bulk of their population does not live with hunger and economic insecurity. This has led to a gradual shift: Needs for belonging, esteem and intellectual and aesthetic satisfaction become more prominent" (1998, p. 17). The growth of this new class has occurred in step with post-Fordist extensions of consumer culture. The new elite co-exists in the space of past industrial powers with a spatially concentrated underclass; however social groups are sharply segregated, and urban elites often cordon off the underprivileged from the re-valorized zones of consumption (see Davis, 1990 on "Fortress LA"), generating the gated neighborhoods also common to many underdeveloped countries.

The shift from industrial, Fordist capitalism is accompanied by increased attention to aestheticized consumption practices, indeed even the "aestheticization of everyday life" (Featherstone, 1991). This trend has been observed by numerous and ideologically varied cultural theorists. Daniel Bell argued in the 1970s that the locus of individual identity has shifted from the subject's position in the productive apparatus. "As the traditional social class structure dissolves, more and more individuals want to be identified not by their occupational base (in a Marxist sense), but by their cultural tastes and lifestyles" (1976, p. 38). That Bell considered this trend as a "cultural contradiction" of capitalism still suggests that he underestimated the depth of the changes emerging from his forecasted post-industrial society. Flexibility in production, characteristic of the postindustrial economy, finds its analogue in flexible consumption. Harvey argues that post-Fordist economic strategies of "flexible accumulation" motivate production of desires based on the sign value of objects. "Flexible accumulation has been accompanied on the consumption side, therefore, by a much greater attention to quick-changing fashions and the mobilization of all artifices of need inducement and cultural transformation that this implies" (1989, p. 156). The tastes of Stendhal's (1824, 1985: Chap. 4) *dilettanti*, or the "conspicuous consumption" which Thorstein Veblen (1899) identified as informing the privileged classes, have vastly shifted in social meaning. They have so expanded as now to include a majority or residents in some locations, and nearly everywhere an increasing proportion of social participants – as education, media access, and Internet-linked technologies grow. The magnitude of this change is so great that it becomes a near revolution for social theory. The new forces of production stand the old on their heads in an historic conjuncture: they encompass such large and growing portions of the entire society that they can no longer be dismissed as "mere elitism".

"Conspicuous consumption" practices extend even to the American underclass, where status may be measured by the display of gold jewelry or exorbitantly priced, meticulously maintained athletic apparel. The subtleties of taste and consumption practices of lower-status African-Americans are inscribed in the lyrics of popular music enjoyed across class and geographic borders. Yet consumption patterns are not uniform. Tastes and dispositions of social groups vary, not coterminously with their economic positions. Perhaps Bourdieu (1984) sought an opening in this direction by choosing the term *habitus* – which has the flexibility of Weber's status rather than Marx's class – to help interpret aesthetic practices and dispositions. This concept helps us to understand the importance of artists in the city as key contributors to the Entertainment Machine. Artists, whose attraction to urban districts has been evident since the Parisian bohemia, have been a vanguard population in processes of urban gentrification, as educated professionals mimic their residential practices (Zukin, 1982; Mele, 2000; Lloyd, 2001). Occupations in the arts have been growing at high rate in the past decade, consistent with the development of the Entertainment Machine and the increase in demand for aesthetic products. As Markusen points out, arts occupations also evince "siginificant cross fertilization with other sectors and occupations" (2000, p. 2). In the San Francisco Bay Area, local artists are finding increased employment opportunities in the technology sector, demonstrating how the new economy enhances opportunities to valorize creative skill sets. Artists often share educational profiles with employees in other postindustrial industries, and this element of the habitus helps to explain the similarities between the preferences of artists and a larger proportion of the urban middle class. The city becomes a spatial arena for the realization of intellectual and aesthetic concerns led by a talented and mobile workforce. Among these talented workers are a privileged strata of urban residents that help inform both the political agenda and economic activity patterns.

Political elites enact strategies and entrepreneurs start businesses that respond to new consumer preferences. The emergence of this "new class" we argue heightens the importance of aesthetic concerns for urban fortunes; politicians appear to agree. Mayor Richard M. Daley, presiding over Chicago during Fordism, would scarcely recognize the aesthetic bent driving his son's policies. While Daley I bragged of pouring concrete, Daley II confidently claims to have planted more trees than any other mayor in the world (Clark, 1999). In many cases, aesthetic public goods may be shared by all urban residents, not just elites. The distribution of such amenities is still not uniform. Moreover, public encouragement of private amenities, as in Chicago's liberal use of Tax Increment Financing, when it motivates upscale commercial strips, can abet gentrification trends which price out less advantaged residents and prevent them from enjoying the local improvements. Additionally, arts agendas are increasingly promoted by the local bureaucracy; Chicago's Department of Cultural Affairs has experienced significant increase in its funding and mandate over the past decade.

3. GENTRIFICATION, THE NEW CLASS, AND THE ENTERTAINMENT MACHINE

Gentrification is a key aspect of the Entertainment Machine, creating amenity-rich neighborhoods for affluent urban residence. It is spurred by the residential patterns of individuals with distinct attributes, often a group identified by the popular shorthand Yuppies. Yuppies presumably share: (1) Relative youth; (2) High education; (3) The absence of children, and; (4) Relatively high disposable income. Such individuals are disproportionately employed in growth sectors of the global economy. However, the defining characteristic of yuppies in the popular imagination is not occupation, but consumption habits. Says Suttles: "the term "Yuppie" most obviously applies to young singles, who are heavily preoccupied with their nightlife, exploring the new reaches of consumerism, and staying abreast of the trends" (1990, p. 97). Smith concurs: "Apart from age, upward mobility and an urban domicile, yuppies are supposed to be distinguished by a lifestyle of inveterate consumption" (1996, p. 92).

The absence of children suggests that Yuppies will not be particularly interested in local schools or churches as relevant amenities. Rather, they are excited by opportunities for recreation, like along Chicago's refurbished north shore lakefront, with its bicycle paths, beaches, and softball fields; and by opportunities for up-to-the minute consumption in the hip restaurants, bars, shops and boutiques abundant in restructured urban neighborhoods. Heavily gentrified areas like Chicago's Lincoln Park neighborhood contain a heady mix of such public and private goods. If they are *machers* driving the Entertainment Machine, these new class urban residents have hardly been hailed as urban saviors by academics or in popular discourse. Ethnographic work in gentrifying neighborhoods often suggests that even Yuppies don't like Yuppies. Suttles points out that in contrast to urban groups like blacks or homosexuals, "Yuppies you can openly discredit, and in Chicago that term has achieved exceptionally wide usage" (1990, p. 98). The standard ideological assault on Yuppies takes them to task for callously destroying the community fabric of once vibrant poor and working class neighborhoods, leading to wanton cultural and economic displacement.

And, "some of the Yuppies are gay" (Suttles, 1990, p. 95), which creates a small quandary for left-leaning sociologists, who may favor Yuppie bashing but abhor gay bashing. Whatever one's likes or dislikes, in the post-Stonewall decades, gay men have served as a prominent vanguard population in urban gentrification, along with another group favored by the left, the swelling population of self-identified artists (Lloyd, 1996). Urban gay males face obvious constraints on having children, either through biological partnership or adoption. This frees up substantial discretionary income as well as free time that may be directed toward consumption of urban amenities. Additionally, urban gay males evince relatively high education. While common theories of gay locational patterns stress political concerns, and the amicability of the city to "subcultural" variations (see Fischer's general "subcultural theory of urbanism," (1975) or more specific treatments by Chauncey, [1995]), Black, Gates, Sanders and Taylor incorporate an explanation more in line with the Entertainment Machine thesis. They argue that the sorting of gay men into amenity-rich urban neighborhoods can be largely explained by their specific form of affluence: "If 'local amenities' are a normal good, gay men will disproportionately sort into high-amenity locations like San Francisco" (2000, p. 3). Significantly within their findings, the presence of gays in the city also correlates strongly to the strength of the high technology sector in the urban region.

Other young, childless, and affluent professionals have similar priorities. In fact, these authors maintain: "As earnings increase with education and as the number of children falls with earnings in developed countries, it is likely that households with more educated members will also disproportionately locate in high-amenity cities" (2000, p. 4). While these amenities are broadly public goods in being formally open to anyone who chooses to consume them, and thus consistent with Samuelson's (1954) usage of the term, they can still often be niche-specific. The lakefront neighborhoods on Chicago's near North Side provide exemplary cases, where public goods such as the lake and Lincoln Park coexist with health clubs, bars and restaurants supported by Yuppie residents. But consider specific niches: The Boy's Town area around Belmont Street and Halsted is lined with the bars defining much of Chicago's gay night life. Other niches include low culture venues like sports bars, a Chicago staple, proliferating in Lincoln Park and Wrigleyville alongside trendy multi-ethnic restaurants, the North Side's huge live theater scene,[5] and a recently opened multiplex devoted to trendy "arthouse" films. Other cities from Frankfurt, Germany to Mexico City are experiencing similar changes.

The diversity of entertainment opportunities in these neighborhoods, mixing mass spectacle like sports with more "bohemian" flavored urban fare, reflects not only Yuppie affluence, but also education. Despite the revolutionary ascendance of the "informational economy," just over twenty percent of all Americans over 25 have a college degree; however in an elite urban district like Lincoln Park, the figure approaches 70%, with 27.5% of all residents possessing a post graduate degree (1990, census). Employment is disproportionately in services and in the elite FIRE (Finance, Insurance, and Real Estate) sector. 70% of all residents in 1990 were between 18 and 45 years old, and 90% of the households were childless, conditions for a substantial nightlife demand; a walk down the neighborhood's main commercial streets shows that this demand is amply met.

4. "GRIT AS GLAMOUR" AND OTHER MOTIFS OF THE ENTERTAINMENT MACHINE

Although changes in transportation and communication technology make possible the dispersal of major industries, cities offer distinct cultural advantages which appeal to key members of the postindustrial workforce and resist deconcentration. This motivates firms, particularly in creative enterprises, to retain metropolitan addresses. The advantages of the city are found both in the breadth of cultural offerings, i.e. cultural diversity, and their depth, i.e. cultural history. Our approach demonstrates the limitations of the "Disneyfication" thesis, which argues that the city is becoming a homogenized collection of theme park-like attractions with little room for innovation, cultural distinctiveness, or culture on a small (non-corporate) scale (Sorkin, 1992). Sassen and Roost are correct in pointing out that "major cities, in addition to being centers for finance and business services, are strategic sites for the coordination of global entertainment conglomerates" adding that "the same cities that produce entertainment also consume it, giving rise to a new form of urban tourism, one that ... uses the city itself, especially the global city, as an object of consumption – the city as theme park" (1999, p. 143). But it is critical to recognize that large conglomerates do not replace smaller, more local, and more idiosyncratic cultural offerings.

Large corporate interests in fact co-exist with local providers of culture and entertainment, and these may complement one another. Many who find employment as service workers in tourist bubbles also produce culture elsewhere in the city of a smaller, less profitable sort, contributing to ongoing cultural diversity. The growth of the Entertainment Machine produces large scale, spatially contained, consumption driven developments like Navy Pier, Inner Harbor or Ghirardelli Square, however it also includes alternative art galleries, poetry readings, independent theater and film, and a proliferation of funky bars and restaurants. These offerings are less concentrated, though they follow a spatial logic in the city. They are not themselves highly profitable, however they are linked to gentrification and facilitate other types of economic development. Cultural diversity is an urban amenity, central for cosmopolitan identification even if a given individual uses only a small fraction of the available culture. Size, density and heterogeneity (Wirth's classic urban variables), advantage the city for the provision of such cultural diversity over competing social environments. Moreover, the cultural diversity of the city does not only attract human capital with a cosmopolitan amenity profile; ethnographic work suggests that exposure to such urban diversity and cultural vitality enhances creativity in a broad sense – from computer programming to financial transactions to artistic styles (Lloyd, 2001). Understanding the operation of the Entertainment Machine requires not limiting analysis to only its most visible manifestations. A more sensitive analysis would look for the empirical interactions between cultural offerings at different scales.

Urban culture is distinctly characterized by its "cumulative texture" (Suttle, 1984); the specific historical palimpsest of past industrial powers is a crucial aspect of the Entertainment Machine. Modern cities were never only locations for industrial production, but have long included arenas for cultural production and consumption (albeit limited). When George Pullman created a community on Chicago's South Shore in the late nineteenth century to house workers for his railroad car factory, he was not indifferent to the role of consumption in the fabric of urban life. He included shopping arcades modeled on the Parisian example in his local design. The Entertainment Machine signals the extension rather than the invention of many consumption motifs within the contemporary city. Consistent with this, Paris, Benjamin's "Capital of the 19th Century," has long been a model for political elites as they enact new strategies of urban development designed to respond to aesthetic concerns. Paris was the most important single inspiration for Mayor Daley when he brought back ideas like grand boulevards and wrought iron fences for Chicago. Parisian motifs flourish in cites worldwide. However, the Entertainment Machine is characterized by more than the quantitative increase in restaurants, shops and other cultural offerings. Increasingly elements of the city whose functions were considered instrumental (use value) are being valorized through aesthetic concerns (sign value). Spaces of the gritty industrial past become mined for their aesthetic potential, like bars or theaters locating in former steel plants. Even Pullman's factory is being rehabilitated as a potential tourist destination. These strategies do not treat anachronistic elements of the city as a drag on new development, but as potential resources to heighten the aesthetic significance of urban fortunes.

The production of sign value, central to extensions of consumer culture and penetration into previously uncommodified spheres of everyday life, creates a self-referential system indifferent to traditional associations and instrumental logics of production. The sign values of objects are central to the logic of fashion, in which individuals are encouraged to contemplate objects on their aesthetic dimensions. As Benjamin recognized: "Newness is a quality independent of the use value of the commodity. It is the source of that illusion of which fashion is the tireless purveyor" (Benjamin, 1939, [1999] p. 22). This striking analysis anticipates Baudrillard's later assertion that the sign value of commodities has replaced their objective use value (1981). Production oriented to this concern both responds to and produces possibilities of consumer preference. Fashion foregrounds the design intensivity of production over assemblage, which is deskilled and displaced from elite urban spaces. The city as Entertainment Machine extends practices of fashion circulation and display recognized by Benjamin. But in contrast to the fixation on the new and modern in 19th century Paris, the fashionable in the postmodern urban landscape is often the recycled urban tropes and spaces combined with a "new" use and correspondingly multi-tonal aesthetic resonance. In the competition for tourist dollars and elite residence, a city may trade on what Zukin termed its place. "Place in this sense is a form of location rendered so special by economy and demography that it instantly conjures up an image: Detroit, Chicago, Manhattan, Miami" (1991, p. 12). The images of cities in the minds of potential consumers do not develop overnight nor are they imposed by any particular strategist. In Chicago, such an image might include the stockyards, Al Capone, the Water Tower, the Magnificent Mile, the Blues, Sears Tower, Michael Jordan, and, recently, individually painted cows. If images of this sort are not the product of strategic actors to sell the city, they are nevertheless available tools in their repertoire. Thus the former economic functions of the city find a second or third life by donning a new outer attire.

"Downtown developers derive a theme from former economic uses – the harbor, the marketplace, the factory – and offer consumers the opportunity to combine shopping with touristic voyeurism into the city's past" (Zukin, 1991, p. 51). The Inner Harbor development in Baltimore trades on past industrial glory by recreating the harbor as a simulacrum of industrial space, in fact realized entirely as a space of consumption. Baltimore's Camden Yards, home to its Major League Baseball team, has similarly been hailed as an unqualified success, combining "the best of the old and new." Among its "quaint" features, the right field wall is the side of an adjacent restored industrial warehouse. By utilizing such tropes of the industrial past, *instant tradition* is created, in contrast to the sterility perceived to characterize many modern ballparks like Chicago's "new" Comisky Park, an economic and popular failure. In 2000, Comisky's residents, the Chicago White Sox, were soundly outdrawn by the cross town Cubs despite the fact that the White Sox won their division. The Cubs, who hovered near the bottom of the standings, play in the beloved Wrigely Field, a classic venue nestled among popular bars and restaurants. This suggests that the quality of the venue outweighs the quality of the product for many Chicago sports fans – or, more to the point, in this case the venue is the product.

Not all cities are equally successful in producing a bankable image. Oakland continues to be dogged by Gertrude Stein's famous proclamation that "there is no there there." That her quote has been taken out of context is of no consequence; Oakland's ongoing identity problems have made it a longtime poor sister to its storied crossbay rival San Francisco. Natives complain that the toll system on the Bay and San Mateo Bridges bespeaks the relationship: the toll is paid on entry to the San Francisco side of the Bay. The conclusion is inescapable – in San Francisco you pay to get in, in Oakland to get out. Recently Oakland's fortunes have turned better with the election of high-profile Mayor Jerry Brown bringing it a new trope – that of the legendary Brown political family. The massive wealth of the San Francisco-Oakland-San Jose triangle, the premier zone of advanced technological enterprise in the country, has finally penetrated Oakland, transforming once poor, predominantly black neighborhoods into gentrifying spaces with substantial "quality of life" amenities, such as in the area around Lake Meritt, where "Dot-com" households multiply. The rehabilitation of Oakland's Jack London Square has followed the consumption oriented directions characterized above.

Highly sanitized and regulated consumption centers like the Disney creations or Navy Pier, Chicago's number one tourist destination, are described by Judd as "tourist bubbles" within the city (1999). They produce substantial profits, however this tells only part of the story of the Entertainment Machine. Elite consumers also respond to urban tropes foregrounding a grittier milieu, particularly enjoying the affectations of bohemia, with the ambiance of youth and creative energy. Zukin notes: "It is inconceivable that 'living like an artist' would have appealed to segments of the middle class if significant changes in the social position of art and artists had not taken place since the end of World War II. From a marginal and often elitist aesthetic concern, art moved into a central position in the cultural symbolism of an increasingly materialistic world" (1989 p. 82). Spatially concentrated communities of artists and fellow travelers, self selecting into relatively derelict urban neighborhoods, have been well documented in the old industrial city Zorbaugh described "Towertown" as such an enclave in *The Gold Coast and the Slum* [1929]), however they were far less important in driving economic dynamics. The intervention of large numbers of artists in space evincing depopulation and postindustrial decay often precedes and provides a condition for increased capital investment and more advanced gentrification.

The role of the artist in constituting the culture of the city has been recognized enough by political leaders that their residential position is often reinforced through public subsidy. The spatial practices of artists contribute to the revalorization of old industrial space, including the re-imagination of abandoned warehouses and factories as highly desirable "residential loft living" (Simpson, 1981; Zukin, 1989).[6] The appeal of these practices reflects changes in the economy, in providing more income for the affluent if not for everyone, combined with the aesthetic dispositions of the "new class" elaborated above. The symbolic expressive content pervading the informational economy decreases the social distance in terms of taste between the urban "mainstream" and its bohemian fringe. The division between bohemia and the bourgeoisie (the "artist" and the "Yuppie") is belied by the extent to which Yuppies turn to aesthetic practices of artists for their cultural cues. Indeed, in the 1990s bohemian-themed gentrification was only one aspect of popular culture's fascination with an imagined underground. Controversial fashion motifs, Heroin "chic," were part of an aesthetic turn leading a *Time* magazine cover story to pose the question "Is Everybody Hip?" in 1994. While urban artists and intellectuals are among the most vocal critics of the Yuppie,[7] they in fact share with Yuppies similar educational profiles, and the same preoccupation with nightlife and cultural consumption.

Bohemian districts resembling New York's Greenwich Village in the 1960s and San Francisco's North Beach in the 1950s could be found in virtually every large U.S. city in 2000. Their gritty charms undergird trendy entertainment destinations for "sophisticated" young urbanites. As a neighborhood lays claim to the mantle of local Bohemia, entrepreneurial efforts spring up, and art galleries, night spots, restaurants and associated businesses proliferate to constitute the neighborhood as a distinct entertainment space. The rehabilitation of SoHo and the East Village in Manhattan provide the model; it is broadly followed in cities nationwide.

Creative pursuits clearly benefit from the urban innovative milieux; as one of our informants, a successful sculptor, succinctly put it: "I came to Chicago because that was where the conversation was." At the same time, artistic work is rarely self-supporting. The growth of bohemian entertainment destinations allow would-be artists to market themselves as service workers, in places where the aesthetic self work they perform heightens the marketable ambiance of "hipness." The growing technology sector also provides occupations that valorize artistic and creative dispositions, enhancing the importance of artistic communities to the new urban economy.

Additionally, extraordinarily affluent young participants in highly valorized urban sectors like Madison Avenue and Wall Street patronize "glamour zones," arenas for conspicuous consumption and the display of high fashion. These spatial practices of the urban intellectual and financial elite stand in stark contrast to conditions in the corollary space of the dual city, the space of devastation.

Their restaurants and nightclubs, whose ambiance is improved by patrons for whom being beautiful is a full time job,[8] are not the property of ordinary tourists. The exclusiveness is part of the package, with New York's Studio 54 of the late 1970s providing the model, consigning undesirables to lines that would never move. In Chicago, many new destinations for the glamour circuit have emerged in the former warehouse districts and barrios of the near West side, where the glamour is underscored by the grit of surrounding structures in the unevenly redeveloped zone. This juxtaposition is thematized in the Entertainment Machine; as in New York's East Village, uneven development shades into the avant-garde, and provides patrons with a sense of their own urbane sophistication.

The link between the city and aesthetic innovation is not new. As Harvey points out, cultural modernism has since the middle 19th century been "very much an urban affair phenomenon, that it existed in restless but intricate relationship with explosive urban growth" (Harvey, 1989, p. 25). The world exhibitions, including Chicago's Colombian Exposition of 1892 (coinciding with the founding of the nearby University of Chicago), demonstrate that elites recognized the importance of culture and spectacle to advancing urban centrality even during the past periods of industrial growth. But such concerns were not central in past years. As the classic variables of transportation, housing for blue-collar labor, and fixed capital in the form of factories decline in importance with new information and transportation technologies, however, cultural elements generally rise in power. The new class of talent workers, employed in an increasingly aestheticized economy, is so much in demand that it can make increased and highly differentiated niche demands on the city's aesthetic dimensions.[9] The ongoing advantage of central cities in the new economic geography hinges heavily on its cultural advantages, so that both public and private sector leaders are enacting strategies to enhance their aesthetic profiles. Often these new strategies run counter to the growth machine ideology which Molotch argued quasi-automatically informed elite urban decision making. Future study of cities cannot brush aside the cultural concerns; to an ever larger extent, culture has become the work of the city.

NOTES

1. Judd advances the concept of the "tourist bubble," (1999) regulated spaces in the city given entirely to consumption. Examples include sports stadiums or districts like Times Square. We examine the tourist bubble, but argue this does not exhaust the Entertainment Machine concept.

2. As has been well documented (Castells, 1989; Garreau, 1991; Sassen, 1994; Zukin, 1982) routinized administrative tasks, "back office" work, is mostly displaced to "edge

city" locales; however, the most intellectually intensive forms of administration and producer services still concentrate in urban cores, suggesting agglomeration benefits for advanced intellectual production.

3. Tax Increment Financing retains revenues in the TIF geographic area that derive from economic growth in that area. Hundreds of small neighborhoods across Chicago have become TIFs. This strategy is used by cities nationwide, but under Daley Chicago has become "the TIF capital of America" (Lehrer, 1999).

4. Cosmopolitan means, literally, "citizen of the world." The elite cosmopolitanism discussed above unfolds in the aesthetic realm however, seldom joined to any responsibility of good citizenship toward the immiserated workers in Third World sweatshops, or even toward the disenfranchised members of proximate underclass communities.

5. In addition to large companies like Stepenwolf and the Goodman, the League of Chicago's Theaters has over a hundred members with budgets below one million annually, many of whom alternate in venues disproportionately located on the North Side

6. This practice began in New York's SoHo, and has been aped nationwide. Where such conversions have been exhausted or appropriate space never existed, "lofts" have been built from the ground up for residential use.

7. A local performance poet in Chicago for example delivered the following evocative dis: "Well, they might smell good to real estate developers and personal trainers and the producers of Ally McBeal and Friends. But no amount of designer fragrance from Calvin Klein will ever mask the heinous stench of self satisfied bullshit from ME" (Shappy, 2000).

8. "Club kids" have mastered the social codes of privileged entertainment spaces so thoroughly they are paid by owners in free admission and other perks. They are human props. Whatever they think however, they are not the point of these enterprises, even if the enterprises have become the point of them.

9. We must mildly dissent with Florida (2000) in his generally sophisticated and multi-methodological study. He omitted the largest U.S. cities, which provide the strongest examples that we cite. If his informants more often stressed sailing and skiing than theater and nightlife, this may be in part due to his selective focus on primarily mid-sized and smaller cities.

REFERENCES

Atkinson, R. D. (1998). Technological Change and Cities. *Cityscape, 3*(3), 129–170.
Benjamin, W. (1939). *Paris, Capital of the Nineteenth Century in The Arcades Project.* Belknap Press: Cambridge MA: 1999.
Baudrillard, J. (1981). *For a Critique of the Political Economy of the Sign.* New York: Telos.
Bell, D. (1976). *The Cultural Contradictions of Capitalism.* Basic Books: New York.
Bennett, L. (1999). *The New Style of U.S. Urban Redevelopment: From Urban Renewal to the City of Liesure.* Presentation at the Annual Meeting of the Urban Affairs Association.
Bourdieu, P. (1984). *Distinction.* Cambridge: Harvard University Press.
Black, D., Gates, G., Sanders, S., & Taylor, L. (1997). Why do Gay Men Live in San Francisco?
Castells, M. (1989). *The Informational City.* Oxford: Blackwell.
Chauncey, G. (1994). *Gay New York.* New York: Basic Books.
Clark, T. N. (2000). *Trees and Real Violins: Building Post-Industrial Chicago.* Book Manuscript.

Clark, T. N., & Ferguson, L. C. (1983). *City Money: Political Process, Fiscal Strain and Retrenchment*. New York: Columbia University Press.
Clark, T. N., & Inglehart. R. (1998). The New Political Culture: Changing Dynamics of Support for the Welfare State and other Policies in Postindustrial Societies. In: Clark & Vincent Hoffman Martinot (Eds), *The New Political Culture*. Westview: Boulder CO.
Clark, T. N., & Lipset, S. M. (1991). Are Social Classes Dying? *International Sociology*, 4, 397–410.
Clarke, S., & Gaile, G. (1998). *The Work of Cities*. Minneapolis: University of Minnesota Press.
Cruze, T. (2000). *Area Tech Star Rising: Region lauded for turnaround*. Chicago Sun-Times.
Davis, M. (1990). *City of Quartz: Excavating the Future in Los Angele*. London: Verso Press.
Featherstone, M. (1991). *Consumer Culture and Postmodernism*. London: Sage.
Fischer, C. (1975). Toward a Subcultural Theory of Urbanism. *American Journal of Sociology*, 80, 1319–1353.
Florida, R. (2000). *Competing in the Age of Talent: Quality of Place and the New Economy*. Report Prepared for The R. K. Mellon Foundation, Heinz Endowments, and Sustainable Pittsburgh.
Friedman, J., & Wolff, G. (1982). World city formation. An agenda for research and action. *International Journal of Urban and Regional Research*, 6(3), 309–404.
Gans, H. (1962). *The Urban Villagers*. New York: The Free Press.
Gottdiener, M. (1985). *The Social Production of Urban Space*. Austin: The University of Texas Press.
Hannigan, J. (1998). *Fantasy City: Pleasure and Profit in the Postmodern Metropolis*. London: Routledge.
Harvey, D. (1990). *The Condition of Postmodernity*. Cambridge: Blackwell Press.
Jameson, F. (1992). *Postmodernism, or the Cultural Logic of Late Capitalism*. Durham: Duke University Press.
Judd, D., & Fainstein S. (1999). *The Tourist City*. New Haven: Yale University Press.
Judd, D. (1999). Constructing the Tourist Bubble. In: D. Judd & S. Fainstein (Eds), *The Tourist City*. New Haven: Yale University Press.
Kornblum, W. (1974). *Blue Collar Community*. Chicago: University of Chicago Press.
Lacayo, R. (1994). Is Everybody Hip? *Time*, Aug. 3.
Lash, S., & Urry, J. (1994). *Economies of Signs and Space*. New York: Sage.
Lehrer, E. (1999). *The Town That Loves to TIF*. Governing Magazine.
Lloyd, R. (2001). Grit as Glamour: Neo-Bohemia and Urban Change. *Journal of Urban Affairs*, forthcoming.
Markusen, A. (2000). *Targeting Occupations Rather than Industries in Regional and Community Economic Development*. Presented at the North American Regional Science Association Meetings, Chicago.
Mele, C. (2000). *Selling the Lower East Side* (Minneapolis: University of Minnesota Press).
Mollenkopf., & Castells. (Eds) (1991). *Dual City: Restructuring New York*. New York: Russell Sage Foundation.
Molotch, H. (1976). The City as a Growth Machine American. *Journal of Sociology*, 82(2), 309–330.
Nevarez, L. (1999). Working and Living in the Quality of Life District. *Research in Community Sociology*, 9, 185–215.
Reich, R. (1991). *The Work of Nations: Preparing Ourselves For 21st Century Capitalism*. New York: Alfred A. Knopf.
Samuelson, P. A. (1969). Pure Theory of Public Expenditure and Taxation. In: J. Margolis & H. Guitton (Eds), *Public Economics*. New York: St. Martin's Press.
Sassen, S. (1991). *The Global City: New York, London, Tokyo*. Princeton NJ: Princeton University Press.

Sassen, S. (1994). *Cities in a World Economy*. London: Pine Forge Press.
Sassen, S. (1998). *Cities: Between Global Actors and Local Conditions*. 1997 Lefrak Monograph College Park MD: Urban Studies and Planning Program.
Sassen, S., & Roost. F. (1999). The City: Strategic Site for the Global Entertainment Industry. In: Judd & Fainstain (Eds), *The Tourist City*. New Haven: Yale University Press.
Shappy, N. (2000). *Stinky Little Book of Ass*. San Francisco: Kapow Press.
Simmel, G. (1971). The Metropolis and Mental Life. In: D. Levine (Ed.), *Georg Simmel on Individuality and Social Forms*. Chicago: University of Chicago Press.
Simpson, C. R. (1981). *SoHo: The Artist in the City*. Chicago: University of Chicago Press.
Smith, N. (1996). The New Urban Frontier: Gentrification and the Revanchist City Routledge: New York.
Sorkin, M, (Ed.) (1992). *Variations on a Theme Park*. New York: Hill and Wang.
Stendhal, H. B. (1824) in French, (1985) English translation. *The Life of Rossini*. London: John Calder.
Suttles, G. (1984). The Cumulative Texture of Local Urban Culture. *American Journal of Sociology*, *90*, 283–304.
Suttles, G. (1990). *The Man Made City: The Land Use Confidence Game in Chicago*. Chicago: University of Chicago Press.
Veblen, T. (1899). *The Theory of the Leisure Class*. Prometheus: New York.
Wilson, W. J. (1987). *The Truly Disadvantaged*. Chicago: University of Chicago Press.
Wirth, L. (1938). Urbanism as a Way of Life. *American Journal of Sociology*, *44*, 1–24.
Zorbaugh, H. W. (1929). *The Gold Coast and the Slum*. Chicago: University of Chicago Press.
Zukin, S. (1982). *Loft Living: Culture and Capital in Urban Change*. Baltimore: Johns Hopkins University Press.
Zukin, S. (1991). Landscapes of Power. Berkeley: University of California Press.

THE "DISNEYFICATION" OF TIMES SQUARE: BACK TO THE FUTURE?

Bart Eeckhout

> 42nd Street Now! calls for the restoration of New York's quintessential entertainment district, our most democratic good-time place. The renewed 42nd Street will be an enhanced version of itself – not a gentrified theme park or festival market. The focus of the renewed 42nd Street will be theaters and all that goes with them: restaurants and retail establishments related to entertainment and tourism. Once again 42nd Street will be able to take its rightful place among the world's great urban entertainment destinations.
>
> Robert A. M. Stern Architects and M & Co.

> It would have been less ambiguous if Times Square had died a "natural" death, if it had fallen victim to some other necessary expansion – but the leap from sick but energetic authenticity straight into the embalmed cheer of Disney has an intolerable perversity. It is as if the transition from the harmful to the innocent offends a sense of urban dramaturgy, as shocking as a movie suddenly played backward. A coalition of moralists, planners, and a nostalgia-driven entertainment giant expelling, as if in some Biblical scene, the unwanted from the city . . . it hardly seems a good omen for Manhattan's continuing relevance in the twenty-first century.
>
> Rem Koolhaas

INTRODUCTION

One of the most poignant paradoxes of New York City's Times Square at the turn of the millenium is that it is both one of the most singular, instantly recognizable, and inimitable places in the United States and a site that embodies a great many of the most generic characteristics commonly associated with redeveloped shopping and entertainment destinations in contemporary American cities. Many of the issues that are at the core of current theories about the redevelopment of American central cities have also been raised in recent discussions about Times Square, the symbolic heart of New York City. These issues range from the construction of a themed fantasy city that is geared towards pleasure and profit, or the transition from an industrial to a postindustrial service and information economy, to the hegemony of middle-class consumerism, globalized cultural tourism, and the spectacular commodification of cityscapes; the commercial simulation and manipulation of history; the importance of preservation- and arts-based redevelopment; the wish to provide edutainment and vicarious forms of risk-taking; the privatization and even militarization of public space; and, finally, the deeply ingrained American tendency towards noncoercive forms of social segregation.

It is precisely the tension between specificity and generality, between the uniqueness of a place called Times Square and the forcefield of its wider historic and socioeconomic determinants, that gives purpose and meaning to the following extended inquiry. For it remains of the utmost critical importance in urban studies at once to distinguish between the specific and the general and to get a purchase on the manifold, frequently indivisible interrelationships between the two in reality. We do not learn from what is radically particular, idiosyncratic, or local – except that it is such and must be recognized and respected as such. By contrast, we do manage to learn a great deal from what appear to be larger underlying structures and processes that apply across a wide range of similar situations – except if we forget that these structures and processes are necessarily entangled with, and complicated by, the particular, the idiosyncratic, and the local. In the case of a densely constructed, constantly changing, variously used, and symbolically overdetermined space like Times Square, the complicating entanglements are indeed such that they can only be approached through a thoroughly multidisciplinary analysis. In what follows, I will attempt such an analysis, combining, among other things, extended historical research (by way of an opening), micro- and macroeconomic theory (for my subsequent reading of the local Business Improvement District and its emblematic role in a postindustrial service economy), the debate in especially architectural theory and urban sociology on the so-called Disneyfication of cityspaces (which takes up a long central section), and an investigation into the effects of recent redevelopments on the class, race, gender, and sexuality of visitors to the Times Square area. In an afterword, I will zoom out once more to pose a few remaining questions about the wider meaning of Times Square's remarkable transformation during the past decade.

Although much of what is to be found in the present essay has been gathered from other published academic sources like Taylor (1991), Zukin (1991, 1995), Wallace (1996), Hannigan (1998), Delany (1999), and especially Reichl (1999), these sources have been mixed throughout with articles from *The New York Times* and other journalistic media, and have been expanded and corrected by means of personal fieldwork, the most recent of which happened during the summer of 2000. If one of the main purposes of the present essay, then, is to offer a synthetic overview of the transformation of Times Square and its attendant critical discourses during especially the 1990s, I nevertheless propose at the same time a number of reconsiderations that should help to focus and deepen future debates. Central to these reconsiderations is the need to historicize recent redevelopment efforts by taking a sufficiently large view of Times Square's twentieth-century history as well as of its wider socioeconomic context. Only thus, I would suggest, are we able to separate out continuities and discontinuities within the transformations that have been taking place and that frequently enough produce the illusion of radical novelty.

A similar wish to be more precise in identifying the nature of recent changes also underlies my central attempt at unpacking the term "Disneyfication." Even if many of the eight factors I discuss under that heading have been the object of earlier analysis, they have never been taken apart and brought together in this systematic way to explain their relevance to the current Times Square. What is more, I follow up on my analysis of the Disney Company's material and symbolic influence by devoting a long supplementary section to an issue that, although at times implied and tangentially touched upon in the debate on Disneyfication, nevertheless threatens to be obfuscated or displaced by that debate: the question of the extent to which the redevelopment of Times Square has altered the social profile of its visitors and the social uses of its space. Here again, many of the facts and figures are derived from published sources, but they have been augmented by personal research and observation and organized in a systematic format that should contribute towards a more focused and balanced discussion of what are by their very nature hotly contested issues. I will demonstrate in particular that the social space of the new Times Square differs from the old one in several respects: (1) by offering a consumption and entertainment district primarily catering to the middle classes at the expense of the earlier working classes; (2) by being constructed on the basis of racial fears for the earlier low-income minority groups using the area; (3) by its attraction of a group that was largely missing from the earlier population of visitors: middle-class women (and their children); and (4) by virtually disabling the survival of a wide variety of sexual subcultures and practices to which Times Square, for more than three quarters of a century, owed at least part of its fame.

SYMBOLIC CROSSROADS OF THE WORLD

Times Square, we might begin by simply reminding ourselves, is not really a square. It is a fluid area in midtown Manhattan centered around the diagonal slicing of 7th Avenue and Broadway – a slicing that stretches out over five blocks, between West 42nd and 47th Streets. The heart of the area is the so-called "bowtie" open space surrounded by neon supersigns and other "spectaculars," which derives some of its original and lasting appeal precisely from the fact that it breaks up the monotony of the Manhattan grid pattern and constitutes a more complex and open space than the grid elsewhere tends to provide. The name "Times Square," however, extends beyond the southern half of this elongated crossroads (to which it officially refers) to include all of West 42nd Street roughly between 6th and 8th Avenues, as well as the surrounding streets and blocks especially to the north that are home to the theater area collectively described by the synecdoche Broadway.[1] By its very location and outlook, Times Square has always carried considerable symbolic value. As the cultural and popular heart of America's prime city it differs essentially from its European counterparts (see also Rybczynski, 1995, pp. 15–34). It does not offer the prototypically European large rectangular square filled with monuments and bordered by the civic or religious emblems of the class societies that shaped the nineteenth-century and pre-nineteenth-century European city. No palaces here, no parliament, no big statues, no churches or cathedrals, no museums or libraries. Times Square has from the beginning been typical of those pragmatic, egalitarian, and mercantilist aspects characteristic of the twentieth-century American metropolis – and if one may be so sweeping, of the twentieth century overall. Typically, one of the most popular nicknames of Times Square is that of Crossroads of the World.

A HISTORY OF MATERIAL AND SOCIAL PERMUTATIONS

To fully understand Times Square at the turn of the millenium, we should recall at least some of the history of the place. What immediately strikes us then is how radical some of the area's material and social permutations in the course of a single century have been. This is a site that would seem to corroborate Richard Sennett's provocative prediction (1994, p. 360) that "in a hundred years people will have more tangible evidence about Hadrian's Rome than they will about fiber-optic New York." American city centers, with their general lack of a long-standing historical residue and their constant drive for vertical as well as horizontal expansion, have always been particularly susceptible to cycles of *tabula rasa* and reconstruction. Manhattan's history and geography as the nation's most densely populated "city" and an island faced with the strongest possible natural limits for horizontal expansion have arguably only strengthened this tendency to produce the most extreme instances, one of which appears to be Times Square.

The strong mutability of this area is one important reason why, as Marshall Berman (1997, p. 78) has argued, "So much of the talk about Times Square seems to be both driven and crippled by nostalgia." Indeed, the redevelopment of Times Square in the 1980s and 1990s has consistently depended for its political survival as well as public support on a careful manipulation and exploitation of such nostalgia.[2] Since nostalgia is a form of imaginative indulgence that most naturally feeds on what is most irredeemably lost, the feeling surfaces almost spontaneously when we look at pictures from a century ago, when the area was still known as Long Acre Square and was famous (or infamous) for its horse stables and silk-hat brothels. The same applies to pictures taken four years into the century, when the area's further history was marked by an early instance of the kind of commercial "branding" that has returned to become such a hot topic in economic theory today. As soon as the original 25-story Italianate Times Tower had been erected by *The New York Times* (in a somewhat unlikely spot and wedge-like shape), the square on which it bordered was renamed Times Square. The construction of the sleek but unusually tall tower, the second-highest building in the city at the time (and the highest if one included the several levels below ground), was symptomatic in more ways than one, since it also presented a classic, even literal Marxist story of economic infrastructure and cultural superstructure: the newspaper offices were built right on top of an important North-South junction in the city's first subway system that was under construction at the same time. Thus, the newspaper was able to avail itself among other things of immediate access to the transportation network needed for local distribution.

Within less than a decade, Times Square managed to become not only New York's foremost media district but also its prime theater district – an evolution that may be viewed as part of a more general shift in the American industrial metropolis. In the early years of the twentieth century, historian David Nasaw reminds us, the industrial city was slowly becoming "as much a place of play as a place of work" (quoted in Hannigan, 1998, p. 15). It did so, moreover, along more democratic lines than before. For it was only by the end of the nineteenth century that urban entertainment zones which sought to span the entire length of American society began to spring up. No more than twenty years earlier, for instance, the realms of entertainment and leisure in the city were still strongly segregated along class lines, and many of them were completely inaccessible to large groups of women and non-whites (Hannigan, 1998, pp. 15–16). Oscar Hammerstein's turn-of-the-century Olympia, a multifunctional and grand entertainment complex in Times Square, was one of the first and most spectacular buildings to bear witness to the newly emerging, would-be democratic and pleasure-seeking urban culture; it was also one of the first major buildings in the area to be demolished. Topics that have come to be associated with postmodern urban theory, like theme parks and the production of phantasmagoric spaces, were equally applicable to this era with its mushrooming urban amusement parks. The Paradise Roof Garden on top of the Victoria Theater, for instance, contained not only a windmill with cottage, but also a stream, ponds filled with live swans, and a rock grotto inhabited by monkeys – all of this covered by a glass dome (Stone, 1982, p. 40). More popularly even, the Hippodrome, which took up an entire block at 6th Avenue and 40th Street, offered a gigantic, spectacular "plaything for the masses" at a wide range of prices (Reichl, 1999, p. 53; Register, 1991). Impresarios like Oscar Hammerstein were the driving force behind the development of what made Times Square famous in the early decades of the century: theater, vaudeville, the music business of Tin Pan Alley, dance halls, cabarets, and the would-be democratic mix of upper-to-lower classes that patronized these sites of leisure. Times Square became New York's first real "public space" in a city whose street grid had been laid out for purely pragmatic and commercial reasons and had not so far invited a crowding and clustering of functions and social classes (Sandeen, n.d., p. 6). The symbolic heart of this entertainment area was to be found on the single block of 42nd Street between 7th and 8th Avenues, so much so even that theaters built on 41st and 43rd Streets made sure to have their entranceways built through the block in order to have a marquee also on 42nd Street (Reichl, 1999, p. 44).

Two grand hotels, both prime examples of the Beaux Arts style promulgated by the City Beautiful Movement, took up a central iconic position in the square. They were the Knickerbocker (now landmarked, renovated, and converted into condominiums) and the Astor Hotel (demolished in the 1960s). Both hotels stood as magnificent emblems of how by the 1920s large-scale tourism and business travel had developed, and how these phenomena in turn played a prominent part – together with the decline of religious and moral strictures and the concomitant rise of sensuous excitement as a form of self-sufficient pleasure – in drawing a rapidly increasing number of visitors to the area (Taylor, 1992, pp. 95–96; Harris, 1991). The nightscape these visitors found was an ode on electricity and neon that went by the moniker of Great White Way (and that went back to at least the 1880s). With the rise of standardized mass production and consumption in the industrial metropolis had come the need for mass-oriented advertisements, and Times Square's entrepeneurial, originally undercapitalized low-rise environment had been one of the first urban spaces to fully embrace the potential for creating commercial outdoor spectacles to a fun-loving audience. In only a few years' time, the area had come to offer the largest concentration of advertising space in the world (Reichl, 1999, p. 52), which proved trendsetting in its stimulation of a new "commercial aesthetic" (Leach, 1991).[3]

Nostalgia has often dictated that the late 1910s, early 1920s be seen as the area's golden age, which is then supposed to have ended with Prohibition (Erenberg, 1991, pp. 164–170) and, more fatally, with the Great Depression. But as Alexander Reichl (1999, p. 49) has noted, the nostalgic picture of a golden age has frequently been lopsided and romanticized, eliding among other things the fact that even then the Times Square area was also famous for its high levels of crime and prostitution (which had only temporarily been moved indoors) and that it bordered on one of the unsafest districts in the entire nation, Hell's Kitchen. Even so, by the 1930s, Times Square did get caught in an economically downward cycle that would drastically change the commercial and social use of its buildings and practically halt new construction for almost half a century: one by one theaters and grand movie palaces were downscaled to burlesques or began to show sexually explicit films and non-stop B-movies; cheap restaurants, peep shows, penny arcades, and dime museums entered the neighborhood, which also began to offer a reincreasing array of commercialized sex (including the sociologically new arrival of male hustlers, often as the result of unemployment), thereby recharting the tradition of Long Acre Square's blossoming sex industry in the 1890s and around the turn of the century. The relative slump, which to some commentators nevertheless involved a socioeconomic broadening of the audience (McNamara, 1991; Senelick, 1991), continued into the Second World War, despite the fact that a first clean-up operation prohibited the production of burlesque shows by the early 1940s and even if the 1945 celebrations of the war's ending have continued to register with many New Yorkers as the acme in the square's history of collective enchantment. After the war, with sailors and soldiers returning home to find themselves hard pressed for housing and hot for company, the area did not really manage to pick up its former glory again. It gradually lost its glamor still further as the process of suburbanization weaned many white middle-class citizens away from downtowns to the shopping malls in the periphery and the media stronghold shifted from the movies and newspapers to the television industry. Already in the late 1940s, Times Square began to act as a magnet on drug-using outcasts, the early rebels of an emerging youth subculture. The Beat poet Allen Ginsberg was one of them; in a poem called "'Back on Times Square, Dreaming of Times Square,'" composed in July of 1958, he could already reminisce about a dead friend and how he "lay in / gray beds [at the Globe Hotel] and hunched his / back and cleaned his needles" while Ginsberg himself slept "many nights on the nod/from his leftover bloody cottons

/ and dreamed of Blake's voice talking." "The movies took our language," the poem goes on to recall, "Teen Age Nightmare / Hooligans of the Moon / But we were never nightmare / hooligans but seekers of / the blond nose for Truth / Some old men are still alive, but / the old Junkies are gone –" (Ginsberg, 1984, p. 188).

In this 1958 poem, Ginsberg still quarreled with the way he and his friends were being represented as "teen age nightmare hooligans," especially in sensationalist B-movies, but the protestations were in vain. The image of Times Square and its residents had not yet reached its nadir. Through the 1960s and 1970s, when the tax base of cities was further eroded through the ongoing "white flight" to suburbia, American metropolitan downtowns became more and more synonymous with physical deterioration, escalating crime, racial tension, drug abuse, pornographic shops, and sexual vice, all of which were again emblematized by Times Square, more particularly by West 42nd Street. Rundown, dismal, dreary, dangerous: those were some of the words that came most readily to mind in conjuring up Times Square in this era. For some time, the area acquired the reputation of "sleaziest block in America" (Hannigan, 1998, p. 71) and harrowing portraits were provided by dystopian films like John Schlesinger's *Midnight Cowboy*, in which a simple-minded Texan boy is taught a number of reality lessons in the local world of prostitution, and Martin Scorsese's *Taxi Driver*, in which the psychotic protagonist Travis Bickle famously lashes out verbally as well as physically at the "scum" of "whores, skunk pussies, buggers, queens, fairies, dopers, junkies" peopling the sidewalks and sleazy hotels. The image that was created in the public imagination through such films was again – as Allen Ginsberg had previously protested – overblown: the narrative of urban decline that portrayed "the area's devolution from the Great White Way to the Dangerous Deuce" and that became dominant among especially white observers drew on myth as much as on reality (Reichl, 1999, p. 44). Nevertheless, Times Square did manage to become the most crime-ridden district in all of New York City and the ambivalent attractions of danger and lawlessness, more than before, became one of its characteristic features.

The perceived social threat posed by the area was possibly still enhanced in the 1980s through an "increase in homelessness and the invention of crack as a means of marketing cocaine to low-income users" (Reichl, 1999, p. 44). Certainly to white middle-class outsiders and infrequent visitors, the image of the area became dominated by sex shops, porn movie theaters, hustlers, con men, small-time crooks, pickpockets, drug users, homeless people, and menacing youngsters from minority groups. It was this dominant image also that informed political and media discourses at the time and came to support a local redevelopment agenda – an agenda that depended on "constructing an idealized past around the symbolism of the Great White Way that would both highlight the exceptional severity of the contemporary problems and provide a grand vision to guide redevelopment, and [on] defining the area in the 1970s and 1980s as without any redeeming social functions and home to only a depraved subculture so deeply entrenched that it would repel all but the most determined assault" (Reichl, 1999, pp. 44–45). Times Square came to be seen by pro-growth politicians and business elites as a "cancer" that needed to be cured by making deep cuts in the urban fabric. Although, as Alexander Reichl (1999, p. 68) has insisted, the area was in fact still a thriving (if risky and disturbing) multi-purpose entertainment center and a popular tourist destination in the 1980s, a coalition of political and business leaders under the Koch administration launched a massive redevelopment plan that would practically raze all of 42nd Street between 7th and 8th Avenues (including the southern part of the bowtie) and, under the pretense of helping to restore a number of theaters, amount to a transformation of the area into a new midtown high-rise office district.

Development incentives provided by the Special Midtown Zoning District resulted in a series of new skyscrapers that drastically altered the face of the bowtie during the 1980s, but failed to affect 42nd Street in time before the economic and cultural climate changed again. By the early 1990s, the string of sex shops, cheap stores, and porn movie theaters along 42nd Street had finally been seized by eminent domain (after many years of procrastination through challenges in court). What remained of these was no more than a series of permanently closed shutters painted in bright colors and lending to the street "the aura of a ghost town" (Reichl, 1999, p. 143). Yet the collapse of the market for office space in the early nineties, caused by the economic and real-estate slump of the late 1980s as well as by severe overbuilding of the West Side, combined with a growing opposition to the redevelopment plan from arts and culture aficionados to stall the long-devised megaproject and necessitate an interim plan. This interim plan rather preferred gradual upgrading to massive *tabula rasa* and changed its attendant discourse from a restoration of the Great White Way to a revalorization of Times Square as an essentially popular, honky-tonk entertainment district (Reichl, 1999, pp. 76, 144–157). By the mid-1990s, the outlook of 42nd Street between 7th and 8th Avenues suddenly began to change again from one of desolate, closed shops to a canyon of construction sites on either side of the street. In the final years of the century (and continuing to date), especially the southern part of the Times Square bowtie, the adjacent block of 42nd Street, and several blocks of 8th Avenue have changed and continued to look different – more spectacular and more brilliant – almost by the month.

Typical of Times Square at the turn of the new millenium are a number of once again radically altered sites and views. Most striking and imposing among these are the two sites on the north side of 42nd Street across from One Times Square (as the former and redesigned Times Tower is currently called). To the east of One Times Square (which itself now parades a colorful Warner Bros. Studio Store in which merchandise for kids is sold) rises a silvery 48-story skyscraper known as the Condé Nast Building because it is mainly occupied by the international publishing company Condé Nast, which includes among others *Vogue*, *GQ*, *Wired*, *The New Yorker*, and *Vanity Fair* in its ample magazine portfolio. (The other occupant of the building is the major law firm of Skadden Arps Slate Meagher & Flom.) The new building is, in other words, a corporate office tower for a media giant, and contains not only a gigantic high-tech videoscreen for Nasdaq that has become the new focal point of the constantly multiplying videoscreens around the bowtie, but also flashy neon signs or "spectaculars" on top of the building.[4] Directly across, meanwhile, to the west of One Times Square, a comparable story has been unfolding. On the exact spot where the Rialto Theater used to be (that is, the 1935 Art Deco successor to Oscar Hammerstein's earlier Victoria Theater flaunting the Paradise Roof Garden), we now find the blitzy and sleek Reuters Building, a 32-story corporate office tower mostly occupied by Reuters, the international news and financial services company. It, too, parades a barrage of spectacular signage, bands of LED messages, and videoscreens at street-level and on top. Both towers have indeed been built by the same architects, Fox & Fowle. Characteristically, these architects, too, have tapped into a discourse of nostalgia to defend their work, for the Reuters building is meant to "recall the rhythm of the Rialto facade," as one of the architects claimed, so that "We think it will develop its own historic nostalgia over time" (quoted in Dunlap, 1998, n. p.). Walking further west along 42nd Street, finally, and passing two lavishly restored historic theaters on either side of the street, we are dazzled and dazed nowadays by a long string of new restaurants and eateries (many of them belonging to national chains like Applebee's, Starbucks, Chevy's, and Manchu Wok), theme shops (from The Museum Company Store to a Yankees Clubhouse), high-tech games arcades, a new Hilton hotel, an HMV music store, and two big multiplex movie theaters. All of these come with their own, often restlessly flickering signs – some of these banal and garish, others scintillating (the exceptionally bright, ever-changing Loews sign), and one of them aesthetically subtle even and in its quiet, constantly shifting way quite magical (the phantasmagoric light effects devised by Anne Militello to play along the louvered facade of the rehearsing-space glass tower which now tops the entrance to the American Airlines Theatre, also singled out for architectural praise by Goldberger 2000, pp. 90–92).

The "Disneyfication" of Times Square: Back to the Future?

Fig. 1. Rising up above the new Times Square (and viewed through the construction site for a third new skyscraper, designed by the firm of Kohn Pedersen Fox): the Reuters Building (left) and Condé Nast Building (right), both by the architectural firm Fox & Fowle. August 2000. Photograph by the author.

Fig. 2. With construction works on the south side of the redeveloped 42nd Street largely finished, visitors are greeted by a barrage of colorful signs and billboards, several of which put national and international retail chains on the Times Square map. August 2000. Photograph by the author.

BUSINESS IMPROVEMENT DISTRICTS AND THE POSTINDUSTRIAL SERVICE ECONOMY

"Come take a look at the New Times Square. It's a community changing every moment. It's safer. It's cleaner. It's brighter. And it's more exciting than ever." This is the clarion call with which Internet surfers were until recently welcomed to the website of the Times Square Business Improvement District, established in 1992 by the 404 property owners in the area.[5] Business Improvement Districts or BIDs are non-profit organizations set up by New York State Legislature in 1983 (Zukin, 1995, pp. 33–38). The principle of BIDs is relatively straightforward: property owners in a specific district pay a mandatory assessment collected by the city and returned in full to the local BID. This system has provided the Times Square BID with an annual budget of some $6 million, with which it can then get to work on its various tasks. These include relatively uncontroversial activities like administering tourist services: in 1998, for example, the BID opened a Visitors Center in the restored and landmarked Embassy Theatre.

More conspicuously and importantly, however, the Times Square BID has also assimilated some of the major services ordinarily performed by the city itself. Thus, the BID employs 50 sanitation workers in cherry-red jumpsuits (wielding, among other things, the redoubtable Frankensteamer, a hot-water, high-pressure steamcleaner for scrubbing sidewalks and removing graffiti). The eye-catching cherry-red suits were specially designed by an award-winning theatrical costume designer. The constantly visible sanitation workers, moreover, are supplemented by a no less visible team of 45 public safety officers, a fully trained (albeit unarmed) private security force that is linked by radio to the NYPD. In addition, the BID prides itself on being an active supporter of the Midtown Community Court, a nationally recognized court established to handle misdemeanors swiftly, locally, and visibly by assigning offenders to community service in the area and social services in the Court. To put the procedure at its most graphic extreme: minor offenders in and around Times Square are promptly picked up, publicly tried in a former theater turned courthouse, and given a broom to sweep the sidewalks – a practice that Sharon Zukin (1997, pp. 37–38) has understandably labeled "elitist" and, more caricaturally, "worthy of Dickens."

The power of BIDs may be taken as in some sense symptomatic for the redevelopment of urban downtowns in the 1990s. A few years ago, already some 1,200 of these BIDs were active all over the U.S., and the phenomenon has begun to spread to Europe, the Caribbean, Australia, and South Africa (Hannigan, 1998, p. 139). The success of BIDs thus epitomizes a growing tendency, begun in the 1980s, to privatize and source out the management of commercially prosperous parts of public space. It simultaneously highlights what we may call the hegemony of a growth- and business-oriented economic logic in urban politics: these are quite aptly called *Business* Improvement Districts, not *Neighborhood, Community, Arts,* or *Culture* Improvement Districts, and the outreach program for the homeless which the BID also trumpets appears to have been an afterthought: it was launched only five years after the BID itself was established.

To understand the rise of BIDs, we need to zoom out for a moment and rehearse the macroeconomic story of postwar American cities. It is the story, basically, of how the industrial metropolis lost its manufacturing base in the course of the twentieth century and turned into a so-called postindustrial metropolis rooted in consumption more than production. In order to survive, high-profile cities like New York have had to tap into other economic opportunities, principally those offered by a service and information economy: they have come to capitalize on their function as nerve centers for the worlds of finance, banking, insurance, and real estate; on their aura of cultural producers and art capitals; on the services offered by the restaurant, fashion, entertainment, and media industries; and on their touristic potential. For New York City, and especially Manhattan after the annexation of its four surrounding boroughs (Bronx, Queens, Brooklyn, and Staten Island), this process was already well under way in the late nineteenth century. Already by the turn of the twentieth century, Manhattan had established itself as the nation's foremost commercial hub and clearing-house that provided the necessary services for its industrial hinterland (Taylor, 1992, p. 96). The very emergence of Times Square as the heart of the country's advertising, media, and entertainment industries is itself indistinguishable from the early rise of a service and information economy. Yet in the second half of the twentieth century, arguably, the production of images upon which this economy depends has massively gained in importance and scope, so that, as sociologist Sharon Zukin (1995, p. 7) has argued, a "symbolic economy" devised by "place entrepeneurs" has acquired a central function. "What is new about the symbolic economy since the 1970s," writes Zukin, "is its symbiosis of image and product, the scope and scale of selling images on a national and even global level, and the role of the symbolic economy in speaking for, or representing, the city" (p. 8). Today, the identity of places in a city is established above all by "sites of delectation" tailored to tourists and middle- to upper-income spenders (p. 9). As David Harvey in *The Condition of Postmodernity* (1989, p. 92) notes: "imaging the city through the organization of spectacular urban spaces became a means to attract capital and people (of the right sort) in a period (since 1973) of intensified inter-urban competition and urban entrepreneurialism." The Times Square area currently draws between 20 and 30 million tourists every year. And whatever our objections to current evolutions might be, the touristic success has at least been undeniable: the occupancy rate of hotel rooms in New York City (one fifth of which are in the Times Square area) was 83% for 1998, year-round, with the average room rate reaching an all-time high of $192 a night. No less tellingly, the annual income per hotel room (after expenses) has multiplied from $9,179 in1991 to $32,008 only six years later. New York hotels today are three times more profitable than the average hotel in the U.S. (Holusha, 1999).

"In the late 1990s," explains sociologist John Hannigan (1998, pp. 1–2), "nearly every major multinational entertainment company has established a development team to evaluate, plan and initiate urban entertainment destination (UED) projects. . . . These developments are indicative of a new urban economy which has its roots in tourism, sports, culture and entertainment." In his thoroughly researched study of today's "Fantasy City," Hannigan defines the postmodern metropolis of pleasure and profit in terms of six central features, all of which prove to be central also to the changing status of Times Square. First, writes Hannigan (1998, p. 3), today's successful Fantasy City consists above all of a *themed* environment in which "everything from individual entertainment venues to the image of the city itself conforms to a scripted theme, normally drawn from sports, history or popular entertainment." Secondly, it is "aggressively *branded*" or name-oriented: for those who finance and market these developments a lot depends on the "potential for selling licensed merchandise on site" (p. 3). Thirdly, unlike the suburban shopping mall (which appears to be past its prime), Fantasy City "operates *day and night*," which "reflects its intended market of 'baby boomer' and 'Generation X' adults in search of leisure, sociability and entertainment" (p. 3). Fourthly, it is highly *modular*, that is to say, it consists of "an increasingly standard array of components in various configurations. Typically, an UED project will contain one or more themed restaurants (the Hard Rock Cafe, Planet Hollywood, the Rainforest Cafe), a megaplex cinema, an IMAX theater, record (HMV, Virgin, Tower) and book (Barnes & Noble, Borders) megastores, and some form of interactive, high-tech arcade complete with virtual reality games and ride simulators" (p. 4). Fifthly, Fantasy City is *solipsistic*, a (somewhat infelicitous) hyperbole by which Hannigan wishes to indicate that it is "isolated from surrounding neighborhoods physically, economically and culturally" (p. 4). And finally, it is *postmodern* "insomuch as it is constructed around technologies of simulation, virtual reality and the thrill of the spectacle" (p. 4).

Today, every one of those six features applies to Times Square. In less than a decade, the area has risen from the position of seedy and sleazy underbelly of the city to the prime symbolic embodiment, together with Las Vegas, of this

The "Disneyfication" of Times Square: Back to the Future? 393

Fig. 3. One of the fifty omnipresent sanitation workers (in the specially designed red jumpsuits) who work for the Times Square Business Improvement District. August 1998. Photograph by the author.

Fig. 4. To be able to use the landmarked Empire Theater as a lobby and mezzanine to AMC's 25-screen multiplex on the south side of 42nd Street, the entire theater had to be lifted up, put on wheels, and rolled 170 feet to the west of its original location. August 2000. Photograph by the author.

The "Disneyfication" of Times Square: Back to the Future? 395

Fig. 5. To advertise an exhibition at the Whitney Museum in 2000, the west side of the new Hilton hotel (which turns its blank walls on an as yet undeveloped parking lot awaiting future redevelopment) was decorated by a gigantically enhanced art work by Barbara Kruger reminding us that "It's a small world (but not if you have to clean it)." In its 42nd Street location, the saying seemed fraught with ironies and ambiguities. August 2000. Photograph by the author.

Fig. 6. A shot of the recently built hotel-and-casino in Las Vegas which themes on New York and its skyline, the symbolic counterpart to a Las Vegas theme restaurant on the new 42nd Street. July 2000. Photograph by Steven Jacobs.

American Fantasy City. From a place no honorable entertainment company or media corporation wanted to be associated with, it has evolved into the number one site for corporate self-promotion. A roll-call of power players in the area makes this abundantly clear. Among the investors who have already arrived are the Disney Company, Warner Brothers, Viacom, Virgin (with its largest megastore in the world), HMV (with another music store), Bertelsmann (the biggest publishing company in the world), two media giants already mentioned, Condé Nast and Reuters, MTV (with a second-floor street-front studio drawing huge crowds whenever a show goes live on the air), ABC Television (with conspicuous videoscreens, multicolored bands of LED messages, and yet another street-front studio from which it broadcasts its live show *Good Morning, America*), Livent (the Canadian producers of *Ragtime*, *Showboat*, and other recent Broadway hits), Madame Tussaud's of London (with a new high-tech wax museum), American Multi-Cinemas (AMC) with a 25-screen movieplex, and Sony with another 13-screen multiplex (run under the banner of Loews). There is a new Hilton hotel, an Official All Star Cafe (a themed restaurant owned by celeb athletes like Tiger Woods, Andre Agassi, and Shaquille O'Neal), a 53-story hotel by Planet Hollywood (a chain owned by people with names like Arnold Schwarzenegger, Demi Moore, and Bruce Willis), a magic-theme restaurant devised by the Disney Company in collaboration with David

Copperfield (which should contain levitating tables and disappearing diners), and finally, perhaps most symptomatically of all, a Las Vegas theme restaurant, which brings the exchange of symbolic capital between America's prime urban entertainment destinations full circle, since today we also find a $460 million hotel-and-casino in Las Vegas that depicts a compacted and dwarfed version of the New York skyline.

IT ALL COMES TOGETHER IN THE DISNEY COMPANY

In almost no time, the "Disneyfication of Times Square" has become one of the most hackneyed clichés in journalistic and academic discourse. Predictably, the phrase has been used mostly as a slur. By the same token, it often threatens to become empty of content. Yet it is a phrase worth analyzing and critically unpacking, for it helps to tie together several phenomena that are crucial to the history not only of Times Square but also of other central symbolic sites in American (and even European or Asian) cities. "What's in a name?" asks Juliet of her fatally misnamed Romeo. "That which we call a rose / By any other name would smell as sweet" (*Romeo and Juliet*, II.ii, 43–44). But as Juliet is tragically made to realize, there is a difference between the natural realm of roses and the cultural realm of society. Adamic acts of arbitrary naming are no longer simply available once we enter the symbolic order of culture. What's in the name "Disneyfication"? A whole tangle of possible meanings for which there is no immediate substitute, it would appear.

Perhaps we should start by pointing out what the term clearly should *not* be taken to mean and what it is nevertheless all too easily made to suggest: that the Disney Company is the major investor and power player – even the mastermind – behind the transformation of Times Square in the 1990s. To put the record straight: the redevelopment of Times Square was launched and steered in the first place by municipal and state authorities in a decision-making process that reaches back all the way to the fiscal crisis of the mid-1970s. To this crisis a number of business elites and pro-business politicians responded by focusing on the redevelopment of economically strategic and largely unresidential sites with a high symbolic capital.[6] With the urban renewal projects of the 1950s and 1960s increasingly contested and discredited, local officials and developers in the late 1970s began to realize the possibilities of a new, ostensibly more cultural strategy based on historic preservation and the arts. The Times Square area appeared to offer one of the choicest locations for massive office construction that could be sold to the public by linking investments directly to the restoration of a number of historic Broadway theaters and the cultural life that

Fig. 7. A view from the third floor of the Warner Bros. Studio Store in One Times Square (the former New York Times Building) at the crater that replaced the Rialto Theater in 1998 to make room for the Reuters Building and at the competing merchandise store across the street operated by Disney. August 1998. Photograph by the author.

went with them. To be sure, the tangle of organizations (from publicly controlled to mixed public-private) that have steered the redevelopment efforts begun in the late 1970s is complicated, but it does not include the Disney Company: it ranges from the 42nd Street Development Corporation established in 1976 and the New York State Urban Development Corporation (which entered the field in 1980 and installed the 42nd Street Development Project), to New York City's Board of Estimate (BoE), its Department of City Planning (DCP), its Landmarks Preservation Commission (LPC), its Planning Commission (CPC), and its Public Development Corporation (PDC), as well as The New 42nd Street, Inc. (New 42) and the Municipal Arts Society (MAS).[7] It is these organizations that have set design rules, provided special tax breaks and incentives, clinched deals, overlooked plans, and entered into the unavoidable wheeling and dealing with private investors. The most important among such private investors, moreover – those to have pumped the really large sums of money into the area – were powerful real estate developers, who in the new entertainment complexes on 42nd Street currently under construction or just finished have names like Park Tower Realty, Forest City Ratner, and Tishman Urban Development.[8] In terms of touristic boostering and the active management of streets, as we saw, the area has been in the hands of the local Times Square BID. It is only as one of the numerous multinational media and entertainment companies listed above, then, that the Disney Company has been adding its own share to private investments in the area.

And yet the name of Disney stands out and resounds most strongly among this densely packed field of political authorities, non-profit organizations, investment partners, and commercially competing players. If the shorthand term of "Disneyfication" must in some sense be seen as a reductive and obfuscating hyperbole, its use is nevertheless warranted in at least eight different ways. More than in any single meaning, it is in this unique and impressive octuplicity (so to speak) that the strength of the term resides.

(1) To start with the most obvious and visible level first: at the southwest corner of 7th Avenue and West 42nd Street, across from the glamorously redesigned Times Square subway entrance that spawns millions of visitors a year and strategically located as a kind of fulcrum between the bowtie and the new entertainment strip of 42nd Street, the Disney Company in 1996 opened one of its quickly multiplying Disney Stores. (Four years later, the store was temporarily razed again to make room for the construction of another imposing skyscraper, but it is scheduled to return at the streetlevel of this new behemoth.) The general success of Disney's retail stores has been such that a few years ago there were already more than 500 of them in over eleven different countries (Hannigan, 1998, p. 93). The significance of this particular flagship thus transcends its particularity: the store is itself illustrative of the kind of synergistic "shopertainment" strategies, on-the-spot sales of merchandise, and general image-building that entertainment companies like Disney have devised to expand their markets and raise their profits.

(2) What the Disney Store with its world-famous red letters mimicking Walt Disney's personal signature has been out to achieve for the company's visual image, the refurbished New Amsterdam Theater next door has sought to do for the company's symbolic and cultural capital. Built in 1903 and originally home to the legendary Ziegfeld Follies that contributed a large share towards the popular success of Times Square in the early decades of the century, the Amsterdam Theater, like so many theaters on this block of West 42nd Street, had devolved over the years into a seedy movie house, been closed in the early eighties, and fallen into such disrepair that only a major renovation could save the building (Weber, 1997). When the Disney Company decided to look for a stage for another of its profit-enhancing synergistic expansions (the production of live musicals based on animated

movies), it was very actively wooed by local officials and by the interim redevelopment plan's architectural mastermind, Robert Stern (about whom more in a moment), who did their utmost to persuade the family-entertainment company to land on 42nd Street. Officially started in 1993, the wooing was done at a time when the megaproject for 42nd Street launched under the Koch administration in the early 1980s had completely stalled. Aware of this and of its own symbolic power, Disney preferred not to act the Cinderella but instead played hard-to-get, extracting a deal from project officials that greatly exceeded what other developers under similar circumstances had been able to obtain (Reichl, 1999, pp. 158, 211). Among other things, the company found itself in a position to demand that at least two other entertainment companies that met with Disney's approval would enter the redevelopment of 42nd Street. These turned out to be AMC (with its 25-screen multiplex, excellent for the dissemination of Disney's own film products) and Madame Tussaud's (with its highly Disneyesque venture of memorializing famous people in wax form) (Reichl, 1999, p. 158). The deal was finally clinched and Disney forked out $34 million to restore the Art Nouveau building of the Amsterdam – "the crown jewel of all the theaters in the city" according to the chairwoman of the city's Landmarks Preservation Commission (quoted in Reichl, 1999, p. 157). The New Amsterdam opened in 1997, garnered encomiums from the general public and architectural critics alike (see e.g. the 2000 edition of the *AIA Guide to New York City*), and has been playing to capacity audiences since.

The restoration of the New Amsterdam, like the opening of the Disney Store, was of more than local and anecdotal importance, for as soon as the Disney Company had made its commitment, other investors from the entertainment and media industries came flocking to what was now marketed as the New 42nd Street. Accepting to restore the theater made the family-entertainment company serve as the torch-carrier and final trigger for the whole interim redevelopment plan devised by the mid-1990s. This symbolic weight was arguably still increased by the fact that the project of restoring an old building, as architectural critic Paul Goldberger reminds us, was "really quite un-Disneyesque" (quoted in Hannigan, 1998, pp. 6–7). That the Disney Company should ever have come to consider, leave alone undertake, such a venture, and have met with such wide public approval for doing so, is further evidence for Alexander Reichl's contention (1999, p. 22) that "Historic preservation (and adaptive reuse) is ... the one identifiable and coherent principle guiding urban development in the postmodern era." Disney's successful restoration of the Amsterdam Theater came at the end of some two and a half decades in which preservation activists in American central cities had steadily gained power, tax policies had come to favor preservation-based commercial development, and the built environment in a city like New York had come to be protected and solidified more than ever before (Reichl, 1999, chapter 2).

(3) The goodwill gained by the Disney Company was translated into an important cultural precedent. On June 14, 1997, on the occasion of the world premiere of its animated film *Hercules* in the New Amsterdam, the Disney Company was given the permission to stage a high-octane *Hercules Electrical Parade* along West 42nd Street and up Fifth Avenue. The occasion was a historic first in two ways: never before had a purely commercial parade without any historic or ethnic constituency in the city been organized in New York; and never before had the famous parade of Disney characters been shown outside Disneyland (Purdy, 1997). The effect of this mass spectacle on the public perception was as simple and straightforward as it was intentional: by rolling its allegorical muscles in the figure of Hercules, the Disney Company was demonstrating its presence and centrality in the new Times Square and, by extension, the power of its vision on the reconstruction of American city centers. A quintessentially suburbanophile and family-oriented entertainment company in origin, Disney had finally invaded America's most urban heart and was marching its mock-victorious troops into what it called "the media capital of the world" (Purdy, 1997). For the theoretically reflecting observer, it did not take much to read this spectacle as a classically postmodern and somewhat undecidable mix of half-serious ironies, hyperreal simulations, and self-promotional myth-making.

(4) While the Disney Company's immediately visible presence in the Times Square area has been restricted thus far to its merchandise store and the restored New Amsterdam, there are other plans on the table as well as indirect ways in which the company can be found on the scene. At the northeast corner of the redesigned 42nd Street and 8th Avenue, for instance, at the end of the redeveloped block of new complexes, there is today an internally related network of buildings and buildings-under-construction in which the company has been involved as one of the main developers. A large retail and entertainment complex devised by Disney and its associates (called E Walk) serves as the flashy base of a spectacular, still-to-be-constructed 47-story hotel designed by Arquitectonica. This hotel will not only use laserbeams to simulate the striking of a meteor on its outside but will also provide room for 100 time-share apartments operated under Disney's Vacation Club (Reichl, 1999, p. 5, 159). More centrally located in the Times Square bowtie itself, meanwhile, at Broadway and 43rd Street,

we find the new iconic markers of ABC Television, with its second-floor street front studio, its overarching and jutting-out video displays, and its multicolored, wavily sculpted bands of LED messages. This promotional complex offers one important *indirect* illustration of the many ways in which the Disney Company has invaded the area, for ABC Television itself has for some years now been owned by Disney (and, not so indirectly, does not fail to promote some of the mother company's products on the big videoscreens).

(5) With Universal and Sony, the Disney Company has become one of the three key players in the American entertainment industry today. Its name may thus also stand as a synecdoche for the role of UED projects in the revitalization of metropolitan downtowns already indicated by John Hannigan. Yet Disney is also more than that. Contrary to its two main competitors, the company has had a strong land development connection for many years. "The Disney Company is America's urban laboratory," as a journalist once noted (quoted in Zukin, 1995, p. 55). A bird's-eye history of the company is in place here, and it should start with the father figure of Walter Elias Disney himself, the tirelessly self-promoting, darkly obsessive man who invented Mickey Mouse and created an instant sensation when he launched his invention in the first talking cartoon in 1928. "Cocky, and in his earliest incarnations sometimes cruelly mischievous but always an inventive problem solver, Mickey would become a symbol of the unconquerably chipper American spirit in the depths of the Depression" (Schickel, 1998, p. 76). For almost three decades, Walt Disney devoted himself to founding Hollywood's animation industry and turning his name into a business empire. By the 1950s, however, he threw himself with equal zeal on bringing his two-dimensional cartoons alive in three-dimensional form. In 1955, Disneyland was opened in Anaheim, California. Its impact on the postwar American landscape has been so enormous that to some commentators it even eclipsed Disney's earlier achievements. Thus, Sharon Zukin (1991, pp. 221–222) claims that Disney's "real genius was to transform an old form of collective entertainment – the amusement park – into a landscape of power. All his life Disney wanted to create his own amusement park. But to construct this playground, he wanted no mere thrill rides or country fair: he wanted to project the vernacular of the American small town as an image of social harmony. 'The idea of Disneyland is a simple one. It will be a place for people to find happiness and knowledge,' Disney said." In the middle of the Cold War era, Disney wanted to pacify and reassure people, offering them a landscape of escape, a dreamscape into which they could collectively flee. The idea, as the cultural historian Mike

Wallace (1996, p. 137) notes, was emblematic of larger developments in fifties America: "The dominant culture, seemingly determined to come up with a happy past to match its own contented present, contracted a selective amnesia. Leading academic historians downplayed past conflicts and painted optimistic, even uplifting pictures of the American past." Disney revitalized and reformulated the turn-of-the-century urban amusement park, "a form that beer magnates, real estate developers, and transportation kings had fashioned for the urban working classes in the 1880s and 1890s, and whose culmination was Coney Island. Disney's park was a cleaned-up version, aimed at a middle-class 'family' audience. He quite consciously stripped away the honky-tonk legacies of the carnival. . . . Disney's park erased any lingering traces of rituals of revolt and substituted highly organized, commodified fun" (p. 141).[9]

The popular success of this first postwar American theme park was immediate. "Disneyland succeeded on the basis of [its] totalitarian image-making," claims Zukin (1991, pp. 222–223), "projecting the collective desires of the powerless into a corporate landscape of power. In this way it paralleled the creation of a mass consumption society. . . . Disney's business growth also related to important processes of change in the larger society: notably, the demographic growth of the baby boom, the spread of television, and the increase in domestic consumption. . . . [It] coincided with the expansion of the suburbs and a population shift to the Southwest, the growth of the service sector, and a boom in leisure-time activities, including sales of recreational land and travel." The popular and business success inspired Disney over the years to build a second amusement park, Disney World, near Orlando, Florida. In Disneyland, he had looked back nostalgically to the past, rebuilding the heavily idealized Main Street of his youth on a psychologically shrewd three-fifths scale. In Disney World, by contrast, he looked to the future and boldly sought to go where no-one had gone before. He wanted to build the Experimental Prototype Community of Tomorrow, a.k.a. EPCOT. As Disney himself defined this project:

> It will be a city that caters to the people as a service function. It will be a planned, controlled community, a showcase for American industry and research, schools, cultural and educational opportunities. In EPCOT there will be no landowners and therefore no voting control. No slum areas because we will not let them develop. People will rent houses instead of buying them, and at modest rentals. There will be no retirees. Everyone must be employed (quoted in Zukin, 1991, p. 224).

In Disney's conservative utopian vision, EPCOT was to be a laboratory city with twenty thousand people in it. According to Mike Wallace (1996, pp. 142–143), the vision derived its ambition to a large extent from the fact that

Disney "had been praised extravagantly as an urban planner. James Rouse, master builder of new towns and historical shopping malls modeled on Main Street (Boston's Faneuil Hall, Baltimore's Harborplace, New York's South Street Seaport) told a 1963 Harvard conference that Disneyland was the 'greatest piece of urban design in the United States today.' Architectural critic Peter Blake called the Anaheim park the only significant new town built in the United States since World War II – 'staggeringly successful' – and suggested, only half-humorously, turning Manhattan over to Disney to fix up. All this went to Walt's head, and he flowered into a utopian capitalist."

In the end, EPCOT was not built as Disney conceived of it. Its inventor and would-be absolute monarch died of cancer in 1966, still in the middle of the planning stage. His corporate heir, Walter Elias Disney (WED) Enterprises, Inc. (later to be renamed the Disney Company), backed out and, fearing especially the legal responsibilities, shelved the idea of building a living city with actual people in it. EPCOT was cut down to a technological theme park, a showcase for its multinational corporate sponsors. Despite this adaptation, however, Disney's town planning ideas had already managed by then to establish "a new vernacular image of a postmodern society" (Zukin, 1991, p. 230). As Diane Ghirardo in *Architecture After Modernism* (1996, p. 46) indicates, the Disney model began to exert a major influence on smaller cities across the U.S.: "the approaches to public space, work space, and urbanism embodied in Disneyland and its successors came to appeal both to developers and to architects as a standard against which to assess buildings and public spaces." Succinctly defined, this means that urban environments were adapted to a template that was originally that of a privatized, consumption- and spectacle-oriented theme park intended to combine several ideological functions: to simulate a shared middle-class family culture, to aestheticize social differences, to harmonize history, and to offer a reassuring, fully controlled and preprogrammed environment without arms, alcohol, drugs, or homeless people (Zukin, 1995, pp. 52–55). As one of Disney's "imagineers" confessed: "What we create is a 'Disney Realism,' sort of utopian in nature, where we carefully program out all the negative, unwanted elements and program in the positive elements" (quoted in Wallace, 1996, p. 137). Importantly, the company's "imagineers" (a term coined by Disney himself) came to spread this view on urban planning outside their mother company as well, since several of them made lucrative transfers to other entertainment and real estate companies and projects (Hannigan, 1998, p. 4).

When in the late 1970s, early 1980s, the downtowns of America's bigger cities started to crawl back to their feet after decades of deterioration, the influence of the Disney model began to make itself felt there as well. As Mike Wallace already indicated, the successful inventor of "festival market places,"

James W. Rouse, drew considerable inspiration from Disney's example in creating his consumerist urban pockets of redevelopment organized around deftly manipulated historic themes. The relative success of these festival market places in terms of popular appeal (more than in economic terms) fired off and influenced various types of urban revitalization in the 1980s and especially 1990s, one of which happens to be Times Square. The Disney Company itself, however, jumped on this bandwagon of central-city redevelopment only relatively late. As an idealizer of small-town America with the largest following in middle-class family suburbia, the Disney Company was long hesitant to enter urban downtowns and only began to make the move by the early 1990s. What is more, the company itself had had to effectuate its own revitalization first. In the early eighties, it had been languishing and come under siege from raiders attempting hostile takeovers that caused the company's stocks to plummet. The eventual opening of the fully achieved high-tech paradise EPCOT in 1982 helped put Disney back on the Wall Street map, and a change of management two years later did the rest. Three Hollywood heavyweights were brought in: Frank Wells was piloted in the presidential chair; 31-year-old Jeffrey Katzenberg revamped the Disney film studio and produced a series of blockbusters; and Michael Eisner became the company's CEO (Wallace, 1996, p. 161). Unlike "Uncle Walt," known to have been an anti-Semitic Republican supporter, Eisner was a Jew from New York and a Democrat (who would actively come to support Bill Clinton in both his presidential bids). He decided to break the moulds set by the company's founding father and initiated a major diversification operation, recycling the company's film materials into other commodity forms: from home videos, toys, and merchandise, to video games, vacation clubs, a touring troupe, a publishing firm, a Disney Channel, a hockey team, and a cruise line, all produced under the banner of the Walt Disney Company (Wallace, 1996, pp. 161–162). In no time, the company was booming again and by the end of the 1980s it could envision participating in the rebirth of city centers as well. Not that the portents for this participation were entirely positive: "in its first and most extensive foray into urban planning on non-company property," John Hannigan (1998, p. 8) informs us, "Disney consultants proposed (unsuccessfully) a redesign for the civic center in Seattle which would have made an admission fee inevitable." What drew the company eventually into Times Square was, as we saw, the promise of live theater. By the 1990s, the possibility of a company-owned Broadway theater seemed like another profitable spin-off for recycling old and new movies. In addition, the newest technological innovations "meant that sophisticated special effects could be compressed in time and space; making them more amenable to downtown sites where land was at a premium" (Hannigan, 1998, p. 60).

When local officials and political authorities involved in the Times Square redevelopment plan began to court Disney's CEO to make his company serve as the new symbolic anchor on 42nd Street, they were able to make good use of one supporter in particular: Robert A. M. Stern, the urban historian and practising architect recently named Dean of the Yale School of Architecture. Stern had just been officially appointed to lead the design team for the interim redevelopment plan for Times Square in the early 1990s. His personal ties with Michael Eisner were already strong at the time. Not only did Stern draw up the masterplan for Celebration in Florida (a gated community recently built by Eisner in an attempt to pick up Walt's dream of building an ideal living city),[10] but he was also on the board of directors of the Disney Company, and had personally built a private house for Eisner himself (who has become a major patron of big-name American architects; see Wallace, 1996, p. 173; Dunlop, 1996). At some deep level, then, there *is* a connection between the community-building and town-planning ideals of the Disney Company and the masterplan for Times Square currently taking shape by the minute. But it is at most an *indirect* connection and a connection with the Disney Company *New Style*, which would appear to be an ideologically more progressive (or at least more flexible) institution, and we should not jump to simplistic conclusions.[11] Walt Disney's original "friendly fascism," his mania for control, his radical escapism into an infant's dream world, his harmonizing melting-pot ideology, and his suburban fear of life in the fast lane and on the dangerous edge are not all being simply transplanted onto the new 42nd Street and Times Square. Robert Stern was quick to announce that he was "not doing Disney," that he did not want to make the new 42nd Street "so gentrified that there is no sleaze or sensationalism," and that the area historically requires "a little sense of threat, excitement, derring-do – a sense of adventure" (quoted in Reichl, 1999, p. 151). As the first epigraph to the present essay illustrates, Stern's knowledge of the area's history and awareness of the increasing contestation of the earlier megaproject inspired him to represent his plan as "the restoration of New York's quintessential entertainment district, our most democratic good-time place. The renewed 42nd Street will be an enhanced version of itself – not a gentrified theme park or festival market" (quoted in Reichl, 1999, p. 152). At the same time, however, a certain Disneyesque rosiness and selectivity could be discerned in the way the area's history was being represented, for it relied above all on what Alexander Reichl (1999, p. 153) has called the "cultural imagery of good, clean, honky-tonk fun – Times Square à la 1950s Americana."[12] The picture is thus a complex one. On the one hand, the interim plan has clearly resisted the tendency of modeling itself too strongly on a Disneyfied theme park, and the Disney Company has itself been changing ideologically together with the world in which it lives and the new enterprises in which it engages. On the other hand, a term like "Disneyfication," for want of a more neutral one, does remain particularly grateful for accommodating criticisms both on the impact of consumption- and spectacle-oriented private entertainment companies in shaping the streetscape and on the sanitizing, socioeconomically upscaling effects of a clean-up operation that has for instance resulted in the removal of homeless people and other sidewalk undesirables, or in the crackdown on the sex industry in the area (see also Weber, 1997). And there are still more ways in which the term has a certain legitimacy.

(6) If, as Fredric Jameson (1996, p. 14) has argued, the past decades have seen the rise of "a new 'society of the image' in which consumerism and market frenzy are not the issue so much as consumption by the eyes," then the Disney experience, too, may stand as an emblem of this *"concupiscio oculis."* In *Collage City*, Colin Rowe and Fred Koetter (1978) described Disney's Main Street as a "machine for the production of euphoria" (p. 46) and situated it in a tradition of townscaping which tries "to provide sensation without plan, to appeal to the eye and not to the mind, and, while usefully sponsoring a perceptual world, to devalue a world of concepts" (p. 36). This kind of 1970s critique may bespeak its age somewhat in setting up an antagonism that threatens to diminish the intrinsically seductive and gratifying powers exerted by the senses. Yet one may with some justification wonder about the spectatorial regime produced in Times Square today and how it compares to that stimulated in places like Disneyland and Disney World. We know how the rides in Disney's parks are consciously designed to preclude social intercourse: the scopic focus is systematically pulled towards the spectacular stage sets and props, and visitors are actively discouraged from noticing – much less touching or talking to – other visitors (see Hannigan, 1998, p. 198). What is more, Disneyland and Disney World are not places you go to explore, since "individualized itineraries disrupt the standardized visitor flows which are essential to their smooth and efficient operation" (Hannigan, 1998, p. 82). "At Disney World," notes Sharon Zukin (1991, pp. 227–228), "the solar-powered 'traveling theater cars' that propel blocks of sixty visitors through the sights, sounds, and smells of the exhibits literally hold the public captive to the image. The sensation of consuming has become more intense, but also more ambiguous, so that visitors to Disney World feel that while they can control technology, they are also being controlled." No such strenuous control is exerted in a public streetscape like Times Square, yet the emphasis on a one-sided visual

consumption of spectacles and on crowd control does seem to be growing, which in turn facilitates the association with a visit to a Disney theme park.

The problem with a Foucauldian analysis of the ways in which underlying power structures shape the disciplinary practices in and around Times Square is that it can only be handled in a "wet," mostly unverifiable form – the kind of form Clifford Geertz has called "thick description." Nevertheless, it is a measure of the popular and touristic success of the revamped Times Square that we are reminded more and more of the area's heydays in the 1920s, when the daily practice was to keep the crowds moving as much as possible (see Sandeen, n.d., p. 5). Around the time of my latest visit, in the summer of 2000, Mayor Rudolph W. Giuliani announced that the sidewalks in the area would be widened to counter increasing pedestrian congestion. The continuous flow of people, in combination with the constant monitoring of loitering by-standers by private-security and police patrols, may incline visitors to keep walking on cruise control. A visit to the area would thus prevent many individuals from standing still and either reflecting upon the place or seeking out contact with strangers. Certainly, the streetscape – brighter and more restless than ever and increasingly dominated by giant videoscreens – invites a constant visual immersion that seems interrupted or marked only by those moments when visitors are sucked away from the endless pedestrian flow into megastores, interactive virtual-reality arcades, movie multiplexes, or theaters. In addition, the omnipresence of the conspicuously attired and equipped sanitation workers exerts its own surreptitious mental corrections on passers-by. The cherry-red outfits, in combination with the highly visible as well as symbolic activity of cleaning up, have understandably caused critics to wonder about the Disneyesque tendency to associate theming, style, and bright colors with the idea of social order (Zukin, 1995, p. 66).

Against such criticisms, however, it should be said that there is no apparent, direct pressure on pedestrians to keep on moving: shoppers and tourists do stop and gaze, take pictures, and cut out their own itineraries, and this may well be more often the case now than it used to be one or two decades ago, since the atmosphere in and around the square, and certainly along 42nd Street, has become noticeably more relaxed. One important subconscious determiner of this more laid-back attitude could indeed be olfactory: the smell of the place nowadays is more likely to be that of honey-roasted nuts or incense sticks than of the earlier urine-stained pants and rotting garbage. At the same time, the current Times Square is (or remains) not a place for hanging around and socializing but for no-holds-barred impression management. It invites visual consumption, but not social exploration. It does not, in other words, seem to serve that quintessentially urban role of stimulating serendipitous encounters and hardly acts as the kind of public space where, in the French thinker Roland Barthes's utopian words (1967, p. 96), "subversive forces, forces of rupture, ludic forces act and meet." In the late 1960s, Barthes loved to depict city centers as spaces of erotic play and of a constant confrontation with, and embodiment of, the other. Although a soft form of erotic play and a confrontation with the other are both what Times Square admittedly still provides, these urban attractions are characteristically made available now at one remove from immediate, involved contact: they tend to be displayed as the objects of advertising images *about* such urban qualities, the virtualized ideological subject matter for publicity and corporate images that bears little or no relation to the interaction of people at streetlevel or in the intimacy of local buildings. Times Square as a *publi-city* does not do much to promote a more *public city* (the opposition derives from Rutheiser, 1996, p. 10). To what extent recent redevelopment efforts have indeed "sacrificed the provocative, raw energy produced by the friction of different social groups in close interaction for the stultifying hum of a smoothly functioning machine for commercial consumption" (Reichl, 1999, p. 179) is a question I will need to come back to.

(7) Yet another phenomenon to which the expression "the Disneyfication of Times Square" may be taken to point is that of a macroeconomic trend in those multinational media and entertainment industries which have collectively come to print their stamp on the area's outlook. The term "Disneyfication" then stands in for the phenomenon of megamergers, takeovers, and synergies typical of economic thinking in the 1990s. As John Hannigan (1998, pp. 61–62) explains: "In a series of mergers and takeovers which stunned financial analysts in the mid-1990s, Walt Disney acquired Capital Cities/ABC in 1995 for $19 billion, thereby becoming an owner of one of the original three television networks; Viacom bought Paramount Communications Inc. and, with it, Blockbuster Entertainment in 1994 for a combined $17.4 billion giving it control of a variety of entertainment properties: video rentals, a movie studio, a book publisher and MTV, the music video channel; and Seagram, the Canadian liquor and orange juice giant, bought MCA for $5.7 billion thereby acquiring a movie studio, a theme park and a theater chain." This trend is continuing to this very moment. More recently, for example, Viacom closed an even bigger deal with CBS, which it acquired for $37 billion, thus propelling itself to the rank of third largest media-entertainment conglomerate, behind Time Warner and the Disney Company (Hillenbrand, 1999). "The deal reflects a decade-old business doctrine that size does matter in the media and

entertainment industries. Megacompanies like Viacom [or Disney, or Time Warner, which, in the biggest deal of all, has been seeking to merge with AOL] are able to weave intricate product tie-ins and spin-offs across markets and media. Thus a best-selling book begets a movie, which begets recordings, which beget an Internet website – all of which is promoted on television and radio outlets until the film is endlessly rerun on cable systems and flogged in neighborhood video shops around the world. The consumer can hardly escape the hype and the company can hardly fail to see the profits roll in" (Hillenbrand, 1999). Understandably, this trend is viewed with severe critical skepsis by cultural commentators fearful of a *Gleichschaltung* or leveling that is inclined to stake too much on the lowest common denominator in matters of cultural production and that has an eye almost only for the concerns of the mainstream middle classes with the highest spending powers.

Translated into the Times Square cityscape, the evolution implies that the anarchic gaggle of come-and-go stores and small, low-priced retail outlets once typical of the area has had to make room for global entertainment moguls monopolizing the area and peddling their hyped-up wares that no longer bear the slightest link to the concrete location in which they are being advertised. "Above all, an urban lifestyle is about choice and opportunity," suggests Hannigan (1998, p. 9), and one may wonder whether the current Times Square is fully able to embody that principle – whether it still creates the most desirable conditions for fostering cultural and social variety and heterogeneity. The idiosyncratic local "leisure merchants" and impresarios that shaped Times Square during the first decades of the century have been displaced with generic multinational chains. These chains operate on a global market; they parachute themselves anywhere in central cities – from New York to Tokyo or Berlin's Potsdamer Platz – and contribute largely to the production of a Western type of that "generic city" so provocatively etched by the architect Rem Koolhaas in his eponymous essay of 1994. The result is a monoculturalization of sorts: a reduction to modular supplies and an aesthetic form of globalization of the streetscape that is the visual corollary of economic processes of globalization. As Steven Johnson (1999, p. 4) has suggested, the more demanding and critical urbanite who desires individuality and surprise may thus feel increasingly short-changed and disappointed by places like the new Times Square, "not because they're tacky, or commercial, but because they're boring. And boring because they are everywhere the same. . . . That is a change worth regretting, if only because it threatens to diminish a sense of place that the market formerly amplified. Markets once helped give Amsterdam its canals, and Istanbul its architecture; their multinational descendants now bestow the same Virgin Megastores and IMAX theaters on every locale, or eliminate locale altogether by building the massive synthetic environments of the West Edmonton Mall or the Mall of America. It's a fantasy city, all right, but it looks like every other fantasy on the market." The new 42nd Street and Times Square admittedly escape this tendency to some extent – by a return in the interim development plan to a more variegated streetscape, for instance, or by the individual efforts of global corporations and imagineers to stand out in this competitive environment and display their most creative forces here – yet it cannot be denied that the dulling effects of generic recognizability are at times palpable.

(8) Finally and most evidently, the term "Disneyfication" also manages to speak to evolutions in and around Times Square by illustrating the growing importance of "branding" and "naming rights" in today's fantasy city. This phenomenon is by no means new, of course, since Times Square derives its very name from an early instance of such corporate branding.[13] But in a postmodern symbolic economy, brand names and their associated images appear to be increasingly what is being sold. As Kevin McNamara (1999, p. 190) notes, paraphrasing an argument by architectural historian M. Christine Boyer, today "the visually acquisitive experience of consumer society ... dissolves the distinction between the object of use and symbolic attributes as well as between the commodity and its setting." "While such public places as Grand Central Station [sic] or Piccadilly Circus are not up for grabs just yet," adds Hannigan (1998, p. 3), "a kind of precedent was set recently when Continental Airlines became the first national corporate sponsor of New York's theater district, soon to be rechristened 'Continental World.'" If the synecdoche "Broadway" comes under siege from the promotional flag "Continental," one may legitimately wonder how much further big-name corporations will seek to extend their attempts to appropriate profitable urban environments like Times Square. "Disney," Hannigan (1998, p. 194) tells us, "has a history of being reluctant to allow authors the right to reproduce photos of their theme parks, especially if these are thought to be in any way critical of the Disney organization ... Disney World, Michael Sorkin quips, 'is the first copyrighted urban environment in history, a Forbidden City for postmodernity' ... It also signifies the triumph of market over place ... as brand identity and protection is extended from manufactured goods to the spaces and places of the city." Nobody is able to predict at this point whether some day global corporate players will be able to buy up entire urban districts as promotional spaces and attach their names to them. Nor can we foresee whether they will then decide to close off

streets and charge entrance fees, as the Disney Company dreamed of doing in Seattle. But behind the screens we may be sure that people are investigating the possibility and working to achieve that end. A term like the Disneyfication of Times Square would then be more than descriptive of recent changes: it would also contain a possible scenario for the future.

THE NEW TIMES SQUARE: BACK TO WHOSE FUTURE?

A back-to-the-future ideology in urban (re)development may take various shapes. In one of its more conservative guises, it for instance informs the nostalgic cityscape of an idealized smalltown Main Street developed by Walt Disney already in the 1950s. With many modifications, moreover, a similar ideology may be felt to underlie the rising success of the historic preservation movement since especially the late 1960s (though its real roots for a city like New York, as historian Max Page [1999, pp. 111–143] has argued, reach back much farther to at least the turn of the twentieth century). "Back to the future" appears to be the recipe also of other architectural and socioeconomic trends from the last three or four decades, from gentrification (based on the restoration of derelict houses and spaces for adaptation to modern upper-income usage) to the dissemination of festival market places organized around local historical themes and other themed or semi-themed commercial areas in contemporary cities. As Alexander Reichl (1999, pp. 167–168) claims: "Historic preservation allows pro-growth forces to depict development as a means to restore romanticized images of the good ol' days, whether the glorified past is packaged in the form of an elegant theater district, a lively place of popular entertainment, a bustling waterfront market, a friendly main street, or some other reassuring symbolic place."

The "Disneyfying" redevelopment of Times Square thus partakes in a much wider movement that has wrought changes in the urban fabric, paradoxically, by striving to revive the past. From the beginning of the current redevelopment epoch of Times Square in the late 1970s and early 1980s, the policy of local developers (officials as much as private investors) has been to legitimate their efforts by mobilizing historic images and redefining and repackaging the area's cultural history. The adaptation to modern uses has consistently relied on a simulation of (selective parts of) the area's past. This selectivity and simulated quality is only as to be expected: history, after all, is a social construction that, except for the most disinterested academic investigation, has traditionally been produced to legitimate and consolidate the powers that be. Nor should we be surprised, in such a context, that many of the political, economic, and symbolic processes that have been steering the redevelopment of the greater Times Square area turn out to copy processes that were also at work in the early history of the neighborhood. Several of the issues that have come to the fore in recent critical debates are not as novel as they may seem. This is the case, for instance, with complaints about how the area has fallen victim to a moral clean-up operation (which is not so new: it equally happened in the Prohibition era or in the late Depression years); with complaints about the predominance of real-estate interests in the decision-making process behind the screens (and the fact that these interests have in reality been more marked by an economic than any moral or social logic); with observations on the stalling of the large-scale redevelopment project because of an interfering economic downturn (which applies to the wake of 1929 and 1974 as much as to the post-1987 era); with an insistence on the appropriation and invasion of the area by media and entertainment industries (also applicable to the era in which *The New York Times* and the early amusement complexes served as local growth machines); or with strictures about the crass and slick commercialization of this symbolic space and its economic boosting through upper-income tourism (both of these phenomena having a long pedigree).

There are, in other words, a great many constants in the twentieth-century history of Times Square, many of which have merely been replayed (to a greater or lesser extent and in a postmodern form) during the final two decades. What appear to be the most glaring discontinuities at the surface often hide surprisingly tenacious continuities underneath. Once these underlying continuities have been observed, however, it becomes possible to be more precise about what constitute actual discontinuities and uncharacteristic shifts that cannot be presented as typical of the area's history. These discontinuities and shifts, it appears upon closer inspection, have less to do with the spectacularly altered and altering *material* outlook of the area than with the production of Times Square as a *social* space. Alexander Reichl (1999, p. 2) is clearly right when he claims that "the essence of Times Square is found as much in its history as a contested social space as in its dramatic physical characteristics." For all the useful discussion about the Disneyfication of the area, there is the constant danger that this label and its attendant topics will obfuscate or sideline a more thorough consideration of alterations in the social production of Times Square's space – alterations that may not follow altogether clearly or directly from the principally architectural, iconographic, symbolic, and economic debate that tends to surround the name of Disney. As a supplement to the previous analysis, then, I will undertake to identify the most crucial variables in the social alteration of Times Square. These variables are, as far as I can see, to be situated above all at the levels of class, race (and ethnicity), gender, and sexuality. In all of these respects, important and striking discontinuities in the profile of visitors to the area *may* be observed. Zooming in on such alterations, moreover, proves to be of more than local importance again, for the direction in which the changes point tells us a lot about the tension between progressive and regressive social forces with which many American central cities are currently grappling.

(1) *Class*: Although any discussion of socioeconomic class in a U.S. context is inevitably inflected by race and ethnicity, it is still useful to keep these categories apart. If, as Alexander Reichl (1999, p. 53) has contended, the Times Square of earlier in the century "represented a breakdown of traditional class-based cultural boundaries," then we should wonder whether that class-mixing heritage has been acknowledged and respected in the back-to-the-future project of the recent redevelopment. And here the answer seems to be that it has not – or only very secondarily. Active social engineering of the kind that seeks to bridge class barriers has never been on the redevelopment agenda. The latter has been powered especially by a mix of economic and cultural-artistic goals. These goals were themselves inevitably inflected by class interests, but did not explicitly foreground these. If we pan out again to look at the wider urbanistic trend towards preservation- and arts-based redevelopment in which the recent history of Times Square partakes, we notice how "The source of the move from renewal [in the 1950s and 1960s] to preservation lies in a class-based political conflict over urban space. ... Urban renewal, guided by its cultural vision of urban modernism, failed to construct a new cityscape that appealed to middle- and upper-income people. ... Rallying behind Jane Jacobs's assault on the planning principles of modernism – and the 'marvels of dullness and regimentation' that they produced – higher-income city dwellers began asserting their commitment to the value of neighborhood preservation" (Reichl, 1999, p. 11). In residential neighborhoods, this led to the class struggles known as gentrification, but it also had its effect on a number of potentially profitable commercial areas like Times Square. There it meant that politicians could strengthen their ties with middle- to upper-income voters by undertaking historically themed redevelopments. "Whereas urban renewal increasingly became the focus of political conflict," Reichl (1999, pp. 12, 19) argues, "preservation- and arts-based development has tended to enjoy broad public support. ... The discourse on Times Square was also a discourse on urban decline that drew on deeper anxieties in mobilizing support for redevelopment among middle- and upper-income whites, groups that were crucial to the endurance of [Mayor Edward I.] Koch's regime" in the early 1980s as well as to the survival of Mayor Rudolph W. Giuliani's political powerbase through much of the 1990s (p. 190).

The effects of serving principally a middle- to upper-income agenda have been unmistakable: with respect to class membership, the social profile of visitors to the new Times Square has considerably altered. Among other things, the single-room occupancy hotels for which the area was also famous have been torn down or converted to permanent upscale residences. Although at first this merely increased the number of homeless people in the area (Reichl, 1999, pp. 75, 200), the eviction of low-income residents has since been supplemented by a variety of measures that made sure to remove the homeless from the local streetscape. Likewise, the typical small-business bric-à-brac stores "featuring everything from T-shirts and souvenirs to discount electronics and scuba gear" or "selling cheap electronics, jewelry, and martial-arts supplies" (Reichl, 1999, pp. 143, 182) have effectively been ousted, as have the several forms of inexpensive entertainment like cheap movie theaters. As a result, consumption possibilities in the area now represent primarily the upper end of the continuously growing income gap between rich and poor in the city.[14] Various non-affluent visitors that were lumped together in the category of "loiterers" by the media, politicians, and project apologists in the 1980s and that were represented as particularly menacing, are no longer made to feel welcome. Drug users, down-and-out alcoholics, peddlers, homeless people, con men, bohemians, and various members of marginal, alternative youth subcultures: they all seem to have disappeared largely from the area.

The simple but efficient politics that have underlain the clean-up operation and transformation of Times Square into a city block of unmitigated consumerist fantasy and fun-making are those of an "out of sight, out of mind" politics. Whether the area ever offered a full cross-section of New York's sociopolitical spectrum is doubtful, but today it obviously falls short of doing so by a considerable margin. The apologists of the new Times Square do not seem to have been concerned with restoring the mix of classes that inspired a 1930s musical to hail a place "Where the underworld can meet the elite / naughty, bawdy, gaudy, sporty, 42nd Street" (quoted in Reichl, 1999, p. 55). Next to the New York subway, Times Square for much of the twentieth century used to offer the city's most heterogeneous social landscape, with tuxedoed patrons of Broadway shows walking side by side with three-card-monte players and stoned youngsters. Today's visitors, however, despite their being perhaps more numerous than ever before, have clearly become more homogeneous also. And the hegemony of the visible is such that what cannot be immediately perceived by people – in this case by the middle- to upper-income majority looking for "shopertainment" – becomes hard to push also as a sociopolitical issue. As Kevin McNamara (1999, p. 189) contends: "Without any visual evidence of poverty and injustice, shoppers need not calculate the degree to which forms of oppression underwrite their own contentment." Or in Alexander Reichl's words (1999, p. 9), condensing an argument by David Harvey: "spectacular places and public festivals serve to obfuscate underlying class conflicts and inequalities." To be sure, we should beware of the masculinist kind of romanticizing in which a few mid-1980s sociologists could still indulge when they pleaded for Times Square as "a place where the laws of conventional society are suspended, where people come to seek adventure, to take risks in dealing in the fast life" (quoted in Frieden & Sagalyn, 1989, p. 242). Yet we must face the possibility that a majority of postmodern urbanites, despite their unprecedented average level of education and the unprecedented amounts of information with which they are being bombarded, prefer not to be too visibly and directly confronted with their own socioeconomic prerogatives. For political scientist Alexander Reichl (1999, p. 177), "the ideal authentic public space is one where the physical environment supports a diversity of uses and users, thus creating an area of genuine, or relatively unrestricted, social interaction." The redevelopment of 42nd Street and its surrounding area, by contrast, "has proceeded with disregard for these principles. Before its redevelopment, the Times Square area, although an imperfect place, was a public space of genuine diversity that served a wide range of racial, ethnic, and income groups." Novelist and critic Samuel Delany, a long-time regular in the area, has formulated the contrast even more starkly (1999, pp. 159–160): "The old Times Square and Forty-second Street was an entertainment area catering largely to the working classes who lived in the city. The middle class and/or tourists were invited to come along and watch or participate if that, indeed, was their thing. The New Times Square is envisioned as predominantly a middle-class area for entertainment, to which the working classes are welcome to come along, observe, and take part in, if they can pay and are willing to blend in."

(2) *Race and ethnicity*: As Alexander Reichl's above formulation about a reduction of the "range of racial, ethnic, and income groups" suggests, lower-class visitors to the Times Square area had become, by the 1960s and especially 1970s, to a very large extent ethnically non-white. A study by sociologists from the City University of New York about the situation in the late 1970s noted, for instance, how "42nd Street between 6th and 8th Avenues is probably one of the racially most integrated streets in the city. Particularly in the daytime, the racial and ethnic breakdown of pedestrians approximates the proportions of white, Black, and Hispanic persons in Manhattan's residential population" (quoted in Reichl, 1999, p. 62). At the same time, and contrary to widely held beliefs, "studies found that whites constituted a 'numerically dominant group' of persons on the street at almost all times of day. It was only after midnight that the number of African-Americans might exceed that of whites" (Reichl,

1999, p. 62). In spite of these facts, local officials and politicians like Edward Koch repeatedly pleaded in favor of redevelopment by surreptitiously cashing in on white fears and resentments and representing young minority males as somehow by definition threatening. The New York State Urban Development Corporation drew up an influential environmental-impact statement whose "implicit conclusion was that African-Americans, Latinos, and whites cannot share the same entertainment district because non-Latino whites are unable to distinguish the criminal element from among a crowd of African-Americans and Latinos" (Reichl, 1999, p. 63). Since the equation with crime was often implicit in this kind of discourse and reasoning, it may be useful to recall that the relatively high levels of crime for the area in the late 1970s and early 1980s lost much of their peak values when compared to the enormous number of people using the area. As the sociological CUNY report again noted: "Given the vast numbers of tourists and commuters who pass through the Port Authority Bus Terminal and the 8th Avenue subway station, [the] numbers of felonies and felonious assaults, while disturbing in themselves, reflect extremely low rates of incidence" (quoted in Reichl, 1999, p. 64).

During the 1970s and early 1980s, "Times Square was probably a relatively safe entertainment district for residents of low-income neighborhoods that were plagued by crime and drugs and lacked basic attractions, such as movie theaters," suggests Alexander Reichl (1999, p. 134). At the time redevelopment plans were initiated, then, "the city's low-income minority populations . . . stood to lose an easily accessible, centrally located district of inexpensive fast-food outlets and discount first-run movie houses" (p. 134). Even middle-class African-Americans appeared to come under siege as a result of redevelopment, for by the late 1980s preservationists discovered that the Times Square area had for a long time also accommodated a thriving network, hitherto unnoticed by white New Yorkers, of mostly family-owned bars and restaurants stemming from the heydays of the theatrical district and catering to an audience of middle class African-Americans (Reichl, 1999, p. 68). Subsequently, there was a growing worry that the redevelopment plan might be "inadvertently racist" (quoted in Reichl, 1999, p. 69). But the insight and sensitivity came too late: the ball had been set rolling and evictions were nearly completed. By the mid-1990s, both this hidden world of leisure for an African-American middle class and the much more visible streetlife of younger low-income African-Americans and Latinos had effectively been made impossible. Not that the area has come to be shunned therefore on strictly ethnic grounds. Personal observation inclines me to say rather that ethnic minority members in search of entertainment (whether in the games arcades or as patrons of the movieplexes) are both numerous and fully at ease. Yet it seems also clear that a large part of these visitors are middle-class consumers. The class factor, in other words, has been more influential in redefining the area's social space than that of race and ethnicity – at least, if we take the trouble to remember that a majority of low-income people who no longer frequent Times Square are African-American and Latino. Nor should we close our eyes to the constant and careful monitoring of the sidewalk population. Any form of collective "loitering" that could be perceived as menacing tends to be nipped in the bud. On the handful of occasions when I visited the area in recent years, I have noticed police officers interfering with groups of young African-Americans more than once, entering into altercations and removing youngsters from the scene. It seems safe to presume, given the official policy of "zero tolerance" that has marked the Giuliani administration and the omnipresence of private BID as well as public NYPD officers, that latent ethnic and racial tensions in the city are painstakingly prevented from breaking out in and around the redesigned, ever-populous Times Square.

(3) *Gender*: This factor is probably the least noted, most underestimated and understudied aspect in the history of Times Square's twentieth-century alterations. In his invaluable historical-political study, even Alexander Reichl (1999) has a tendency to underplay the gender side of the story, although he does acknowledge it repeatedly. Partly because its solid academic grounding is so recent, moreover, an analysis of the gender distribution in the area cannot fall back on reliable comparative statistics. Still, it suffices to look at pictures from the early decades of the century to be reminded of the frequently unequal distribution of men and women in the street and as independent consumers of entertainment. The role of women in Times Square during the first half of the century was all too often that of prostitute or burlesque dancer. An early feminist like Gloria Gould Bishop, who managed the lavish and exclusive Embassy Theatre (now restyled as the Times Square Visitors Center), was the proverbial exception proving the rule. Yet even her positive discrimination of women in the hiring of staff was somewhat offset by the accompanying demand that all women be between 17 and 21 years old and have impeccably white teeth. At the latest by the 1930s, 42nd Street was "taken over as a male domain . . . and the grinders, arcades, bookstores, and novelty shops continued to cater to a predominantly male clientele. World War II brought new crowds of soldiers and sailors into Times Square, reinforcing the tendency toward a male-dominated cultural space and bringing back a market in commercial sex" (Reichl, 1999, p. 56). Unsurprisingly, during the postwar era of suburbanization (which relied to a considerable extent on a gender division between rough, male cities and safe, feminized "bedroom communities" [Saegert, 1981]), many women continued to shun the sleaze of Times Square and only frequented it when they were either

actively involved in the sex business or belonged to the working class and performed as barmaids, store clerks, or ticket takers (Delany, 1999, p. 160). The report drawn up by CUNY sociologists about activity in the area during the late 1970s reconfirmed that "the real pattern of dominance was based on gender with men greatly outnumbering women on the street" (Reichl, 1999, p. 62). Predictably also, the large majority of patrons for the local sex businesses appeared to be white-collar, middle-class males (Reichl, 1999, p. 66).

Although gender arguments never managed to play a significant role in the political and commercially motivated discourse that promoted redevelopment from the early 1980s onwards, we should not be surprised to find that one of the civic organizations entering into an alliance in favor of redevelopment at the time of the public hearings in 1984 was the Women's City Club of New York (Reichl, 1999, p. 124). For political-historical reasons, the needs of women had never really been factored in during discussions about the social use of Times Square, and in the masculinist worlds of urban planning, architecture, large-scale real estate, and politics, they were still receiving short shrift at a time when emancipatory values elsewhere were beginning to percolate. It is symbolically not inappropriate, though, that by the mid-1990s the three most powerful local nonprofit organizations overseeing redevelopment efforts were headed by women: Gretchen Dykstra chaired the Times Square Business Improvement District, Rebecca Robertson the 42nd Street Development Project, and Cora Cahan The New 42nd Street, Inc. (Weber, 1996). No less symbolic and telling is the fact that the first theater to be restored on 42nd Street (the Victory) was turned into a children's theater. For any discussion of gender in a context of entertainment and consumption is intimately tied up with that of motherhood. Times Square has indeed managed to become an area that draws in massive numbers of families. Not only do (middle-class) women seem to be no longer afraid of the area, they also bring their children with them. For the first time in history, then, Times Square seems to have become a family entertainment destination. That this remarkable shift has been achieved to a considerable extent through the image-building effect of the Disney Company's presence goes without saying. If there is one way in which the "Disneyfication of Times Square" signals a radical change not only in the iconography of the area but also in the social constellation of its visitors, it is precisely in the great numbers of middle-class women and children currently participating in the local worlds of consumption and entertainment.

(4) *Sexuality*: The apparent corollary of Times Square's strong regendering has been the disappearance of a sexually vibrant culture. This, too, constitutes a novelty in the history of the area, for sexual transactions have always been part and parcel of that history, even if the apologists for preservation-based

redevelopment have consistently downplayed or ignored this aspect of the story. Historically, theaters and brothels operated side by side in 19th-century and early-20th-century New York (as in many other American cities) and prostitution was a central part of the entertainment economy (Reichl, 1999, p. 49; Buckley, 1991). By the turn of the century, for instance, more than 130 addresses out of which prostitutes worked were identified in the blocks around Long Acre Square and commentators "reported that on Broadway there were ten to twenty prostitutes on every block, forming a 'two-mile parade of prurient commerce'" (Reichl, 1999, p. 50, quoting Gilfoyle, 1991, p. 300). A few decades later, by the 1920s and especially 1930s, the reach of the sex industry expanded still further with the emergence of a homosexual subculture in the neighborhood that among other things attracted numerous male prostitutes, from gender-bending "fairies" over so-called "rough trade" (Reichl, 1999, p. 57; Chauncey, 1991). Through the war years and after, Times Square remained the most prominent sexual center of New York. This epoch culminated in the pre-AIDS era of sexual emancipation and countercultural experimentation – the late sixties through early eighties – when "consumers could choose from a smorgasbord of books, magazines, films, peep shows, private one-on-one peep booths (usually, but not always, divided by a glass partition), stripteases, and even live sex performances in the Times Square establishments. Prostitution was thinly veiled behind the facades of these and other enterprises, such as hotels, topless shoe-shine stands, and, especially, the many massage parlors of Times Square" (Reichl, 1999, p. 57). At the height of such activities, in 1978, some 121 sex-related businesses were identified in the area, and by 1983, there were still an impressive 18 of these just on the single block of 42nd Street between 7th and 8th Avenues that has since been Disneyfied and completely cleaned up (Reichl, 1999, pp. 59, 65).

It pays to list the whole sexual menu of these practices and establishments, if ever so briefly, for such a list helps to make the contrast with the current situation all the more palpable. A few sex shops and peep shows admittedly survive on the outskirts of the area (most notably on 8th Avenue below 41st Street), despite the tough zoning laws implemented by the Giuliani administration. But even they can no longer display any of their sexual wares in their windows. In the bowtie of Times Square itself, meanwhile, not a single XXX-movie theater can be spotted anymore and potential patrons for a few peripherally located striptease bars must be lured by way of flyers. No street prostitution of any significance seems to have survived in the core area (though some apparently has moved to the inside of derelict buildings to the west). The local drag queens have gone and no developed gay community life remains; the single male striptease venue that could still be found by the summer of 2000 was anonymously stacked up a staircase that did not bear the slightest sign at streetlevel to indicate the purpose of the establishment. In some sense, then, the redeveloped, "sanitized" Times Square perfectly embodies the striking shift over the past decade from a sex industry that was spatially organized to facilitate direct sexual encounters to one that has been virtualized through the internet (whether for passive consumption or as an interface to set up sexual contacts). Yet surely the radicality of the change in Times Square embodies also more than such a virtualization of sexual transactions: it also tells of a moral backlash generally that has taken no interest in salvaging the material institutions able to underpin a wide diversity of sexual practices.

One colorful and strongly argued example serves to illustrate the surprising diversity that has been lost in the process, the more so since the example ties together the four categories of class, race, gender, and sexuality that summarize the most striking social transformations in and around Times Square. In *Times Square Red, Times Square Blue* (1999), the leading African-American science fiction writer and cultural critic Samuel Delany reports on his own decades-long experience in the local porn movie theaters. The world he recalls and portrays at the point of its very demise is that of men engaging in casual and often anonymous sex with each other – a world considerably more complex than outsiders would tend to imagine, for in Delany's description it gave rise to a relatively affable (though in most cases transient) system of interaction and contact across the boundaries of class, race, and sexual preference. However perverse and marginal to the discussion this particular kind of practice may have seemed to all those involved in redeveloping the area, it at least embodied an instance of an active type of interaction, in a semi-public space, between people of different colors, classes, and sexualities. And it is precisely this kind of heterogeneous interaction that, as Delany argues, is missing from the current environment. Yet Delany's narrative – which goes out of its way to avoid romanticizing the experiences described – at the same time also highlights how strongly masculine this world was and how it failed to accommodate contact across the boundaries of gender as well. Only once do we hear of an inquisitive and daring female friend who accompanied the author on a visit to get an inside view at this sexual subculture (Delany, 1999, pp. 25–31). Aware of this shortcoming, yet still believing in the need for institutional sexual outlets that the current New York is almost completely without, Delany (1999, p. xv) indeed prefers to present his analysis as one whose "polemical thrust is toward conceiving, organizing, and setting into place new establishments – and even entirely new types of institutions – that would offer the services and fulfill the social functions provided by the porn houses that encouraged sex among the audience. Further, such new institutions should make those services available not only to gay men but to all men and women, gay and straight, over an even wider social range than did the old ones. ... The polemical passion here [i.e. in Delany's book] is forward-looking, not nostalgic, however respectful it is of a past we may find useful for grounding future possibilities." The little bit of controversial history Delany has rescued from oblivion offers no more to him than an embryonic and imperfect example of a kind of social practice that may seem provocative and increasingly utopian now but is not so easily dismissed as irrelevant for the future.

AFTERWORD: TAKING THE PULSE OF POSTMODERNITY

Although the transformation of Times Square during the 1990s has been most eye-catching and spectacularly visible at the material level, the area has also been radically reconfigured as a social space. From a place that uncomfortably mixed heterogeneous, at times conflicting uses, it has changed (or been changed) to one that more comfortably accommodates more homogeneous, less combustible uses. Its target audience by the turn of the millenium is clearly that of fun-loving visitors who have the spending power to satisfy their consumerist desires and to participate in the various types of mainstream entertainment offered by Broadway theaters, high-tech games arcades, and national or global chains. Women as well as families with children are welcome, and so are members of all ethnic groups, but they will only be inclined to return and participate in the local life on a regular basis if they are in no immediate financial distress. To people with reduced budgets (and in the greater New York City these are still preponderantly non-white), the area no longer offers the opportunities it used to provide. Those whose interest is primarily sexual will stand to be disappointed and fall on hard times.

Does this profile also correspond to the patrons of Disney's amusement parks? To a certain extent it clearly does, though with some important modifications: Disney's parks are obviously still more children- and family-oriented; they strictly demand consumptive participation; and they attract fewer adult adventurers in search of a modicum of metropolitan excitement. What is more, Disney's parks are considerably more coercive, since as privatized spaces they are practically able to keep out all undesirable visitors and to prohibit all undesirable types of behavior. And yet, a space like Times Square, in its more complex and more appealing ways, does seem to share at least a tendency with Disney's amusement parks – a tendency that extends beyond either type of idiosyncratic and phantasmagoric space to be part of a larger and perhaps more enduring sociological trend. According to John Hannigan (1998, p. 7), the contemporary fantasy city embodied by Times Square "is the end-product of a long-standing cultural contradiction in American society between the middle-class desire for experience and their parallel reluctance to take risks, especially those which involve contact with the 'lower orders' in cities." Borrowing a term from the cultural commentator Russell Nye, Hannigan (1998, pp. 71–72) talks of a desire for "riskless risk": "The 'riskless risk' so evident in the themed environments of Fantasy City is part of a wider trend in which various foreign cultures and domestic subcultures are appropriated, disemboweled and then marketed as safe, sanitized versions of the original. ... [I]f Ulrich Beck [in his influential study of 1986, *Die Risikogesellschaft*, translated in 1992 as *Risk Society*] is correct in his assertion that in the contemporary 'risk society' we are increasingly subject to an escalating barrage of global-generated risks over which we have little control, then such mildly exciting but essentially harmless activities as virtual reality skiing, or skydiving the Grand Canyon in IMAX 3-D, may provide just the right measure of 'reassurance.'" Hannigan (1998, p. 74) has noted how contemporary suburbanites are attracted by the affective ambience of light noncommitted interaction in crowds that is missing from suburban life, yet it is only possible to bring these suburbanites round to visiting a place like Times Square if the urban crowds there are sufficiently defused, carefully monitored, and so to speak mallified in order to suppress the dangers and random contacts with all-too-different people traditionally shunned by suburbanites. Hence the frequently heard, if at times reductive complaints that Times Square is becoming a "sanitized," "domesticated," "disemboweled," "suburbanized" open-air mall and that the place is trying to gloss over the productive necessity of conflicts in a democratic society, blandly erasing tensions in an unwillingness to let these conflicts interfere with the profit-making agenda of private companies. Public spaces like today's Times Square testify to what Kevin McNamara (1999, p. 201) has called "a fear of conflict that marks a failure of urbanity, the ability to value the diversity and unpredictability that has characterized the life of cities. ... [T]he greater challenge, at least in the United States, is to reassert the value of the public sphere and a democratic culture in which people are responsible to others whom they do not know and may not like or even understand." A populist and essentially infantile iconography like Disney's, one may doubt, will not do much to help reassert this value of the public sphere. Disneyfied cityscapes pose aesthetic and political problems perhaps above all by insufficiently *resisting* visual and intellectual assimilation. They do not sufficiently unsettle or disturb, and thus fail to teach us how to negotiate complexity in an intrinsically complex and ever more interconnected world. Simple childish delights in spectacular images, although they have their undeniable place and use, preclude the development of the kind of critical sense on which a democracy depends for its effective functioning. It may be true that Times Square has never been a real political agora and has always been primarily a place of fantasy, spectacle, and escape. Yet we should still be willing to question what gets lost when profit-oriented developers and megacorporations destroy more complex worlds and simplify their histories. Walt Disney's approach to the past was always to "improve" it, for instance by recasting the Grimm brothers' gothic horror tales in the form of zany and uplifting cartoons (Wallace, 1996, pp. 136–137). Social critics have for decades been wary of the way in which "his work ground off the rough, emotionally instructive edges of the folk- and fairy-tale tradition on which it largely drew, robbing it of 'the pulse of life under the skin of events,' as one critic put it" (Schickel, 1998, p. 77). The pulse of life under the skin of Times Square may well have quickened today, as local business boosters and political leaders would have it, but does it also race with more than a consciously induced and monitored consumption fever for the affluent?

NOTES

Parts of this essay were first developed as a lecture entitled "Sign-of-the-Times Square: The Spectacle of Urban Transformation in the 1990s," delivered at a conference on "The American Metropolis: Image and Inspiration" organized by the Netherlands American Studies Association in Middelburg, June 3, 1999. The original paper, which provided a number of historical and factual backgrounds recycled in the above essay, will be published as part of the conference proceedings by the VU Press of Amsterdam. A first reworked and expanded version of the original materials appeared as "The Disneyfication of Times Square: What's in a Name?" pp. 39–73 in *Spieghel Historiael*, Vol. 39, 1999. The present essay has been further revised, expanded, and updated through extra research and fieldwork. Many thanks are due to Drew Whitelegg of King's College, London, to Carl Smith of Northwestern University, to Themis Chronopoulos of Brown University, and to Kevin R. McNamara of the University of Houston-Clear Lake, Texas, for providing me with comments and pointing me to additional research materials; to my fellow members of the Ghent Urban Studies Team, Judit Bodnár, Kristiaan Borret, Steven Jacobs, Bart Keunen, Dirk De Meyer, Kristiaan Versluys, and Trui Vetters, for their intellectual support and inspiration; to Gert Morreel, who for many years co-hosted me during my regular trips to New York; and to Mikako Kumagai, who graciously put me up during my most recent visit.

1. On the various ways of delimiting the Times Square area, see Taylor (1992, pp. 97–98).

2. This is one of the master narratives in Alexander Reichl's excellent political analysis of redevelopment efforts in *Reconstructing Times Square: Politics and Culture in Urban Development* (1999), which has been one of the major sources and inspirations for this essay.

3. This local embracing of an oversized, in-your-face billboard aesthetic stands in marked contrast with the attitude of property owners and entrepeneurs on Fifth Avenue, which began to establish itself around the same time as the exclusive "Spine of Gotham" by among other things adamantly resisting the arrival of commercial signs; see Reichl (1999, p. 52), and, for an excellent overall case study of Fifth Avenue in those years Page (1999, pp. 21–67).

4. In what appears like an ironic comment on Jean Baudrillard's claim (1988, p. 17) that a humanizing eco-architecture would mean the death of New York, the skyscraper also presents state-of-the-art ecological or "green" architecture: among many other things, it has photovoltaic cells in its outer wall, which convert sunlight into electricity without generating pollution (see Holusha, 1998).

5. These and the following facts and figures have been derived in part from the BID's own website at <http://www.timessquarebid.org/web/index.htm> (accessed 2 March 1999), as well as from personal fieldwork. More recently, the website has toned down its triumphalism at the same time as it has raised its interactive interest by including the newest gadget of live webcams that allow you to view the square at any time of the day.

6. For synoptic versions of this historic shift in public urban policies and its effect on the redevelopment of Times Square, see Frieden and Sagalyn (1989, pp. 291–300), Huxtable (1991), Gratz (1998, pp. 68–77). The long and short of the redevelopment story from the 1970s onwards is beautifully and painstakingly detailed in Reichl's *Reconstructing Times Square* (1999), especially chapters 4–5.

7. These organizations are all briefly explained in Reichl (1999, pp. 185–186) and their specific roles worked out in the rest of his study.

8. The role of Tishman Urban Development is discussed at length in Hannigan (1998, chapter 6, especially pp. 114–119).

9. The importance of class difference between the Coney Island and Disneyland models of popular entertainment found an interesting historic reflection during the 1930s at Times Square, when the area got caught in a downward spin. Whereas before the Depression, Times Square had been considered a place reaching out to a great mix of socioeconomic classes, the subsequent downgrading inspired critics to decry "the transition of Times Square to a new 'Coney Island,' with a 'carnivalesque' and 'cut-rate, amusement park ambiance'" (Reichl, 1999, p. 55).

10. For two fascinating book-length reports on the Celebration project by, on the one hand, a *New York Times* journalist and his family and, on the other, a New York University cultural studies professor, who have all gone to live inside this "town of the future," see respectively Frantz and Collins (1999) and Ross (1999).

11. A political-deconstructionist analysis of the many ways in which the Disney Company has adapted to changing social and political constellations in its recent film production is to be found in Byrne and McQuillan (1999).

12. New Yorkers would of course resist any strong association with Disneyland or Disney World, as Stern must have realized. When in the late 1970s a first privately sponsored plan for massive redevelopment in Times Square was proposed to Mayor Edward I. Koch, he typically shot it down because of its too close resemblance to Disney's theme parks, arguing that "New York cannot and should not compete with Disneyland – that's Florida. . . . We've got to make sure [visitors to Times Square] have seltzer instead of orange juice" (quoted in Reichl, 1999, p. 89).

13. For the direct and indirect role of *The New York Times* in supporting redevelopment efforts in the Times Square area since the late 1970s and through the 1990s, see Reichl (1999, pp. 81, 86, 104–109, 151–152, 155).

14. At the time of the public hearings before the Board of Estimate in 1984 (which had to decide on the fate of the massive redevelopment plan under the Koch administration), a group of residents of the adjacent low- and moderate-income Clinton neighborhood stood up to dispute the way in which the redevelopment project "worked like traditional urban renewal by displacing the poor for the benefit of the wealthy" (Reichl, 1999, p. 129). This resulted in a counteroffer by Governor Cuomo and Mayor Koch that provided $25 million to support low- and moderate-income housing resources and other community services in the Clinton neighborhood, which effectively silenced local opposition (Reichl, 1999, p. 133).

REFERENCES

Barthes, R. (1986). Semiology and the Urban. In: M. Gottdiener & A. Ph. Lagopoulos (Eds), *The City and the Sign: An Introduction to Urban Semiotics* (pp. 87–98). New York: Columbia University Press.
Baudrillard, J. (1988). *America*. Reprint. London: Verso, 1993.
Berman, M. (1997). Signs of the Times: The Lure of 42nd Street. *Dissent*, *44*(4), Fall, 76–83.
Buckley, P. (1991). Boundaries of Respectability: Introductory Essay. In: W. R. Taylor (Ed.), *Inventing Times Square: Commerce and Culture at the Crossroads of the World* (pp. 286–296). Reprint. Baltimore: The Johns Hopkins University Press, 1996.
Byrne, E., & McQuillan, M. (1999). *Deconstructing Disney*. London: Pluto Press.
Chauncey, G., Jr. (1991). The Policed: Gay Men's Strategies of Everyday Resistance. In: W. R. Taylor (Ed.), *Inventing Times Square: Commerce and Culture at the Crossroads of the World* (pp. 315–328). Reprint. Baltimore: The Johns Hopkins University Press, 1996.
Delany, S. R. (1999). *Times Square Red, Times Square Blue*. New York: New York University Press.
Dunlap, D. W. (1998). Where Raunch Met Excess; Demise of Art Deco Temple of Less-Than-High Art. *The New York Times on the Web*, 22 September.
Dunlop, B. (1996). *The Art of Disney Architecture*. New York: Harry N. Abrams.
Erenberg, L. (1991). Impresarios of Broadway Nightlife. In: W. R. Taylor (Ed.). *Inventing Times Square: Commerce and Culture at the Crossroads of the World* (pp. 158–177). Reprint. Baltimore: The Johns Hopkins University Press, 1996.
Frantz, D., & Collins, C. (1999). *Celebration, U.S.A.: Living in Disney's Brave New Town*. New York: Henry Holt.
Frieden, B. J., & Sagalyn, L. B. (1989). *Downtown, Inc.: How America Rebuilds Cities*. Reprint. Cambridge, MA: MIT Press, 1994.
Ghirardo, D. (1996). *Architecture After Modernism*. Singapore: Thames and Hudson.
Gilfoyle, T. J. (1991). Policing of Sexuality. In: W. R. Taylor (Ed.), *Inventing Times Square: Commerce and Culture at the Crossroads of the World* (pp. 297–314). Reprint. Baltimore: The Johns Hopkins University Press, 1996.
Ginsberg, A. (1984). *Collected Poems, 1947–1980*. New York: Harper and Row.
Goldberger, P. (2000). Busy Buildings. *The New Yorker*, 4 September, 90–93.
Gratz, R. B., & Mintz, N. (1998). *Cities Back from the Edge: New Life for Downtown*. New York: John Wiley.
Hannigan, J. (1998). *Fantasy City: Pleasure and profit in the postmodern metropolis*. London: Routledge.

Harris, N. (1991). Urban Tourism and the Commercial City. In: W. R. Taylor (Ed.), *Inventing Times Square: Commerce and Culture at the Crossroads of the World* (pp. 66–82). Reprint. Baltimore: The Johns Hopkins University Press, 1996.

Harvey, D. (1989). *The Condition of Postmodernity: An Enquiry into the Origins of Cultural Change.* Oxford: Blackwell.

Hillenbrand, B. (1999). An Eye for a Media Deal. *Time, 154*(12), 20 September, 60.

Holusha, J. (1998). Commercial Property/The Reuters Building at Times Square; A Corporate Headquarters Next to Bugs and Mickey. *The New York Times on the Web,* 6 September.

Holusha, J. (1999). After a Roller-Coaster Year, Optimism. *The New York Times on the Web,* 10 January.

Huxtable, A. L. (1991). Re-Inventing Times Square: 1990. In: W. R. Taylor (Ed.), *Inventing Times Square: Commerce and Culture at the Crossroads of the World* (pp. 356–370). Reprint. Baltimore: The Johns Hopkins University Press, 1996.

Jameson, F. (1996). Space Wars. *London Review of Books, 18*(7), 4 April, 14–15).

Johnson, S. (1999). Welcome to the Pleasure Dome. *The Village Voice Literary Supplement.* February, 1–6. <http://www.villagevoice.com/vls/160/johnson.shtml> (12 April 1999).

Koolhaas, R. (1994). The Generic City. Office for Metropolitan Architecture. In: Rem. Koolhaas & Bruce. Mau, *Small, Medium, Large, Extra-Large* (pp. 1239–1264). Rotterdam: 010 Publishers, 1995.

Leach, W. (1991). Commercial Aesthetics: Introductory Essay. In: W. R. Taylor (Ed.), *Inventing Times Square: Commerce and Culture at the Crossroads of the World* (pp. 234–242). Reprint. Baltimore: The Johns Hopkins University Press, 1996.

McNamara, B. (1991). The Entertainment District at the End of the 1930s. In: W. R. Taylor (Ed.). *Inventing Times Square: Commerce and Culture at the Crossroads of the World* (pp. 178–190). Reprint. Baltimore: The Johns Hopkins University Press, 1996.

McNamara, K. R. (1999). CityWalk: Los(t) Angeles in the Shape of a Mall. In: Ghent Urban Studies Team (Eds), *The Urban Condition: Space, Community, and Self in the Contemporary Metropolis* (pp. 186–201). Rotterdam: 010 Publishers.

Page, M. (1999). *The Creative Destruction of Manhattan, 1900–1940.* Chicago: The University of Chicago Press.

Purdy, M. (1997). Disney Parade About to Turn New York Goofy. *The New York Times on the Web,* 13 June.

Register, W. W., Jr. (1991). New York's Gigantic Toy. In: W. R. Taylor (Ed.), *Inventing Times Square: Commerce and Culture at the Crossroads of the World* (pp. 243–270). Reprint. Baltimore: The Johns Hopkins University Press, 1996.

Reichl, A. J. (1999). *Reconstructing Times Square: Politics and Culture in Urban Development.* Lawrence: University Press of Kansas.

Russ, A. (1999). *The Celebration Chronicles: Life, Liberty, and the Pursuit of Property Value in Disney's New Town.* New York: Ballatine Books.

Rowe, C., & Koetter, F. (1978). *Collage City.* Reprint. Cambridge, MA: MIT Press.

Rutheiser, C. (1996). *Imagineering Atlanta: The Politics of Place in the City of Dreams.* London: Verso.

Rybczynski, W. (1995). *City Life: Urban Expectations in a New World.* New York: Scribner.

Saegert, S. (1981). Masculine Cities and Feminine Suburbs: Polarized Ideas, Contradictory Realities. In: C. Stimpson et al. (Eds), *Women and the American City* (pp. 93–108). Chicago: The University of Chicago Press.

Sandeen, E. (n.d.). Signs of the Times: Waiting for the Millennium in Times Square, pp. 1–12. <http://www.nottingham.ac.uk/3cities/sandeen.htm> (12 May 1999).

Schickel, R. (1998). Walt Disney. *Time*, *152*(23), 7 December, 74–77.
Senelick, L. (1991). Private Parts in Public Places. In: W. R. Taylor (Ed.), *Inventing Times Square: Commerce and Culture at the Crossroads of the World* (pp. 329–353). Reprint. Baltimore: The Johns Hopkins University Press, 1996.
Sennett, R. (1994). *Flesh and Stone: The Body and the City in Western Civilization.* New York: W. W. Norton.
Stone, J. (1982). *Times Square: A Pictorial History.* New York: Collier Books.
Taylor, W. R. (Ed.) (1991). *Inventing Times Square: Commerce and Culture at the Crossroads of the World.* Reprint. Baltimore: The Johns Hopkins University Press, 1996.
Taylor, W. R. (1992). Times Square as a National Event. In: *In Pursuit of Gotham: Culture and Commerce in New York* (pp. 93–108). New York: Oxford University Press.
Wallace, M. (1996). *Mickey Mouse History and Other Essays on American Memory.* Philadelphia: Temple University Press.
Weber, B. (1996). In Times Square, Keepers of the Glitz. *The New York Times on the Web*, 25 June.
Weber, B. (1997). Disney Unveils Restored Theater on 42nd Street. *The New York Times on the Web*, 3 April.
White, N., & Willensky, E. (2000). *AIA Guide to New York City* (4th ed.). New York: Three Rivers Press.
Zukin, S. (1991). *Landscapes of Power: From Detroit to Disney World.* Reprint. Berkeley: University of California Press, 1993.
Zukin, S. (1995). *The Cultures of Cities.* Cambridge, MA: Blackwell.

REDEVELOPMENT FOR WHOM AND FOR WHAT PURPOSE? A RESEARCH AGENDA FOR URBAN REDEVELOPMENT IN THE TWENTY FIRST CENTURY

Kevin Fox Gotham

INTRODUCTION

> Constant revolutionizing of production, uninterrupted disturbance of all social relations, everlasting uncertainty and agitation, distinguish the bourgeois epoch from all earlier times. All fixed, fast-frozen relationships, with their train of venerable ideas and opinions are swept away, all new-formed ones become obsolete before they an ossify. All that is solid melts into air, all that is holy is profaned, and men at last are forced to face with sober senses the real conditions of their lives and their relations with fellow men (Marx & Engels 1948, p. 12).

In the *Manifesto of the Communist Party*, Karl Marx and Friedrich Engels describe modern society as a society of chaos and turmoil, where class conflict is endemic, and social stability is a fleeting and ephemeral condition. For Marx and Engels, the modern city concentrates and expresses most vividly the peculiarities of capitalism including, for example, the antithesis between wage labor and capital, the valorization of exchange-value, and proletarization of the populace. Yet the city is also an important birthing area for class consciousness and the development of revolutionary social movements that can challenge the exploitative conditions of capitalist social relations. As Engels notes, this "double tendency" of capitalist urbanization makes visible and intensifies the problems of poverty, unemployment, and deprivation that, in turn, provoke agitation and calls for reform and "redevelopment." Local attempts to eliminate the slums in one area of the city, however, mean their simultaneous appearance and growth in another part of the city. As Engels (1969b) states in his essay on urban housing, "the bourgeoisie has only one method of settling the housing question ... The breeding places of disease, the infamous holes and cellars in which the capitalist mode of production confines our workers night after night are not abolished; they are merely *shifted elsewhere*! The same economic necessity which produced them in the first place produces them in the next place also!" (Quote appears in Saunders 1981, p. 23). In particular, Engels observes that in modern cities, the rich and poor live in segregated spaces, neighborhoods and land-uses are in a state of flux, and movement, migration, and "everlasting uncertainty" are the defining characteristics of urban life (1969a).

Marx and Engels' vision of the modern city as a city of upheaval, fragmentation, and discontinuity sheds light on the nature of urban redevelopment and the transformation of metropolitan space and life at the beginning of the twenty first century. Traditionally, urban scholars have connected the term "redevelopment" to the central city, although rural towns and suburbs have become increasing locked in a competitive struggle to redevelop themselves and attract capital. Since the 1960s, urban areas in the United States have been subjected to a series of unprecedented socio-spatial changes, including a decrease of population and employment in the "rustbelt," redevelopment of the inner core of some older metropolitan areas, increase in minority populations in central cities, rapid growth of "edge cities" on the metropolitan fringe, and the economic decline of older suburban communities, among other changes. Moreover, urban politics has moved away from the days of generous federal funding through fiscal austerity to the present period of limited resources, privatization of services, and combined public/private partnerships in pursuit of growth. Today, in virtually all cities, policy-makers and elites have perceived their economic base as endangered from competition from other cities and have struggled to develop various programs, fiscal policies, and other subsidies to attract businesses. The contributors have outlined and discussed these concerns in detail throughout the preceding twelve chapters. At this stage, I wish to bring together common themes for making some preliminary suggestions as to the way forward for a research agenda in urban redevelopment in the twenty first century.

All of this volume's contributors have discussed the significance, characteristics, and consequences of urban redevelopment on urban life and change. Yet a satisfactory and agreed upon definition of "redevelopment" is elusive since the term assumes that someone or something is "undeveloped." What and who is need of "redevelopment" needs spelling out: is it the tax base? Downtown property values? Neighborhoods where poor people and racial minorities live? Other questions inevitably arise: What are the positive effects and negative consequences of redevelopment? Which groups and interests benefit or suffer from redevelopment? How has the nature, pace, and character of urban redevelopment changed throughout the decades? How are these changes connected to macrostructural changes in the state and the economy? The contributors to this volume have addressed these and other important questions in an effort to shed empirical and theoretical light on novel and enduring features of urban redevelopment. To this end, I propose identifying five areas in which the contributors here have identified as key points of concern which may help to propel research on urban redevelopment into our new century. I will then discuss these substantive concerns more broadly under the umbrella of recent developments in theory and method in urban sociology. We can therefore identify the five points of concern arising out of the chapters collected here: growth coalitions and growth conflicts; globalization; gentrification and the transformation of urban space; culture and history in urban redevelopment; class, gender, and race issues.

GROWTH COALITIONS AND GROWTH CONFLICTS

Most of the contributors comment on the various dimensions and meanings of economic "growth" in their studies of urban redevelopment. Indeed, what connects the empirical work in this volume with the various growth-theories of urban redevelopment – for example, the growth machine, the growth regime, the entertainment machine, and so on – is the contention that city leaders seek to create or sustain economic growth and private profits to maintain government services, fiscal health, and project an image of a "good" business climate (Imbroscio 1998). In their seminal work, *Urban Fortunes: The Political Economy of Place*, Logan and Molotch (1987) argued that the formation of coalitions in pursuit of growth permeates all facets of local life, including the political system, as well as local utility companies, unions, media, and cultural institutions such as professional sports teams, theaters, symphony orchestras, and universities (Molotch, 1976, 1999; Logan & Molotch, 1987). The contributors to this volume note that local politics and the pursuit of "growth" is shot through with conflict at multiple levels: from struggles over the use of public monies to finance large-scale redevelopment projects to the role of local government in engineering new growth or containing the negative consequences of past growth initiatives. Not all pro-growth coalitions use the same strategies to influence local politics and policy and the balance power in any community is contingent upon a variety of local and external conditions (Gotham, 2001a). The theoretical and empirical literature recognizes that different cities can embrace different "growth" strategies (pro-growth, managed growth, slow-growth, no-growth, anti-growth) and not all support the same kinds of growth (population growth, manufacturing versus tourism-based growth, downtown revitalization versus outlying residential development, and so on) (Purchell, 1999; Hamilton, 1999; for an overview, see Logan, Whaley & Crowder, 1997).

Over the last two decades, much research has investigated the role of progrowth coalitions in the formulation and specific character of local redevelopment drives, influence of economic elites in local politics and decision-making, and the diverse growth strategies used by elites to promote their agenda (Cummings, 1988; Feagin, 1988; Jonas & Wilson, 1999; Squires, 1989). Who promotes growth and local economic development are central to urban theory, the politics of cities, and places in the global economy. While scholars acknowledge the interconnectedness of local, national, and global forces in the dynamics of urban growth, scholars do not agree on whether research should "start with the local details and build up" (Molotch, 1999, p. 255) or eschew "bottom-up methodology" and examine wider economic and political forces and processes to understand local conditions (Jessop, Peck & Tickell, 1999, pp. 142, 144; Imbroscio, 1998). Most empirical research using growth machine theory embraces the former, more inductive and agency-centered focus to concretely situate actions, events, and processes in broader historical contexts. Future research should relate growth issues – meanings of growth, growth coalition formation and dissolution, mobilization of opposition to growth, and so on – to wider economic and political contexts. In the arena of local redevelopment strategy and policy, we should also focus on how "growth" is itself contested terrain, a battlefield of contending interests with losers and winners. While empiricism and local study is important, "the big story remains beyond the local horizon," according to Flanagan (1993, p. 164). While case studies add descriptive depth, richness of analysis, and contextualization of explanations, we should also strive to be explicit about the frame of reference guiding the research. Moreover, the focus on "the case" should not divert us from being precise about the definition and logic of connection between concepts, and the obligation to attend to the theoretical aspects of our empirical investigations.

The chapters by Aguirre and Brooks, Bures, Crowley, and Lloyd and Clark bring needed attention to the important role that political factors play in shaping urban redevelopment and mediating struggles and conflicts over growth. Some scholars argue that local governments are dependent on the national and global economy (Dear & Clark, 1981, pp. 1278–1283; Clark & Dear, 1984) whereas others argue that local and political institutions remain critical for promoting growth (Friedland, Piven & Alford, 1977). According to Logan, Whaley, and Crowder (1997), scholars have typically viewed the local state as a "limiting force" in the effectiveness of no-growth or anti-growth forces, or a "stabilizing force" for pro-growth coalitions, "maintaining a predictable tilt in favor of development proposals despite minor shifts in the governing coalition" (see also Horan, 1991; Calavita, 1992). Yet this view of the local state has been subject to criticism from scholars who argue that the growth machine approach has failed to consider the efficacy of local political structures and formal politics (Logan et al., 1997), and neglected to specify the connections between the local state and the national state (Kirby & Abu-Rauss, 1999, pp. 214–215; Jessop, Peck & Tickell, 1999, p. 159; Nelson & Foster, 1999). Much research has examined how a growth coalition can unify otherwise disparate groups of investors and capitalists through the ideology of "growth" but few studies have focused on the role of the state arrangements and agencies in fomenting discord within a growth coalition. For Kirby and Abu Rass (1999, p. 215), "the local state is in large measure defined as the locus of growth and of the political issues that surround it." As the chapters in this volume show, the local state plays a crucial role in unifying and stabilizing a growth coalition (Crowley), facilitating the adoption of some growth strategies rather than others (Lloyd & Clark), and mediating conflicts over meanings and growth and redevelopment (Aguirre & Brooks).

Urban scholars will continue to disagree over the dimensions "growth" and the efficacy of growth machine theory, as Lloyd and Clark point out. Although empirical research is beginning to recognize variations in the influence and effect of progrowth coalitions on local redevelopment initiatives, few studies have documented the locally – and historically – specific sources of variation in the composition and operation of a growth coalition. Moreover, few scholars have clearly identified the specific political, organizational, historical, and cultural factors that explain how and why cities pursue some growth strategies rather than others. The chapters by Crowley, Bures, Lloyd and Clark, and Eeckhout offer some suggestions for future research and they acknowledge, in various ways, the necessity of pushing research toward examining the role of culture and history as growth strategies, the internal workings of growth coalitions, the various roles played by their different members, the tactics these actors employ, and the conflicts that develop among them. In particular, scholars should avoid totalizing views and overgeneralizations about the so-called "functions" the local state (or national state) performs, concentrating instead on charting state actions and consequences of actions in a grounded reflexive fashion. Moreover, we clearly need more historical and comparative studies that connect different growth strategies at the local level with the changing character of both U.S. economic and political structure. Other important issues such as territorial or spatial conflict, spatially based power relations, rise and fall of different fractions of capital (property, finance, industrial, etc.), community resistance as a component of political economy, and the consumption and production of space need more attention.

GLOBALIZATION

One debate that has significant implications for the transformation of cities and which appears to be increasingly high on urban scholars' agenda overall is that of globalization. In many respects, urban redevelopment appears to be a global phenomenon (with the requisite synonyms and metaphors) – as Paquin's case study of Seoul, South Korea illustrates. Citing the growth of transnational corporations and the enhanced mobility of goods, services, and money, proponents of the globalization thesis argue that the transition to a globalized economy has created new forms of power and influence that are beyond the reach of federal regulatory agencies, national labor unions, and other forms of democratic decision making and citizen action. Defined in this manner globalization has undermined the scope and role of the welfare state, depressed wages and living standards, and intensified the salience of place-bound identities (Cox, 1997; Wilson, 1997; Amin, 1994; Sassen, 1991, 1998; Featherstone, 1990; for an overview see Riain, 2000). The problem here, as many scholars have noted and several contributors to this volume show, is that the study of globalization suffers from conceptual ambiguity and lack of specificity. Urban scholars have criticized proponents of the globalization thesis for portraying "globalization" as a monolithic and inexorable process following its own internal logic, for downplaying the significance of political variables, and for de-emphasizing the impact of national state arrangements in mediating local impacts of economic change (Logan, Whaley & Crowder, 1997, pp. 84–85; Logan, 1993, 1991; Logan & Swanstrom, 1990; Molotch, 1999, pp. 260–262). Others have argued that the term is a buzzword and a cliche that ignores history and hides the enduring features of capitalism (Giddens, 1996, p. 61; Harvey, 1989; Cox, 1997; Abu-Lughod, 1989; Gottdiener, Collins & Dickens, 1999, p. 7; Silver, 1993).

Despite these conceptual and analytic problems, the concept "globalization" has a key role to play in the study of urban redevelopment throughout the world, whether they are cities in Europe or the United States or cities in less-developed nations (Sassen, 2000). As such, balancing the sometimes contradictory juxtaposition of local history, political and cultural practices, national state arrangements, and the globalizing tendencies of multinational corporations is very important, as Paquin and Indergaard note. Jan Lin's (1995, 1998) work on redeveloped ethnic places as "polyglot honeypots" and Christopher Mele's (1995) discussion of the redevelopment of the Lower East of New York draw attention to the contradictions and conflicts, as well as opportunities presented by globalization. According to Cox and contributors (1997), Wilson and colleagues (1998), and Van Kempen and Marcuse (1997, p. 285), globalization entails different processes that produce different spatial consequences including, for example, loss of local autonomy, displacement, socioeconomic polarization in and between places, racial and ethnic conflict, among others. The growing feeling that – despite conflicting definitions of urban "redevelopment," "globalization," and the like – we all share the same globe and face some form of the balance between the homogenizing tendencies of capitalism and the determination of local people to affirm their unique character has massive implications for the future of cities and communities. The ways in which cities manifest the tensions between global process and local actions should therefore represent a key concern of scholars into urban redevelopment, in the United States and elsewhere.

GENTRIFICATION AND THE TRANSFORMATION OF URBAN SPACE

Urban scholars have long discussed central cities in terms of decay, poverty, and fiscal crisis. Yet in recent decades, new images have been articulated, envisioning central cities with new investment, rejuvenation of central city neighborhoods, and a so-called back-to-the-city movement of affluent whites. Traditionally, scholars have described gentrification as a process of residential development in central city areas that involves renovation for middle- and upper-income people and the displacement of poorer residents. However, Neil Smith (1996, p. 70) astutely observes that "Gentrification is a back-to-the-city movement all right, but a back-to-the-city movement by capital rather than people." This comment brings attention to the notion that key actors and organized interests of real estate and finance capital have much at stake in enhancing the exchange value of urban space in redeveloping the city. Therefore, in order to understand the causes and consequences of gentrification and how it affects socio-spatial organization, we should first examine the role of capital in the process of gentrification, proceed to examine the conflicts between residents and powerful economic and political actors in redeveloping the city, and move towards an analysis of the social and cultural aspects of creating gentrified spaces. Indergaard draws attention to the impact of new media technologies in creating "commercial gentrification" in New York City's Silicon Alley, and Bures discusses the racial conflicts over gentrification and the control of residential space. More broadly, Wyly and Hammel show that debates about gentrification link recent socio-spatial changes in our cities to a number of different issues including, for example, policy discussions over fair lending, racial discrimination in urban housing markets, urban design and growth strategies, anti-poverty policy, privatization, and the devolution of certain federal housing programs to states and cities.

Several contributors bring needed attention to the role of federal policy, housing programs, and national regulatory arrangements in the transformation of urban space generally and public housing specifically. Wyly and Hammel's chapter connects with an embryonic evaluation literature that has documented the progress of the redevelopment of public housing funded through the HOPE VI program. A recent series of reports from HUD maintain that HOPE VI is reducing isolation by providing opportunities for employment and education and engaging residents in the life and prospects of the community (Department of Housing and Urban Development (HUD), 2000, 1999, 1996). Other accounts are not so optimistic. According to a 1998 report from the General Accounting Office (GAO, 1998), from 1993 to 1998, only eleven public housing sites around the nation had developed new units for residents and none of the sites' capital improvement plans had been completed by the middle of 1998. In addition to slow implementation, cost-overruns and resident opposition to demolition and displacement have plagued HOPE VI revitalization efforts in many cities, including Cincinnati, Boston, and Charlotte, North Carolina (Pitcoff, 1999). Gayle Epp (1996) found considerable local variation in the success and implementation of HOPE VI funded comprehensive social and community support services for residents in Seattle and Indianapolis. More recently, Jerry Salama's (1999) comparison of three public housing developments in Atlanta, Chicago, and San Francisco showed that HOPE VI programs were only marginally successful in meeting their goals due to lack of standards for improved physical design and resident involvement. Recent research recognizes the complexity and trade-offs of trying to meet the diverse and overlapping goals of the HOPE VI legislation (Vale, 1997). Yet lack of data at the project level, anecdotal data on indicators of success at the PHA level, and lack of comparability of data between PHAs, make the definition and evaluation of redevelopment "success" problematic (Epp, 1996, p. 567–568; Salama, 1999; Vale, 1997, 1996; 93).

My chapter with Shefner and Brumley, and the chapter by Wyly and Hammel suggest that the redevelopment of public housing is the newest form of state-sponsored redevelopment that seeks to transform public housing authorities into public/private redevelopment agencies. Here the effort is to eradicate the blight of public housing and transform these derelict urban spaces into privatized sites of middle-class residential consumption. The rhetoric and language is that of renewal, progress, and revitalization of poverty-plagued neighborhoods. Yet behind the vernacular of urban progress lies the efforts of private developers and state actors to exclude, segregate, and make invisible the poor and homeless. These disadvantaged groups typically represent negative externalities, endangering the exchange value of space, and threatening the ability of upscale settings to deliver style, distinction and exclusivity, and culture. Thus, as we note, the redevelopment of public housing is a form of "exclusive" redevelopment that is designed to exclude the very poor from the revitalized spaces and render them safe for resettlement by the wealthy and affluent. Not surprisingly, one of the major sources of conflict and delay in the implementation of public housing revitalization around the country has been persistent resident opposition to demolition and relocation (General Accounting Office, 1998; Epp, 1996; Pitcoff, 1999). In an article on the relationship between public housing and urban redevelopment in New Orleans, Cook and Lauria (1995) argued that public housing was an obstacle to urban redevelopment and that the political component of the city's downtown redevelopment strategy was the containment of low-income people within public housing. My chapter with Shefner and Brumley and the chapter by Devine and Sam-Abdiodun suggests that this strategy of spatial containment may be giving way to a strategy of spatial dispersal and dislocation. Whether this strategy is occurring in other cities remains to be seen. Nevertheless, detailed fieldwork is a necessary step in documenting the impact of public housing redevelopment on cities.

The key point here is that the widely praised downtown resurgence and latest attempt to "reinvent" public housing is yet another chapter in the long term process of uneven development across cities, a process with winners and with losers. Yet, all too often, policy analysts take for granted that gentrification and redevelopment benefits everyone. What are the consequences of gentrification when political officials and other elites deny access to space? How far does the denial of that access determine people's experience of city life? Again, the tendency in most urban research is to discuss gentrification in an overgeneralized fashion, with little attempt to contextualize the socioeconomic experience of those who suffer from gentrification. In this respect, research into urban gentrification should tackle the issues of spatial segregation, inclusion and exclusion, and ask how redevelopment in one area of the city can mask the decline and disinvestment of other areas of the city. Wyly and Hammel correctly point out that the analysis should focus on the ongoing deregulation of the nation's real estate and financial institutions, and recent changes in federal regulatory policy and mortgage financing as starting points for understanding the causes and consequences of gentrification. Critical urban scholars have called into question the claim that consumer desires or individual preferences are forces to which capital merely reacts (Feagin, 1998). Consumer taste for gentrified spaces is, instead, created and marketed, and depends on the alternatives offered by property capitalists who are primarily interested in producing the built environment from which they can extract the highest profit. As Wittberg (1992, p. 23) noted, "the invisible hand postulated by urban ecology and neoclassical urban economics to shape the patter of urban land use is an 'ideological smoke screen' hiding the power of those who stand to benefit by such obfuscation."

A number of chapters pick up on themes of gentrification, regulation, and surveillance of everyday life in the city. Proponents of the "fortress city" thesis relate the "militarization of space" to the proliferation of new forms of segregative planning, security, and control to keep out the urban underclass from the gentrified spaces of the affluent (see, for example, Davis, 1992; Christopherson, 1994; Judd, 1995). As Aguirre and Brooks observe in their chapter, the increasing use of war rhetoric accompanies the stigmatization of the homeless by urban officials as they decry the encroachment of the very poor upon the public spaces of the city. Yet the connection between urban redevelopment, gentrification, and the "end of public space" is only one side of the story. In addition to attempts at enhancing police control of "deviant" populations, local officials publicly proclaim the importance of promoting the "diversity" and vitality of urban spaces, revealing a much more complex and ambivalent relationship between gentrification and the city than notions of the "fortress city" consider. Nevertheless, Aguirre and Brooks connect the rhetoric of crime and the reclaiming of public space with the privatization of public space as commercial imperatives increasingly define what is "normal" behavior and attempt to impose their own image of a "good" city on the built environment. One challenge for future research is to explain how city officials imagine urban space, carry out their conceptions, and rationalize their plans and developments by invoking the rhetoric of "progress" or conversely the rhetoric of "safety." At the same time, research should focus on instances of collective resistance to these dominant images and representations of space, asking how and why such resistences appear, and about the classed, gendered, and racialized identities of the peoples involved in these acts of resistance.

CULTURE AND HISTORY IN URBAN REDEVELOPMENT

One of the more interesting lines of urban research looks at how powerful actors and organizations are redeveloping cities as sites of fun, leisure, and entertainment. This scholarly focus reflects a broader interest in the political economy of tourism, the transformation of public spaces into privatized "consumption" spaces, and the latest attempts by urban leaders to provide a package of shopping, dining, and entertainment within a themed and controlled environment – a development that Eeckhout and other scholars have called the "Disneyification" of urban space (Sorkin, 1992). Scholars have examined the commodification of history and culture in the marketing of cities (Boyer, 1992; Reichl, 1997, 1999; Strom, 1999), the use of imagery, theming, and other signifying practices (Kearns & Philo, 1993; Gottdiener, 1997), and the role of "cultural representers" and "urban imagineers" in engineering urban redevelopment (Short, 1999; Holt, 2000). In many cities, political and economic elites are attempting to commodify and market culture and develop various cultural facilities – e.g. art shows and galleries, opera halls, museums, festivals, symphony halls, and so on – to stimulate the local economy by attracting consumers and supporting inward investment (Bassett, 1994; Zukin, 1995). Other cities are emphasizing the aesthetic or historic value of their architecture, redeveloping their river and canal waterfronts, designating areas of the city as artistic quarters, and preserving or reconverting old buildings and archaic technology (adaptive reuse), a process discussed by James Dickinson in his chapter on the "monuments of tomorrow." A common element in these and other "cultural strategies of economic development," according to Zukin (1997, p. 227) is that "they reduce the multiple dimensions and conflicts of culture to a coherent visual representation." Thus, cultural, historical, or architectural materials "can be displayed, interpreted, reproduced, and sold in a putatively universal repertoire of visual consumption" (p. 227).

James Dickinson's chapter points out that many older "rustbelt" cities are increasingly involved in marketing their industrial history and architecture, converting their derelict and abandoned factories and buildings into tourist attractions and places of consumption. During the 1970s, urban areas in the Midwest and northeast that had flourished in the postwar Keynesian boom, entered a phase of chronic industrial decline. Cities such as Kansas City, St. Louis, Milwaukee, Cleveland, Detroit, Philadelphia, Buffalo, and the like, were forced to deal with the triple problems of deindustrialization, falling tax base, and declining public expenditures in an era of intense interurban competition for capital and consumers (Gotham, 1998a). To revive their declining economies, cities have attempted to transform their derelict spaces of industrial production into spaces of consumption such as shopping malls, themed restaurants, casino gambling, mega-complexes for professional sports, and other sites for sales that rely on glitz, hype, and spectacle (Gottdiener, 1997; Gladstone, 1998; Lord & Price, 1992). In the process, redevelopment interests and advertising agencies thematize local traditions, famous buildings and landmarks, and other heritage sights to the point that they become "hyperreal," whereby people lose the ability to distinguish between the "real" and the "representation." (Baudrillard, 1988; Smith, 1996). As far as possible, urban imagineers, cultural representers and other redevelopment interests attempt to remake the industrial city of the past, according to Dickinson, into a "theme park" city (Sorkin, 1992), a city of simulations and visual consumption for entertainment, a point also highlighted by Eeckhout and Lloyd and Clark.

The use of culture and history as marketing strategies in the economic redevelopment of cities has a long history in the United States and elsewhere (Kearns & Philo, 1993; Harvey, 1989; Strom, 1999). However, several chapters in this volume corroborate other studies that have focused on how the increasing dominance of multinational corporations has transformed the marketing of places from a relatively amateur and informal activity into a professionalized and highly organized activity (Holcomb, 1993, p. 133; Judd & Fainstein, 1999). In his recent book on the rise of the "fantasy city," John Hannigan (1998) points out how corporations such as Disney, Universal, Time-Warner, and Sony, among other entertainment giants are becoming the key actors in redeveloping the city. Interestingly, what makes these entertainment companies unique to urban redevelopment is that they control almost every aspect of the redevelopment process, including design and planning, land acquisition and preparation, marketing, and operations. Moreover, as Hannigan (1998, p. 104) notes, urban redevelopment projects today must deliver more than just location, structure, and leases; they must also provide "meaning" and "feeling" to entice the targeted consumer. Thus, large entertainment companies have become highly organized developers, merging culture and capital to "become impresarios, locations their venues, facades and interiors their sets and stages, and retail tenants their characters and chapters in their story" (p. 104).

Several chapters bring needed attention to how local elites and redevelopment actors attempt to market urban images – city myths, localized cultures, identities – to promote the investment potential of cities (Hannigan, 1998; Judd & Fainstein, 1999; Strom, 1999; Sadler, 1993). In particular, Eeckhout and Lloyd and Clark intimate how local institutions have established resource links with the advertising industry to supply products, services, and commodified experiences to tourists. Yet we need more research on how local institutions and elite are inserting "culture" into the "place marketing" strategies of cities as a mechanism of social control to engineer consensus, discourage opposition, and coopt resistance to various redevelopment efforts (Kearns & Philo, 1993). Future research should move beyond the purely descriptive studies and develop sophisticated accounts, informed by political economy, that document the "structural" linkages between the various institutional actors – corporate actors, entertainment companies, real estate developers, retail operators, and public agencies – and the development of the tourist city generally, and the redevelopment of cities specifically. In addition, we need more empirically grounded examinations of the mechanisms that link macrostructural changes in the state and the economy with the development of tourist-led redevelopment in specific cities. Future research needs to examine how "spaces of consumption" are important to the "production of space" of contemporary capitalism.

CLASS, GENDER, AND RACE

Recent years have witnessed a sea change in the way urban scholars conceptualize class, gender, and race and link these categories and processes to theories of power and urban space. Most scholars agree that class, gender, and race are not unilateral or constant but involve different dimensions or "layers" that interpenetrate with each other in complex, multifaceted, and historically changing ways. Patricia Hill Collins' (1991) work shows how a "both/and" reality as opposed to an "either/or" distinction shapes race, class, and gender identities and processes. Urban scholars such as DeSena (2000), Spain (1993), and McDowell (1999), among others have focused on the interlocking dimensions of class, gender, and race, and the connection of these social attributes to spatial arrangements. We can examine spatial arrangements, such as spatial segregation, differential access to public and private spaces, and so on as the context for understanding how class, gender, and race interconnect to shape life experiences, social conflict and action. Segregated spaces and various racialized, gendered, and class-based meanings attached to spaces exist as recurrent patterns of action involving large numbers of people. If we acknowledge that race, class, and gender are sources of inequality and division expressed in space, we have to identify who does what to whom, how do they do it, and how they maintain these spatialized divisions over time. Clearly, we need sophisticated, data-rich analyses that are capable of dealing with material conditions, actions within and coordinated across space, and the "durability" or "fragility" of social structure.

Recent work in urban sociology has explored the use of various overt and covert "gendering" and "racializing" strategies used by urban elites to create maintain segregated spaces (Gotham, 2000; Miranne & Young, 2000). Decisions about urban redevelopment always express class and racial interests and particular groups win and lose, regardless if conflict becomes overt. Indeed, as the history of urban renewal, gentrification, and urban expressway building shows, central city disinvestment and reinvestment are part of the dynamics of urban racial relations, with African Americans and other minorities having to bear the brunt of displacement and neighborhood destablization (Gotham, 2001a, b, 2000, 1998; Squires, 1994; Feagin & Parker, 1990, pp. 145–146; Kaplan, 1999). In *Race, Culture, and the City*, Stephen Nathan Haymes (1995) argues that "in the context of American cities the category 'race' is used metaphorically as a way to juxtapose the different 'social spaces' that make up the urban landscape, describing some as 'normal' and 'order' and others as not." For Haymes, "contemporary urban forms are the spatial expression of racialized values," especially the ideology that equates land value with whiteness, and the residential presence of blacks with neighborhood deterioration and declining property values. As Short (1999) and Beauregard (1993) have pointed out, racial minorities are widely if not always blamed for the deterioration of cities making "urban decline a virtual stand in for race" (Beauregard, 1993, p. 291). Yet "as the focal point of America's social problems, race displaces decline from the political economy of cities and relocates it in the cultural deficiencies of racial minorities" (p. 291). While journalists, politicians, and others may identify victims and actors, most commentators refuse to place them in institutional settings or critically examine the role of social structure.

Much research has focused on the racial and class-based conflicts that accompany urban redevelopment drives but scholars have devoted less attention to the implications and conflicts over attempts to market local racial and ethnic identities as commodifies for tourist consumption. Jan Lin (1998, p. 315) has drawn attention to this issue, pointing out that the redevelopment of the "new ethnic places" in many cities may authenticate "difference" and "diversity" while reinforcing class distinctions, serving as arenas for the investment and further accumulation of capital. Moreover, Zukin (1997) and Wright and Hutchison (1997), among others, have pointed out that local histories and traditions of various racial and ethnic groups have become marketable commodities – resources and strategies to advance the interests of redevelopers and their allies. Smith's (1996) work on the Lower East Side and Hoffman's (2000) case study of the redevelopment of Harlem show how redevelopers package ethnicity and race as culture and art, using frontier motifs and imagery to "tame" a neighborhood, touting images of exotic and benign danger to pull in consumers. This strategy is apparent in Times Square, as Eeckhout points out, where various ethnic and racial groups, foreign subcultures, and domestic subcultures are appropriated, disemboweled, and then marketed as safe, sanitized versions of the original. Hannigan (1998, p. 71) calls this phenomenon "riskless risk" in which "many contemporary leisure and tourist attractions are calculated so as to package our fantasy experiences within a safe, reassuring predictable environment."

The chapters by Bures, Lloyd and Clark, and Eeckhout connect with the work of Zukin (1997) and Reichl (1997, 1999) who note that the marketing of terms like "multiculturalism" and "diversity" allows developers to acknowledge eclecticism, allotting each ethnic and racial group some form of representation, while avoiding ranking oppressed groups in terms of the justness of their claims. Not surprisingly, critics have assailed this strategy for sanitizing and distorting history, ignoring the present reality of racial discrimination and social injustice, and reinforcing social inequalities (Holt, 2000; Hodder, 1999; Eisinger, 2000). Croucher (1997) has focused on how local elites in Toronto construct images of ethnic harmony and diversity to divert public attention away from urban problems, legitimate elite definitions of urban reality, and generate support for some urban growth initiatives rather than others. Yet there are tensions and conflicts in the use of ethnic imagery, community typifications, and local narratives to mobilize support for various urban redevelopment. Indeed, urban redevelopment is always a site of struggle where the powers that be are often forced to compete and defend what they would prefer to have taken for granted. The "undetermined" nature of the imagery, discourses, and typifications surrounding urban redevelopment often allow challengers such as social movements to offer competing constructions of reality in ways that go beyond elite definitions and interpretations of urban reality. Indeed, a nascent literature has begun to identify the various sentimental imagery, nostalgia, rhetorical devices, and various techniques of "framing" that anti-growth activists use to build community identity at the local level to challenge various revitalization initiatives (Gotham, 1999; for an overview, see Gieryn, 2000).

Part of the process of coming to understand the role of race, class, and gender lies in changing the ways in which we understand the city, and the ways in which we translate these understandings into urban research and public policy. Focusing on implications raised by the use of marketing techniques and the racial and spatial consequences of redevelopment means focusing on questions of political accountability, equity, and justice (Merrifield & Swyngedouw, 1997; Kearns & Philo, 1993; Bird, Curtis, Putnam, Robertson & Tickner, 1993; Harvey, 1973). Such an agenda demands a move away from the dead-end of policy "evaluation" and preoccupation with so-called "value-free" "measures" of redevelopment "success" that marketing firms and developers rely on to legitimate their projects. Indeed, constructing an image of the city, as a form of advertising promotion, is an exercise in the manufacture and transfer of meaning. As the chapter by Bodnar points out, the city is not a unitary community nor is it a neutral space that political and economic elites redevelop with all citizens in mind. A downtown is not simply a "downtown", but a social space that caters and privileges some particular segments of a population over other groups. In a world of diminishing spatial boundaries and increasing inequalities, the control of space is becoming more contested and more obviously the source of struggles for power and resources, a point highlighted by Castells and Harvey, among other urban scholars. Theorists of class, gender, and race can benefit from more scrutiny of spatial arrangements while theorists of space should consider how spatial arrangements intersect with class, gender, and race to structure differential access to power and resources.

THEORY AND METHOD IN URBAN SOCIOLOGY

Today, research in urban sociology is in a state of theoretical and methodological flux. Ideas about how to proceed vary tremendously and in somewhat opposite directions. They range from the advancement of strong research programs grounded in political economy (Feagin, 1998; Gottdiener, 1994), and feminist research (DeSena, 2000; Spain, 1992; McDowell, 1999), to interactionist-based theories (Milligan, 1998), discursive analysis and textual deconstruction (Liggett, 1994; Beauregard, 1993), and postmodernist readings of the city (Soja, 2000; Zukin, 1997). In diverse ways, the chapters in this volume borrow something from these very different and contrasting approaches. Some embrace a case study approach and begin at the local/neighborhood level while others focus on the global level, using large data sets and generalized analysis. Some emphasize the cultural dimensions of urban redevelopment while others privilege political and/or economic factors in their research. Indeed, the methodological and theoretical concerns animating this volume suggest that the new urban realities that are emerging do not lend themselves easily to analysis by existing approaches that aim for sweeping generalizations of change or rely on aggregate statistical evaluations to recreate pictures of urban reality. Wyly and Hammel, Devine and Sams-Abiodun, and Aguirre and Brooks take issue with single-method and –data approaches and suggest that research into urban redevelopment should aim for some combination of generalized analysis and ethnographic accounts of cities and urban life. The best kind of methodological strategy relies on multiple kinds of data – quantitative and qualitative – using and weighing them to improve our understandings and explanations.

The study of urban redevelopment is about more than just the study of the redevelopment of deteriorated parts of the city. As the chapters by Crowley and Paquin point out, far from being a spontaneous and isolated phenomenon, urban redevelopment operates within broader socio-cultural, economic, and political contexts and histories. Theories of urban redevelopment connect to theories of the state and the economy (Fainstein & Fainstein, 1986; Zukin et al., 1996; Feagin, 1988; Gottdiener, 1987), modes of consumption (Miles & Paddison,

1998; Urry, 1995), and theories of socio-physical space (Wright, 1997). For instance, one paradigmatic strand of urban research begins at the macro level and describes the functions that cities serve within the large context of capital accumulation (Harvey), or manage crises through collective consumption (Castells). In contrast, a more agency-centered urban sociology begins at the local level to study conflicts over different meanings and conceptions of urban space, including battles between those who seek to maximize the exchange-value of space versus those who value it for its intrinsic use (Logan & Molotch, 1987; Molotch, 1999). One disagreement between the advocates "globalism" and those who emphasize "localism" is whether we can systematically study global level changes and forms of organization without losing sight of local struggles. Is it necessary to eschew depth methods and ground-level scrutiny of action to understand processes at the macro-level? Crowley, Hammel and Wyly, and Paguin do not think so. They all show, in different ways, that the reshaping of urban redevelopment strategies, conflicts, and outcomes under the influence of national and global level restructuring is not a one-way process, and it is not necessary functional or even rational. In fact, when we examine the concrete activities of people and organizations involved in urban redevelopment we find considerable evidence of irrational and contradictory actions, a point highlighted time and time again by the contributors to this volume.

No discussion of urban redevelopment would be complete without acknowledging the growing influence of postmodern urban analysis. According to many urban scholars, we live in a time of the "posts" – postsuburbs, postmetropolis, postindustrialism, postfordism and postmodernism (Soja, 2000; Knox, 1993; Swanstrom, 1993; Watson & Gibson, 1995). The term "post" is a periodizing term that signifies the passing of the old and the advent of the new. Indeed, a number of scholars have written books describing the end of the metropolitan era and the coming of a new postmetropolis era that blurs and makes incomprehensible the boundaries between cities and suburbs, celebrates the demise of class based politics and stresses the centrality of "new social movements." Edward Soja (2000, p. 152), for example, argues that there has been a "restructuring of territorial identity and rootedness amidst a sea of shifting relations between space, knowledge, and power that has given rise to a *new cultural politics* in the postmetropolis, significantly different from the politics of the economy that dominated modernist urbanism" (emphasis in original). Others such as Beauregard (1993) and Liggett (1994) champion new postmodern methods such as textual deconstruction and discursive analysis, problematize the "city" as an empirical referent, and employ "texts' and "stories" as units of analysis. More critical are those who decry a "crisis of representation," pronounce the end of grand narratives and of modern theory, and call for new postmodern theories and politics to deal with the striking novelties of the present (King, 1996; Knox, 1993). The variety of postmodern readings of the city emphasize difference, plurality, fragmentation, and complexity; abandon representational epistemology and unmediated objectivity; and embrace perspectivism, anti-foundationalism, hermeneutics, intertextuality, and simulation (Agger, 1991; Best & Kellner, 1997; Dickins & Fontana, 1994).

Several contributors point to usefulness of postmodernism in highlighting the varieties of culture, difference, and symbolism in cities; the impact of marginalized city residents whom urban scholarship has ignored; and the interconnectedness of discourse and materiality. While often scholars use the term "postmodern" and its most common derivatives "postmodernism" and "postmodernity" in a confusing range of ways, these terms can sensitize us to a series of cultural and economic changes that suggest a more fundamental set of transformations of cities, metropolitan space, and social structures. The contributors also acknowledge the limitations of postmodern urban analysis including its relativism, radical indeterminacy, and lack of normative grounding, among other problems. In particular, Feagin (1998, p. 6) has recently noted that "a postmodern analysis that privileges cultural complexity and diffuseness in cities runs the danger of ignoring or playing down the still central structure-process factors of class, race, and gender." As Ellin (1996) and Best and Kellner (1997) note, too much emphasis on the autonomy of urban subcultures and identities can help hide unequal power relations, downplay race/class/gender antagonisms, and legitimate profound inegalitarian social hierarchies that still shape cities and metropolitan areas. Indeed a lacuna of postmodern urban analysis has been the inability to formulate a critical theory that illuminates the mechanisms of domination in society, the interconnectedness of agency and structure, and the complex mediations among new forms of urban and metropolitan organization, economy, and culture.

Amid the competing debates and controversies over the relative merits of postmodern urban analysis, skeptical epistemology, or textual deconstruction it is vital that urban sociologists not lose sight of the grounding of material life and the centrality of sociological theory in understanding cities and urban life. Sociological theories are not just a description of the past and present, nor an explanation of how and why various institutions, groups, and actors are interconnected. Theories also provide a prescription for social action and change, as Marxism and variants of feminism have done. Henri Lefebvre noted that representations of the city are not politically neutral, nor are they without social implications. Indeed, many dominant representations play down equality, social justice, and issues of power and control in the city.

In our case, urban theories tell us what is urban "redevelopment," why is urban redevelopment undertaken, who are the key actors and organized interests responsible for undertaking urban redevelopment, who benefits and who suffers from various urban redevelopment schemes, and how does urban redevelopment affect cities overall and different groups in particular. The agenda for research on urban redevelopment is to understand how past and present structural arrangements affect cities, redevelopment efforts, and local people. Focusing on the why things are, how they got that way, and how to change them gives us some key answers to some of the most vexing problems in the city today.

REFERENCES

Abu-Lughod, J. (1989). *Before European Hegemony: The World System A. D. 1250–1350*. New York: Oxford University Press.
Agger, B. (1991). Critical Theory, Poststructuralism, and Postmodernism: Their Theoretical Relevance. *Annual Review of Sociology, 17*, 105–131.
Amin, A (Ed.). *Post Fordism*: A Reader. Oxford: Blackwell.
Bassett, K. (1993). Urban Cultural Strategies and Urban Regeneration: A Case Study and Critique. *Environment and Planning A, 25*, 1773–1788.
Baudrillard, J. (1983). *Simulations*. New York: Semiotext(e).
Beauregard, R. A. (1993). *Voices of Decline: The Postwar Fate of U.S. Cities*. Blackwell.
Best, S., & Kellner, D. (1997). *The Postmodern Turn*. New York: Guilford.
Bird, J., Curtis, B., Putnam, T., Robertson, G., & Tickner, L. (1993). *Mapping the Futures: Local Cultures, Global Change*. New York: Routledge.
Boyer, C. (1992). Cities for Sale: Merchandising History at South Street Seaport. In: M. Sorkin (Ed.) *Variations on a Theme Park* (pp. 181–204). New York: Hill and Wang.
Bullard, R. D. J., Grigsby. E., & Lee C. (Eds) (1994). *Residential Apartheid: The American Legacy*. Los Angeles: CASS Urban Policy Series.
Calavita, N. (1992). Growth Machines and Ballot Box Planning: The San Diego Case. *Journal of Urban Affairs, 14*, 1–24.
Christopherson, S. (1994). The Fortress City: Privatized Spaces, Consumer Citizenship. In: A. Amin (Ed.) *Post Fordism: A Reader*. (pp. 409–27). Oxford: Blackwell.
Clark, G. L., & Dear, M. (1984). *State Apparatus: Structures of Language and Legitimacy*. Boston: Allen and Unurn, Inc.
Collins, P. H. (1991). *Black Feminist Thought: Knowledge, Consciousness, and the Politics of Empowerment*. New York: Routledge.
Cook, C., & Lauria, M. (1995). Urban Regeneration and Public Housing in New Orleans. *Urban Affairs Review, 30*, 538–557.
Cox, K. (Ed.) (1997). *Spaces of Globalization: Reasserting the Power of the Local*. New York: Guilford.
Croucher, S. L. (1997). Constructing the Image of Ethnic Harmony in Toronto, Canada: The Politics of Problem Definition and Nondefinition. *Urban Affairs Review, 32*(3), Jan. 1997.
Cummings, S. (Ed.) (1988). *Business Elites and Urban Development: Case Studies and Critical Perspectives*. Albany: State University of New York Press.
Davis, M. (1992). *City of Quartz: Excavating the Future in Los Angeles*. New York: Vintage Books.
Dear, M., & Clark, G. L. (1981). Dimensions of Local State Autonomy. *Environment and Planning, 13*, 1277–1294.

Department of Housing and Urban Development (2000). *HOPE VI: Community Building Makes a Difference*. Washington, DC: Government Printing Office.

Department of Housing and Urban Development (1999). *HOPE VI: Building Communities, Transforming Lives*. Washington, DC: Government Printing Office.

Department of Housing and Urban Development (1996). *An Historical and Baseline Assessment of HOPE VI*. Washington, DC: Government Printing Office.

DeSena, J. N. (2000). Gendered Space and Women's Community Work. In: R. Hutchison. (Ed.), *Constructions of Urban Space*. Research in Urban Sociology, Vol. 5 (pp. 275–297). Stamford: CT: JAI Press.

Dickins, D. R., & Fontana, A. (Eds) (1994). *Postmodernism and Social Inquiry*. New York: Guilford Press.

Eisinger, P. (2000). The Politics of Bread and Circuses: Building the City for the Visitor Class. *Urban Affairs Review*, *35*(3), January, 316–333.

Ellin, N. (1996). *Postmodern Urbanism*. Oxford: Blackwell.

Engels, F. (1969a). *The Condition of the Working Class in England*. St. Albans: Panther Books.

Engels, F. (1969b). The Housing Question In: K. Marx, & F. Engels, *Selected Works*, *Vol. 2*. Moscow: Progress Publishers.

Epp, G. (1996). Emerging Strategies for Revitalizing Public Housing Communities. *Housing Policy Debate*, *7*(32), 563–588.

Fainstein, S. S., & Fainstein, N. I. (1986). Economic Change, National Policy, and the System of Cities In: S. S. Fainstein, N. I. Fainstein, R C. Hill, D. R. Judd & M. P. Smith (Eds), *Restructuring the City: The Political Economy of Urban Redevelopment*. (2nd ed.) (pp. 1–26). New York: Longman.

Feagin, J. R., & Vera. H. (1995). *White Racism: The Basics*. New York: Routledge.

Feagin, J. R. (1988). Free-Enterprise City: Houston In: *Political-Economic Perspective*. New Brunswick: Rutgers University Press.

Feagin, J. R., & Parker. R. (1990). *Building American Cities: The Urban Real Estate Game*. (2nd ed). Englewood Cliffs, NJ: Prentice Hall.

Feagin J. R. (1998). *The New Urban Paradigm: Critical Perspectives on the City*. New York: Rowman and Littlefield.

Feagin, J. R. (1994). A House Is Not a Home: White Racism and U.S. Housing Practices, In: R. D. Bullard., J. E. Grigsby, III, & C. Lee (Eds.), *Residential Apartheid: The American Legacy*. Los Angeles: CASS Urban Policy Series.

Featherstone, M. (1990). *Global Culture: Nationalism, Globalization, and Modernity*. Newbury Park, CA: Sage.

Flanagan, W. (1993). *Contemporary Urban Sociology*. Cambridge: Cambridge University Press.

Friedland, R. F., Piven, F. F., & R. R. Alford. (1977). Political Conflict, Urban Structure, and Fiscal Crisis. *International Journal of Urban and Regional Research*, *11*(October): 447–471.

General Accounting Office (GAO). (1998). *HOPE VI: Progress and Problems in Revitalizing Distressed Public Housing*. Washington, DC: Government Printing Office.

Giddens, A. (1996). *In Defence of Sociology: Essays, Interpretations, and Rejoiners*. New York: Polity Press.

Gieryn, T. F. (2000). A Space for Place in Sociology. *Annual Review of Sociology*, *26*, 463–496.

Gladstone, D. L. (1998). Tourism Urbanization in the United States. *Urban Affairs Review*, Vol. *34*(1), September, 3–27.

Gotham, K. F. (2001a). A City Without Slums: Urban Renewal, Public Housing, and Downtown Revitalization in Kansas City, Missouri. Forthcoming in *American Journal of Economics and Sociology*, *60*(1), January 2001.

Gotham, K. F. (2001b). Growth Machine Up-Links: Urban Renewal and the Rise and Fall of a Pro-Growth Coalition in a U.S. City. Forthcoming in *Critical Sociology*, 26(3).

Gotham, K. F. (2000). Urban Space, Restrictive Covenants, and the Origin of Racial Residential Segregation in a U.S. City, 1900–1950. *International Journal of Urban and Regional Research*, 24(3), 616–633. September 2000.

Gotham, K. F. (1999). Political Opportunity, Community Identity, and the Emergence of a Local Anti-Expressway Movement. *Social Problems*, 46(3), August 1999.

Gotham, K. F. (1998a). Blind Faith in the Free Market: Urban Poverty, Residential Segregation, and Federal Housing Retrenchment, 1970–1995. *Sociological Inquiry*, 68(1), Winter 1998.

Gotham, K. F. (1998b). Suburbia Under Siege: Low-Income Housing and Racial Conflict in Metropolitan in Kansas City, 1970–1990. *Sociological Spectrum*, 18(4), 1998.

Gottdiener, M. (1987). *The Decline of Urban Politics: Political Theory and the Crisis of the Local State*. Newbury Park: Sage Publications.

Gottdiener, Mark. (1997). *Theming of America: Dreams, Visions, and Commercial Spaces*. Boulder, CO: Westview Press.

Gottdiener, M. (1994). *The Social Production of Urban Space*. (2nded.). Austin: University of Texas Press.

Gottdiener, M., Collins, C. C., & Dickens, D. (1999). *Las Vegas: The Social Production of an All-American City*. New York: Blackwell.

Hamilton, D. K. (1999). Organizing Government Structure and Governance Functions in Response to Growth and Change: A Critical Overview. *Journal of Urban Affairs*, 22(1).

Hannigan, J. (1998). *Fantasy City: Pleasure and Profit in the Postmodern Metropolis*. New York: Routledge.

Harvey, D. (1973). *Social Justice and the City*, Baltimore: Johns Hopkins University Press.

Harvey, D. (1989). *The Condition of Postmodernity*. New York: Free Press

Haymes, S. N. (1995). *Race, Culture, and the City: A Pedagogy for Black Urban Struggle*. SUNY Press.

Hodder, R. (1999). Redefining a Souther City's Heritage: Historic Preservation Planning, Public Art, and Race in Richmond, Virginia. *Journal of Urban Affairs*, 21(4).

Hoffman, L. M. (2000). Tourism and the Revitalization of Harlem. In: R Hutchison (Ed.) *Constructions of Urban Space*. (pp. 207–224) Vol. 5 of Research in Urban Sociology. Stamford, CT: JAI Press.

Holcomb, B. (1993). Revisioning Place: De- and Re-constructing the Image of the Industrial City. In: Kearns, Gerry, & C. Philo (Eds). *Selling Places: The City as Cultural Capital, Past and Present*. (pp. 133–44). Oxford: Pergamon Press.

Holt, W. G. (2000). Distinguishing Metropolises: The Production of Urban Imagery. In: R. Hutchison (Ed.) *Constructions of Urban Space*. (pp. 225–252) Vol. 5 Research in Urban Sociology. JAI Press.

Horan, C. (1991). Beyond Governing Coalitions: Analyzing Urban Regimes in the 1990s. *Journal of Urban Affairs*, 13, 119–135.

Imbroscio, D. L. (1998). Reformulating Urban Regime Theory: The Division of Labor Between State and Market Reconsidered. *Journal of Urban Affairs*, 20(3).

Jessep, B., Peck. J., & Tickell, A. (1999). Retooling the Machine: Economic Crisis, State Restructuring, and Urban Politics. In: A. E. G. Jonas & D. Wilson (Eds). T*he Urban Growth Machine: Critical Perspectives Two Decades Later* (pp. 1411–1462). Albany: State University of New York (SUNY) Press.

Jonas, E. G., & Wilson, D. (1999). The City as a Growth Machine: Critical Reflections Two Decades Later In: A. E. G. Jonas & D. Wilson (Eds), *The Urban Growth Machine: Critical Perspectives Two Decades Later* (pp. 3–20) Albany: State University of New York (SUNY) Press.

Kaplan, D. H. (1999). The Uneven Distribution of Employment Opportunities: Neighborhood and Race in Cleveland, Ohio. *Journal of Urban Affairs, 21*(2).

Kearns., Gerry., & Philo, C. (1993). Selling Places: *The City as Cultural Capital, Past and Present*. Oxford: Pergamon Press.

King, A. (Ed.) (1996). *Re-Presenting the City: Ethnicity, Capital, and Culture in the 21st Century Metropolis*. NYU Press.

Kirby, A., & Abu-Rauss. T. (1999). Employing the Growth Machine Heuristic in a Different Political and Economic Context: The Case of Israel. In: A. E. G. Jonas & D. Wilson (Eds), *The Urban Growth Machine: Critical Perspectives Two Decades Later* (pp. 195–212). Albany: State University of New York (SUNY) Press.

Knox, P. (Ed.) (1993). *The Restless Urban Landscape*. New York: Prentice Hall.

Liggett, H. (1994). City Talk: Coming Apart in Los Angeles. *Urban Affairs Quarterly, 29*(3), 454–467.

Lin, J. (1995). Ethnic Places, Postmodernism and Urban Change in Houston. *Sociological Quarterly. 36*(4), 629–647.

Lin, J. (1998). Globalization and the Revalorization of Ethnic Places in Immigrant Gateway Cities. *Urban Affairs Review, 34*(2), 313–339.

Logan, J. (1991). Gambling on Real Estate: Limited Rationality in the Global Economy. *Sociological Perspectives, 34*(4), 391–401.

Logan, J. (1993). Cycles and Trends in the Globalization of Real Estate. In: P. Knox (Ed.), *Restless Urban Landscape* (pp. 33–54). New York: Prentice Hall.

Logan, J. R., & Swanstrom, T. (Eds) (1990). *Beyond the City Limits: Urban Policy and Economic Restructuring in Comparative Perspective*. Philadelphia: Temple University Press.

Logan, J., & Molotch, H. (1987). *Urban Fortunes: The Political Economy of Place*. Berkeley: University of California Press.

Logan, J. R., Whaley, R. B., & Crowder, K. (1997). The Character and Consequences of Growth Regimes: An Assessment of 20 years of Research. *Urban Affairs Review, 32*(5), 6303–6330.

Lord, G., & Price, A. C. (1992). Growth Ideology in a Period of Decline: Deindustrialization and Restructuring, Flint Style. *Social Problems, 39*(2), 155–169, May.

Marx, K, & Engels, F. (1948). *Manifesto of the Communist Party*. New York: International Publishers.

Massey, D. S., & Denton, N. A. (1993). *American Apartheid: Segregation and the Making of the Underclass*. Cambridge, MA: Harvard University Press.

McDowell, L. (1999). *Gender, Identity, and Place: Understanding Feminist Geographies*. Minneapolis: University of Minnesota Press.

Mele, C. (1995). Globalization, Culture, and neighborhood Change: Reinventing the Lower East Side of New York. *Urban Affairs Review, 32*(1), 3–22.

Merrifield, A., & Swyngedouw, E. (Eds) (1997). *The Urbanization of Injustice*. New York: NYU Press.

Miles, S., & Paddison, R. (1998). Urban Consumption: An Historiographical Note. *Urban Studies, 35*(5/6), May.

Milligan, M. (1998). Interactional Past and Potential: The Social Construction of Place Attachment. *Symbolic Interaction, 21*, 1–33.

Miranne, C. B., & Young, A. H. (Eds) (2000). *Gendering the City: Women, Boundaries and Visions of Urban Life*. Lanham, MD: Rowman and Littlefield.

Molotch, H. (1976). The City as a Growth Machine: Toward a Political Economy of Place. *American Journal of Sociology, 82*, 309–330.

Molotch, H. (1999). Growth Machine Links: Up, Down, and Across. In: A. E. G. Jonas &

D. Wilson (Eds), *The Urban Growth Machine: Critical Perspectives Two Decades Later* (pp. 247–266). Albany: State University of New York (SUNY) Press.

Nelson, A. C., & Foster, K. A. (1999). Metropolitan Governance Structure and Income Growth. *Journal of Urban Affairs, 21*(3).

Pitcoff, W. (1999). New Hope for Public Housing? Shelterforce. March/April.

Purchell, M. (1999). The Decline of Political Consensus for Urban Growth: Evidence from Las Angeles. *Journal of Urban Affairs, 22*(1).

Riain, S. O. (2000). States and Markets in an Era of Globalization. *Annual Review of Sociology, 26*, 187–213.

Reichl, A. (1999). *Reconstructing Times Square: Politics and Culture in Urban Development.* Lawrence, KS: University Press of Kansas.

Reichl, A. (1997). Historic Preservation and Progrowth Politics in U.S. Cities. *Urban Affairs Review, 32*(4), March, 513–535.

Sadler, D. (1993). Place-Marketing, Competitive Places and the Social Construction of Hegemony in Britain in the 1980s. In: G. Kearns & C. Philo (Eds), *Selling Places: The City as Cultural Capital, Past and Present* (pp. 175–92). Oxford: Pergamon Press.

Salama, J. J. (1999). The Redevelopment of Distressed Public Housing: Early Results from HOPE VI Projects in Atlanta, Chicago, and San Antonio. *Housing Policy Debate, 10*(1).

Sassen, S. (1998). *Globalization and Its Discontents: Essays on the New Mobility of People and Money.* New York: The New Press.

Sassen, S. (1991). *The Global City: New York, London, and Tokyo.* Princeton, NJ: Princeton University Press.

Sassen, S. (2000). New Frontiers Facing Urban Sociology at the Millennium. *British Journal of Sociology, 51*(1), 143–160, January/March.

Saunders, P. (1981). *Social Theory and the Urban Question.* London: Hutchinson.

Short, J. R. (1999). Urban Imagineers: Boosterism and the Representations of Cities. In: A. E. G. Jonas & D. Wilson (Eds). *The Urban Growth Machine: Critical Perspectives Two Decades Later* (pp. 37-54). Albany: SUNY Press.

Silver, H. (1993). National Conceptions of the New Urban Poverty: Social Structural Change in Britain, France, and the United States. *International Journal of Urban and Regional Research, 17*(3), 336-354, September.

Smith, N. (1996). *The New Urban Frontier: Gentrification and the Revanchist City.* New York: Routledge.

Soja, E. (2000). *Postmetropolis: Critical Studies of Cities and Regions.* New York: Blackwell.

Sorkin, M. (Ed.) (1992). *Variations on a Theme Park: The New American City and the End of Public Space.* New York: Hill and Wang.

Spain, D. (1993). *Gendered Spaces.* Chapel Hill: University of North Carolina Press.

Squires, G. D. (1994). *Capital and Communities in Black and White: The Intersections of Race, Class, and Uneven Development.* State University of New York Press.

Squires, G. D. (Ed) (1989). *Unequal Partnerships: The Political Economy of Urban Redevelopment in Postwar America.* New Brunswick: Rutgers University Press.

Strom, E. (1999). Let's Put on a Show! Performing Acts and Urban Revitalization in Newark, New Jersey. *Journal of Urban Affairs, 21*(4), 423–435

Swanstrom, T. (1993). Beyond Economism: Urban Political Economy and the Postmodern Challenge. *Journal of Urban Affairs, 15*(1), 55–78.

Urry, J. (1995). *Consuming Places.* London and New York: Routledge.

Vale, L. J. (1997). Empathological Places: Residents' Ambivalence Toward Remaining in Public Housing. *Journal of Planning Education and Research, 16*, 159–175.

Vale, L. (1993). Beyond the Problem Projects Paradigm: Defining and Revitalizing Severely Distressed Public Housing. *Housing Policy Debate*, *4*(2), 147–174.

Vale, L. (1996). Public Housing Redevelopment: Seven Kinds of Success. *Housing Policy Debate*. *7*(3).

Van Kempen, R., & Marcuse, P. (1997). A New Spatial Order in Cities. *American Behavioral Scientist*, *41*(3), 285–298, November/ December.

Watson, S., & Gibson, K. (Eds) (1995). *Postmodern Cities and Spaces*. Cambridge, MA: Blackwell.

Wilson, D. (Ed.) (1998). Globalization and the Changing U.S. City. *Annals of the American Academy of Political and Social Science*, *551*, May, 1997.

Wittberg, P. (1992). Perspectives on Gentrification: A Comparative Review of the Literature. Gentrification and Urban Change. *Research in Urban Sociology*, Vol. 2 (pp. 17–46). JAI Press.

Wright, T., & Hutchison, R. (1997). Socio-Spatial Reproduction, Marketing Culture, and the Built Environment. In: R. Hutchison New Directions in Urban Sociology. Vol. 4, *Research in Urban Sociology* (pp. 187–214). Greenwhich, CN: JAI Press.

Wright, T. (1997). *Out of Place: Homeless Mobilizations, Subcities, and Contested Landscapes*. Albany, NY: State University of New Press.

Zukin, S. et al. (1996). From Coney Island to Las Vegas in the Urban Imaginary: Discursive Practices of Growth and Decline. *Urban Affairs Review*, *33*(5), 627–654, May.

Zukin, S. (1997). Cultural Strategies of Economic Development and the Hegemony of Vision. In: A. Merrifield & E Swyngedouw *Urbanization of Injustice* (pp. 233–242). New York: New York University Press.

Zukin, S. (1995). *The Cultures of Cities*. Cambridge, MA: Blackwell.